MW01178524

# Advance Review

For the many of us who read and relished Stevo Julius' first autobiographical volume, which recounted the extraordinary experiences of a young man in war-time Yugoslavia who joined the armed resistance to the Fascist invaders, and who later also actively resisted the Communist regime, this second volume – based on Professor Julius' distinguished career as a physician and medical researcher – has been eagerly awaited.

What could be so exciting about a career in academic medicine? For a start, this book, which is anything but a simple historical narrative, reminds us that to a man like Stevo Julius, driven by deep conviction and passion, so much in life becomes an adventure. For readers themselves in the field of medical research, Stevo's personal history will strike a deep chord. His vivid accounts will trigger their own memories of the elations and frustrations, and especially the competitiveness, of scientific investigation. More importantly, his story reminds us of the visions and hopes that make scientists choose this path in the first place, and how in our daily routine we often forget that the white coats we wear in the laboratory and clinic are really the cloaks of risk takers and explorers.

As well, this book reveals a strong human story. Dr. Julius warmly recalls the leaders, with all their idiosyncrasies, who influenced his early career; and even more warmly, the remarkable number of younger scientists whom he trained and who so often went on to distinguished careers of their own. Stevo shares with us in this book some of the major scientific hypotheses he created that have added substantially to our understanding of human physiology and disease, not to mention community medicine and epidemiology. This information should be satisfying not only to clinical scientists, but also to lay readers.

Except for very rare, dramatic breakthroughs, scientific progress most often emerges from the collective efforts of multiple dedicated researchers. Stevo Julius has shown how highly motivated, innovative leaders can refine and maximize this critical process. His uninhibited autobiographical account gives us a revealing and very personal insight into the professional life of one of this nation's true medical scholars.

Michael A. Weber, MD
Professor of Medicine, SUNY Downstate College of Medicine
Former President, American Society of Hypertension

*Dedicated to the remembrance of the former Division of Hypertension in Ann Arbor and to two pillars of the Division: Andrew Zweifler for over thirty years of loyal support, and to Susan Julius for her help in maintaining the cohesion of our extended family of the Ann Arbor trainees, colleagues and coworkers.*

# About the Author

Dr. Julius is a graduate of the Medical School of the University of Zagreb (Croatia). In 1965 he joined the faculty of the University of Michigan in Ann Arbor where he is presently Active Professor Emeritus of Medicine and Physiology. Dr. Julius is internationally recognized for his contributions to research and education in hypertension.

# Acknowledgements

Many thanks to my old friend, Tony Schork, for editing the book and urging me to persevere in times of creative despair. Thanks is also due to Mrs. Patricia LaCoste for her interest in the book and her hard work in preparing the manuscript for print.

After an incidental chat about a vague plan to write a book on the Ann Arbor research group, and virtually unsolicited, my colleagues at Les Laboratoires Servier in Paris offered to support the first edition of the book and I wish to thank them very much.

# A DOUBLE PREFACE

A preface is usually about the background and the purpose of a book but this is also a preface to the actual writing of the book. A few years ago I started toying with the idea of another book and it has now been about two months since I began typing the first pages of the manuscript. The writing had proved to be much harder than I thought and I wondered why. It took a natural disaster to understand my problem. As part of a lecture tour in Japan, I was scheduled to fly from Yokohama to Fukuoka. A day before departure a strong earthquake shook the island of Kyushu, the lecture was cancelled, I found myself in a luxury hotel with plenty of free time, and I figured out what the problem was. My second book had to be very different from the first.

The first book was a remembrance of how I grew up in rough but emotionally rich times in former Yugoslavia. I wrote it primarily for family and friends, but I secretly hoped that the book might appeal to some of my contemporaries and possibly strike the fancy of some professional colleagues. The book did much better than that. Two soft cover printings of "Neither Red Nor Dead - Coming of Age in Former Yugoslavia During and After World War II" have sold out, and a Croatian translation of the book is also selling well.

Success has its own dynamics. Once you show what you can do, people expect more. Many readers noted that "Neither Red Nor Dead" said nothing about my life in the United States and wondered when I would produce a sequel. Let's be frank, there was no groundswell but sporadic inquiries about the "next one" planted the bug in my mind.

The first book dealt with topics of general interest; war, guerilla movement, Fascism, Communism, human relationships, and how a youngster successfully navigated through hard times. Not to interrupt the flow of the core story, I mentioned only the most relevant specific facts in footnotes. For those interested in understanding the historical context of the conflict I prepared a background chapter at the end of the book. That way the readers primarily interested in what happened and how people interacted during a conflict could enjoy the book. Those wanting to understand the historical background of the conflict could read the last chapter first.

I decided to try the same technique in this book and failed. In the first twenty or so pages of an early draft, footnotes exceeded the length of the main text! Unfortunately, to understand what went on, some mastery of the topics I chose to investigate would be necessary. So, this book has a chronological story line which I occasionally interrupt to insert some basic explanations on how blood circulates in the body and how various systems regulate the circulation. For those who wish to learn more about physiologic underpinnings of my research, I added at the end of the book a chapter entitled "Autonomic Nervous System and Circulation; A Sketch of Underlying Physiology."

Some readers will be more interested in the human side of the story than in science. In the back of the book I have prepared for them a vocabulary of technical terms which throughout the book are denoted in *italics*. A quick reference to that vocabulary will give them the necessary background to understand the elations, excitements, disappointments, doubts, conflicts of ideas, and personality clashes which will be described in the book.

Pursuing science has been great fun and I would love to tell my story to a wider readership. Time will tell whether I succeeded.

# A Truly Different New World 1962-1964

The first time around I came to the United States as my first wife's wife. Mara, a psychologist, had obtained a fellowship to study sociology research methods in Ann Arbor's famous Survey Research Institute. The sponsor was willing to pay the trip for the family but it never occurred to the good people at the Ford Foundation that the primary award recipient might be a woman. So, Mara and her daughter, Ivancica, left for Ann Arbor and back in Zagreb I kept myself busy changing every "wife" on the form into "husband" and generally making sure everybody understood I was a male. Ford Foundation officials either did not read the form or were generous enough to ignore the male chauvinist barbs that I inserted in various places. I thought I was funny but the uncalled for witticisms could have killed my chances to travel to the United States. You don't want to fool around with bureaucrats, anywhere in the world!

Next I wrote to the Dean of the Medical School in Ann Arbor. I informed him of my circumstances, indicating I'd work as a volunteer, and asked whether he knew somebody involved in autonomic nervous system research. The Dean suggested that I contact Dr. Sibley Hoobler in the Hypertension Division of the Department of Medicine. Hoobler responded positively and with his letter in hand, I got an unpaid leave of absence for two years. The Chief of Medicine at the University Hospital in Zagreb, Professor Hahn, wrote a letter of introduction describing me as a hard working Senior Instructor in medicine with a promising academic future. A few months later I was on my way to Ann Arbor.

After he interviewed me, Hoobler decided I could join the clinical team of the Division of Hypertension as an observer. Unfortunately no one in the Division was working on *autonomic nervous system* function, but I could start working in Dr. James Conway's *hemodynamic* laboratory. From him, Hoobler suggested, I could learn about research methods in general. That was fine with me. A gray haired, slender, pipe smoking man in his early sixties, Hoobler was an independently wealthy scion of the Sibley's Shoes fortune. During the Second World War, he served as a medical officer in the Pacific theater. After the war, all the doors were open to him. He was attracted to academic medicine and decided to settle in Ann Arbor where he joined the faculty in the medical school's Department of Medicine. At that time Dr. Max Peet, a neurosurgeon at the University of Michigan, developed a new method of abdominal *sympathectomy* and started to use it in the treatment of hypertension. About 20 % of the patients responded to surgery with dramatic improvement but in the rest the blood pressure wouldn't budge. As it happened, another group in the Department of Pharmacology in Ann Arbor developed the first prototype of a *ganglionic blocking agent*. Because they temporarily blocked the transmission of signals through the entire autonomic nervous system, these drugs caused almost unbearable side effects and could not be used for long-term treatment of hypertension. Hoobler came up with the logical idea of using a short-term "chemical sympathectomy" to predict which patients might respond to Dr. Peet's surgical intervention. His proposal was accepted and Hoobler soon assembled a research laboratory for measurement of intra-arterial blood pressure and *cardiac output* using *dye dilution*. Since the technique required cardiac catheterization, Hoobler became the first person in Michigan to introduce a catheter into a patient's heart. The results were somewhat disappointing. Prior to Hoobler's experiments Ed Freis in Washington and Ed Varnauskas in Sweden had shown that hypertension was a state of generalized arteriolar constriction and that the blood pressure elevation was due to increased *vascular resistance*. Hoobler correctly reasoned that the best candidates for sympathetctomy would be patients in whom excess sympathetic drive caused excessive constriction of blood vessels. He then assumed that such patients would selectively respond to *ganglionic blockade* with a bigger blood pressure lowering due to a larger degree of arteriolar *vasodilation*. Instead, almost everybody's blood pressure decreased and the blood pressure response was found to be associated with a fall of cardiac output with little change in vascular resistance. From this, Hoobler concluded that "something other" than the autonomic nervous system must be involved in the increase of vascular resistance in hypertension and decided to focus on *renin*, the newly discovered pressor agent released by the kidneys.

When it came to science, Hoobler was not given to half-measures. He arranged to take a trip to Argentina where he stayed a few months in Braun-Mendenez's laboratory in Buenos Aires. At that time it was well understood that renin was an enzyme, which acted on a protein substrate to release another pressor agent into the blood. Braun-

Mendenez succeeded in isolating the pressor product of renin, which he eventually called "angiotonin." After his return to Ann Arbor, Hoobler closely followed the work of Irvine Page in Cleveland. His group had also isolated the pressor product of renin which Page called "hypertensin." Hoobler learned in Buenos Aires how to set up rat assay for measurement of plasma renin activity. From that time on, Hoobler lost interest in human hemodynamics and his major publications were in the area of renin. He was a good organizer and in 1957 successfully arranged a conference in Ann Arbor to review the progress in the field of renin. At that conference Page and Braun Mendenez decided to use one term for the peptide they separately isolated and agreed on the new hybrid designation of "*angiotensin*." As we all know, that new name stuck.

**Sibley Hoobler**

Because of his experiments with ganglionic blockade, Hoobler didn't believe that the nervous system played a role in hypertension but he never tried to talk me out of my research fascination with this system. Only once did he mention why he thought that something other than the autonomic nervous system might cause the elevated vascular resistance in hypertension. However, he also understood the limitations of his experiments and readily admitted that he could have used too large blocking doses to draw conclusions about physiological issues. That was all I needed to hear! If there was a chance that he was wrong, I could continue pursuing my fancy. And what I wanted to do was nothing other than fancy. I believed that the autonomic nervous system might be involved in hypertension but could not muster any specific arguments to support the idea. I owe Hoobler a great debt of gratitude for being so supportive. After all, I was only an unknown "walk on" candidate for a position in his team.

They say that if a Chinese wishes to put a curse on you he is likely to say "May you live in interesting times!" Well, Hoobler, the first man I met professionally in the US, was also the most interesting person I encountered in the new world. His father was a pediatrician in Detroit, and his mother taught biology at the University of Michigan. They instilled in him a love of medicine and nature. They also taught him the old-fashioned sense of civic responsibility. Unfortunately, someone else, I suspect from the Sibley Shoes side of the family, conferred on him the love of money coupled with the deep conviction that in everything –charity included – a man is entitled to some return for his investment. Torn between two opposing sets of values, Hoobler evolved into a conflicted and conflict-prone individual. He was parsimonious beyond belief and yet, occasionally he'd give considerable money to charity. But to each gift he attached a spider's net of strings. He gave a special stipend for a woman to finish medical school and, to prevent her from asking for more, stipulated that she should never learn who the donor was. Simultaneously he insisted on regular reports on her progress and attempted to tele-micromanage her schooling. On the private level, one half of Hoobler would succumb to heartfelt generosity but the other half could not help but ask for instant reciprocity. He'd often invite me to his "Camp" in the northern tip of Michigan's Lower Peninsula where he had inherited two square miles of wilderness. Nothing had quite avoided the axe of the ruthless timber barons and presently only a few very small remnants of original forest with tall pine trees could be found in Michigan's Lower Peninsula. Sibley's father got hold of a former logging camp site where, to protect themselves from the scorching summers, the lumber jacks saved six original huge fir trees. There Dr. Hoobler senior built three beautiful but properly Spartan summer vacation cabins. In line with his father's dictum of generosity, Sibley Hoobler would invite us to this piece of paradise. He was an avid dry fly fisherman. I spent many happy hours wading with him through the Pigeon River but as soon as we were back at the camp he'd assign me odd repair jobs. And it was not a "let's do this and that together," I had to do it while he was resting. Later Hoobler sometimes invited me to visit the camp alone, but always handed me a list of things to do. His place was in good shape and most of the jobs were meaningless. That bothered me. If the roof had been leaking, I am sure I would have volunteered with pleasure. Where I came from, a host would never think of asking for reciprocity. Finally, I learned how to cope with Hoobler's peculiarity. Each time I headed up north I'd come up with my own proposal as to what needed fixing. Volunteering was much better than working off my due. On one occasion, our chief laboratory technician Dick Rogers and I built a magnificent outhouse with a panoramic view, a comfortable toilet seat and a window in the shape of a fish. Now, that was fun!

Hoobler was as complex and unpredictable a man as I have ever met. His mood oscillated back and forth from being charming to barely talking with anybody. In one of his expansive days, he offered me a "small stipend" for my work. Sensing the moment was right I informed him that I had met Ernest Harburg, a psychologist, and the two of us could easily set up a study of personality traits in hypertension. I would continue working in Conway's laboratory, and the new project would not interfere with my other responsibilities. Hoobler agreed.

Thrilled, I needed to share the good news with someone. I called up Paul Farkas, my mother's distant relative who emigrated to New York. This kind man still had warm memories of my parents. After my father chose Bela Kuhn's losing side in the Hungarian revolution, he had to leave Budapest in a hurry. He ended up in Prague, graduated from medical school and moved to Koshice in Slovakia to specialize in psychiatry. In the mid 1920's, in the aftermath of the First World War, the business climate changed and father's family back in Panchevo, Yugoslavia, could not continue sending him money. Jews in Europe knew only too well how political shifts quickly could change someone's fortunes. Following the unwritten family rules, Paul's mother who was my mother's aunt took the newlywed Julius couple under her roof in

Koshice. Five decades later, Paul felt he should help me in all matters related to America.

He said, “So they will let you do independent research. That is fantastic. Most people have to wait years on end before they are trusted to work on their own ideas.”

Paul worked in the garment industry, knew nothing about medical research, but intuitively understood I had just appointed him to be my chief cheerleader. He continued to express wonderment and encouragement while I waited for the right moment to pull the ace out of the sleeve.

“That is not all,” I said, “Hoobler must like what I have done so far. He offered me a salary!”

“Terrific! How much?”

“Six thousand dollars!”

“A month? That is sensational!”

“No, a year.”

Stunned, Paul did not know how to respond. “Well, keeping in mind that you showed up on his door and offered your service for free, this is...” Paul was still seeking for the right words, “quite decent of him.”

“This is about 50% more than I make at home,” I said. Nothing could dampen my enthusiasm.

“Yes, but you live in America and six thousand is what I pay our janitor for a half a year’s work,” said Paul. “I presume you will not stay in America. Right?”

“I couldn’t even if I wanted to. We have exchange visitor visas designed to prevent ‘brain drain’ from underdeveloped countries to America. The US Government pledged to return us home after two years in America.”

“You mean you will never again be able to return to America?” asked Paul.

“Not exactly. If I spend another two years in Yugoslavia, I can apply for a visa again.”

“Well, let me then explain something about America, Stevo. It is just as well that you plan on returning home. This low salary could become a serious impediment for your future in Ann Arbor. Generally the next salary always depends on your first one. If a physician works for a janitor’s salary, he automatically admits to not being good enough. And once you start on that road you will never again be appropriately compensated.”

Paul obviously knew how things work in America. He gave me good advice but I never quite learned how to adapt to the new reality. Where I came from, you were not supposed to haggle about the salary lest the person across the table got the idea that you liked the money more than the job. In Yugoslavia your social standing and your income appeared totally unrelated. Before World War II in bourgeois circles in Zagreb, it was considered rude to talk about one's income. My mother ceaselessly emphasized the value of modesty. The postwar Communists opposed just about everything stemming from the Kingdom of Yugoslavia but they knew a good thing when they saw it. The bourgeois ethics of deference to authorities, and the mantle of meekness under which one covered his ambition, was a nice present from the old to the new regime. The Communists added their own touch to these old habits. Just as a good Catholic was expected to atone and ask forgiveness for his sins, a good comrade was supposed to be "self-critical." And just as the church proselytized, the communists kept pushing everybody to join in their confessional ritual. But there was a big difference! A Catholic could recite his sins in the privacy of a confession booth. Not so with the Communists! They humiliated people in specifically organized public "self-criticism" sessions. Regardless of whether you sat in the audience or were chosen for the evening's main menu, you felt diminished by these ugly transgressions of everyone's privacy.

. Back home I rebelled against Mother's insistence on polite humility and I profoundly disliked the Communists' manipulative catechism of modesty. Soon after Paul Farkas suggested that excessive modesty won't do me any good in America, I bought a small collection of "how to succeed" books and embarked on a self-improvement project. Here I come, America! Nevertheless, adjusting to American ways proved much more difficult than I thought. I eventually started to speak out for myself. I got rid of Mother's "don't speak until you are asked" dogma and learned to pose questions from the audience. After a few days of preparation, I could muster enough energy to criticize a younger colleague in a one-on-one session. And at the urging of Harriet "Dusty" Dustan, the great lady of hypertension research, I even got rid of unnecessary introductory apologies.

"Why the hell do you have to say 'I am sorry' or 'Excuse me' before you say what you wish to say," she asked. "You are neither sorry nor do you need to ask for permission. Speak up as a man and do not apologize."

"Or just shut up," she added after a pause. Dusty was a tough critic!

I've learned all those things but to this day I have difficulties in publicly criticizing my colleagues. You could call it aversive conditioning from those uncivil hypocritical self-criticism sessions. Deeply etched in the subcortical crannies of my brain, a Pavlovian reflex waits for its opportunity. I occasionally get the profound urge to give a guy his well-deserved hell, but an invisible force glues me to the chair and, disgusted with myself, I just sit there in frozen silence.

After a few months I started to feel comfortable in James Conway's Ann Arbor laboratory. He was a first rate physiologist, a no nonsense organizer, and an excellent teacher. Conway assembled a laboratory for intra-arterial measurement of blood pressure and dye dilution assessment of cardiac output. Intra-arterial catheter placement is presently a routine procedure but in the early 1960's it was a relative novelty. Now every intern has a choice of neatly packed disposable needles with catheters, in Conway's laboratory we had to sharpen needles and trim the catheters before sterilization. Equipment was hard to come by. Conway had to assemble all components of what is now one single machine for measurement of energy expenditure. He called on the local gas company to acquire a household gas consumption flow meter. He then calibrated the meter and converted the movement of the needle into an electrical signal. From other sources he acquired sensors for online measurement of oxygen and $CO_2$. He fed signals from these sensors, plus the outputs from the dye dilution densitometer and the blood pressure sensor into one recording device. Conway was very meticulous. The outputs from gas sensors were not as steady as expected and he requested that we calibrate anew every device every morning. Later when it turned out that the voltage in our power supply oscillated a great deal, Conway purchased power stabilizers, but we continued the daily calibration routine. Knowing that we always checked every detail gave us a great deal of confidence in our results. Now, when a big white container, which in its entrails hides the proverbial black box, starts producing a stream of neat numbers, I am not quite sure whether to believe the magic.

Conway's lab attracted excellent people. I worked with Runne Sannerstedt from Goteborg, Sweden and Tony Amery from Lueven, Belgium. Both were enthusiastic, intelligent and dedicated young scientists. All of us enjoyed learning new things in the pleasant but hard working environment Conway had created.

Things did not go so well with my clinical activities. I had no problems with the English language. In fact, for somebody who never spent time in an English-speaking country, I spoke remarkably well. But that was colloquial English. Understanding the medical jargon was another matter. I was in for hard times.

Have you ever been the only person not speaking the language during a dinner? While never intending to do so, the happily chatting locals can drive you nuts in many different ways. You'd like to join the fun but your timing is hopelessly off. Just as you think you understand what they are talking about, they change the topic. If somebody decides to act as your translator, his timing is off too. You feel you've become a hopeless nuisance to others. So you withdraw, concentrate on food, drink your wine, and try to look happy. But that does not work either. You cannot help but wonder what they are talking about. Are they poking fun at you? Why else would they look at you when they laugh? What is so funny? If you are robbed of the basic means of communication, it is very easy to grow paranoid.

My reasonable proficiency in everyday English proved woefully inadequate for clinical medicine. Nobody in Zagreb could have prepared me for the American pronunciation of medical terms. As a medical student I spent a few weeks in Germany and Holland where they pronounced Latin terms similarly as we did in Zagreb. But in Ann Arbor I could not understand half of what was said during a patient care conference. Nothing sounded even vaguely familiar as the presenter, with breathtaking speed, worked his way through a sea of terms and abbreviations. How was I to know that "pneumonia," pronounced as "new-monia," was an infection of the lungs. New what? And that sending an ENT referral meant asking the Otorhinolaryngologist to pass judgment?

To add insult to injury, some terms sounded familiar but meant quite different things. A "clinic" in Zagreb stood for a "hospital" and the term "ambulanta" which sound similar to "ambulance" meant an outpatient clinic. I was thoroughly confused but could not interrupt the fast-paced conference to ask what the simplest word meant. When you think about it, all of this was quite funny, but in those initial months I lost my sense of humor. I hated the passive me and, predictably, started to feel a bit paranoid. No, I was not that quiet imbecile they thought I was! However, I held my guns until I finally understood the basic terminology. When I eventually began to talk, no one, including myself, could stop me. Almost overnight, the former imbecile turned into a certifiable nuisance.

There is such a thing as situational behavior. New circumstances will determine how you act. You might anticipate what could happen, decide not to let it bother you, but once you are in the midst of changes they invariably affect you. And the depth of your feelings, as well as the tone of your response, will surprise you. I knew well in advance that I would have to prove myself in Ann Arbor. I planned on working hard and gradually gaining people's confidence. Instead, impatient and eager to assert myself as a good clinician, I became short-fused and argumentative.

**James Conway**

And of all people I chose James Conway to fight with. A Rhodesian in his mid fifties, he grew up in the hierarchical, tough, British educational system. Tender-loving care of junior personnel was not his forte. If he disagreed with something you said, he'd curtly cut you off. I had no problems accepting his judgment in all matters related to research. But the first time when he responded to my assessment of a patient - with his characteristically loud "Rubbish!" - I exploded. Conway was a good clinician but so was I. In Zagreb I learned my trade from absolutely superb clinicians. Most of them were authoritarian but all were unfailingly polite. Conway's rude punch below the belt was more than I could take. Far from being rubbish, I proposed a reasonable diagnostic alternative and I should have stayed with the clinical topic. Instead, I became quite personal. It was not a nice scene. Battle lines were drawn.

I should have met my colleague Vlado Skulj earlier who, before he immigrated to Chicago, was an assistant professor of obstetrics and gynecology in Zagreb. He was a capable teacher and practitioner of his craft. When he realized that in America he'd again have to take a full residency training in Ob-Gyn, Vlado decided to dissimilate. He pretended not to know the first thing about the field. His teachers were amazed at the speed with which he learned and soon Skulj garnered every possible award for being the "best" among his peers.

"I had to turn it all into a big joke, otherwise I'd go mad," he said. "Nobody could change the situation. My teachers had no choice but to run me through the program. Once I accepted the basic premise that my past training did not count I acted accordingly. I considered it as a refresher course and everybody was happy."

At the end of his training he got high recommendations and joined a high quality lucrative practice in Chicago. I wish I had Vlado's wisdom! Unfortunately, during the initial months of my clinical training Conway and I took leave of our senses. We did not argue about the correct diagnosis. We wanted to prove who was smarter. Losing sight of the patient's needs is the worst mistake a physician can make. I am sure we piled up a bunch of unnecessary tests just to show who was right.

Our fiery exchanges could not go on forever. By and by, grudgingly, we developed a mutual respect and managed to conduct discussions with a semblance of academic detachment.

Compared to previous exchanges our last bedside conflict was quite civilized. During my examination of the heart I heard the characteristic opening click - murmur of mitral valve disease. After I presented the case to Conway, he kept listening to the patient's heart for quite a long time. Finally he whimsically looked at me.

"What makes you think the patient has mitral valve disease?"

I'd already reported the history of rheumatic fever and described all clinical signs favoring the diagnosis. What else could I add?

"Actually I half made up my mind that the patient had mitral stenosis before I listened to the heart," I said with more than a bit of bravado.

"And, pray, with what sort of a magic could you conjure the diagnosis from afar?" Conway took the bait.

"He had a typical facies mitralis."

"A typical what?"

"A mitral face."

"Rrr.." Conway stopped himself in time, "Mitral face - my foot! What sort of old wives tales do you believe in?"

It was not an old woman but Professor Ivancic who had taught me what the face of a patient with mitral disease can look like: red, slightly dusky cheeks and lips against the background of a pale face. I'd seen Ivancic diagnosing mitral valve disease right from the doorway of the bedroom. For the first time, in a far away country, I got the knack of it. It felt good.

"I am not saying he doesn't have mitral disease but you should not support your diagnosis with such a flimsy argument," said Conway.

"I thought the classical description of mitral face can be found in all textbooks of cardiology."

"Only in European textbooks."

"England ..."

"Yes, England is in Europe." Conway was quick. "But just because it is in textbooks, it is not necessarily true. A lot of rubbish gets passed down through the generations."

Conway finally found a way to use his favorite word and after a short pause he continued.

"Okay, you wise guy and great diagnostician; what is the chance of the red cheeks being false positive and what is the sensitivity of red cheeks as a diagnostic tool for mitral stenosis?"

Sensitivity? I knew generally what the term meant but Conway seemed to use it in a specific, quantitative way. Instead of admitting ignorance I fell back on a cop out.

"I will look it up."

"And you will not find it. The old time romantic cardiologists described various problematic physical signs but never bothered with statistics. Incidentally, I never said the patient did not have mitral stenosis." Conway fixed his eyes on me and kept looking.

"You mean he also has a bit of aortic insufficiency? I thought that murmur was not significant."

"Yes, that is what I wanted to hear. Including that the murmur was not significant. Congratulations on finally using a statistical term. Maybe you are not as hopeless a romantic as I thought."

Next evening, I looked up the terms "false positive" and "sensitivity" in a textbook of statistics. Conway was referring to what proportion of people who do not have mitral valve disease might have red cheeks (false positive), and if you were to enter a room full of patients with proven mitral disease what percentage would have the typical marks of a mitral face (sensitivity).

Conway's approach was totally new to me. He asked questions never posed by any of my previous teachers. In Zagreb nobody ever bothered with quantitative aspects of medicine. They taught us the art and not the science of medicine. So what if 25% of healthy people have red cheeks and only 10% of all patients with mitral valve disease have red cheeks. It is still useful to know about red cheeks if you understand the hierarchy of signs. Old school clinicians were keen detectives. They loved assembling clues in an organized way - something that we nowadays call "algorithm." They knew the mitral face was a weak sign but they used it in conjunction with other clues. If a patient complained of shortness of breath and had a red face the next question might be "Did you ever have a bout of fever with joint pain?" An answer suggestive of a history of rheumatic fever would further raise the index of suspicion. A cardiac exam compatible with mitral valve disease would trigger a chest X ray and electrocardiogram to confirm the diagnosis. Eventually, if surgery had been contemplated, the patient was scheduled for cardiac catheterization. Nobody was in a hurry and, intellectually, the process was highly satisfying.

For the old diagnostician, the first order of business was to arrive at the correct diagnosis. His responsibility was only to the individual patient. As Conway said, that was the romantic era of clinical medicine. Romantic, but inefficient and possibly dangerous! Dangerous because old doctors had no objective ways of evaluating long-term effects of medications they used. Inefficient because they never sought to determine how many people who considered themselves to be healthy were actually disease-free. Waiting until the disease forces a patient to look up the doctor is not the best way to go.

When Conway wished to know how frequently a test was false positive and what a test's sensitivity was, he asked a rational medical question. The modern health care system seeks to detect the disease before a person feels sick. That is what periodic health exams and routine screening tests are for. A screening test is like a police raid. You throw out a wide net to catch everyone that is suspicious. Because they do not wish to miss the real McCoy, screening tests catch many innocent bystanders. To sift out the true from the false, you next use a highly specific test, which may not be very sensitive but rarely has false positive results. Knowing a test's properties is the precondition for its effective and rational use. Screening tests have been very successful in decreasing mortality from many conditions but nothing is ideal. There is a price to pay. Not by the patient but by practicing physicians. Having worked in "old" and "modern" medicine settings, I know what the doctor's cost is. Gone is the intellectual challenge of the detective work, the creative tension of proving or disproving a hypothesis and, oddly, the closeness with your patient. Present-day doctors rightly complain that they do not have enough time to talk to patients. But I wonder how they would use the time if they had it. These days every interaction with the physician ends up with a lab slip or an order for a diagnostic test or procedure, and a rather noncommittal statement to the tune of "let's see what the tests show." For sure the old timers were misguided in thinking that if they talked with the patient and completed a physical examination, they were saving him from unnecessary tests. They probably failed to find some subtle diseases and may have been less timely in establishing the diagnosis. But, by golly, they did not shy from guessing what could have been wrong and from sharing their thoughts with the patient. They had an intellectual investment in every case and could barely wait for the lab results.

However, these old timers knew very little about drugs they routinely gave to patients. Drugs were used more for their promise, than for their proven efficacy. In my first book, *Neither Red Nor Dead,* I describe how when I contracted typhoid fever my father, who was a physician, summoned a high-powered group of medical advisors. Since they had no means to directly attack the typhoid bacillus, they decided to give me daily painful intramuscular injections of a protein extract to stimulate "nonspecific defense mechanisms" of my body. This amounted to giving me more fever on top of the underlying 103 degrees typhoid temperature. This strategy to increase the defense capacity of the body appeared logical but nobody ever bothered to determine whether that treatment was useful. I suspect that the treatment was more harmful than good and they almost killed me.

Before I came to the United States, I thought I knew which drug to use for what disease. The Americans used digitalis to treat advanced heart failure, we preferred intravenous injections of strophantine. Strophantine had a similar action to digitalis but was considered to be more potent and to act faster than the foxglove extract. A nice theory supported the use of strophantine; given directly into the vein, the drug would reach the heart faster where it would promptly show its magic. My mentors seem to have forgotten the history of strophantine. During his exploration of Africa in 1860s, Livingston watched the natives smearing arrows with a poison derived from seeds of a jungle plant. The poison was strong enough to instantly kill a hippopotamus. Eventually a Scottish pharmacologist isolated strophantine. Well, if it could kill an animal, maybe it was not so good for a patient. In the short term strophantine quickly improved symptoms of heart failure, but occasionally lungs would fill up with fluid (pulmonary edema) and shortness of breath would follow after the strophantine injection. And sometimes the patient would die. We were told that this occurred very rarely and only in patients who have a strong right and a weak left side of the heart. The Americans would never permit such an important issue to linger. They would initiate a study comparing strophantine with digitalis (which worked a bit slower but seemed not to cause edema of the lung). Half the patients would be assigned to strophantine injections, the other half to taking digitalis by mouth, the study would have enough patients to answer the question and eventually we would learn whether one is better than the other.

In the US, every intern seemed to have read the relevant literature and knew exactly what to expect from which drug. Back in Zagreb, Professor Radosevic had seen me filling syringes with strophantine during the weekend. He was a hard-nosed no-nonsense nephrologist.

"Why are you wasting your time on strophantine? This stuff does not work any better than digitalis and in fact, might be dangerous."

"But Professor Ivancic likes us to use strophantine"

"So he does and I don't!" said the cynical kidney man, turned on his heels and disappeared before I could ask further questions.

Had I asked him why he preferred digitalis, Radosevic would have very likely said something to the effect that this was his experience. He had probably seen a patient dying from strophantine related pulmonary edema, whereas his cardiologist counterpart did not. Looking at the same problem, two brilliant clinicians came to opposing conclusions and that was all they could teach us. America was way ahead and, when it came to the usage of drugs, I quickly learned to do it the American way. The realization that American professors teach "evidence based medicine" helped me to come to terms with Conway. Clinically we might have had similar skills but I realized that, just as he did in his lab, Conway meticulously applied scientific methods to clinical practice. There was a lot I could learn from him. By-and-by we buried the hatchet and developed mutual respect. In retrospect I realize that our conflict was not personal, it was a clash of cultures. I was lucky to have had such a tough mentor in the early stages of my US career. Without his shock treatment I would not have changed my ways.

Things went much better with the personality and behavior research. When he approved the project, Dr. Hoobler made it clear that this work must be done on my free time. That was fine with me. I am a nocturnal creature and, as it turned out, so were my colleagues from the Department of Psychology. I met Ernest Harburg during an evening party given by one of my wife's mentors from the Survey Research Institute. I loved these parties and would stay up until the wee hours to chat with fellow insomniacs. The chats in Ann Arbor very much resembled past conversations during the evening get-togethers in the house of Dr. Stefi Steiner in Zagreb. He'd invite a mixture of artists, academicians and physicians for dinner and leisurely after dinner talk. Stefi and his friends were about 20 years older but they never pulled their age or rank on me. The discussions were always interesting, witty and animated. Akin to their counterparts in Zagreb, academicians from the Survey Research Institute in Ann Arbor were knowledgeable and willing to tackle just about any topic. But in one important way the Ann Arbor cast of players was quite different from the intellectuals in Zagreb. In Zagreb, you'd compare abstract ideas and try to win the argument with its intellectual appeal, logic and internal coherence. In Ann Arbor, you had to support ideas with numbers.

In the 1960's, American psychology, sociology, behavioral science and even philosophy started to evolve into quantitative disciplines. People in the Survey Institute in Ann Arbor strongly believed that everything could be measured and that all ideas must be empirically verified. To them, it was a given that qualitative traits such as anger, shyness, or satisfaction could be expressed quantitatively. They were experts in developing appropriate instruments of measurements. The Survey Research Institute preferred to utilize a five point Likert scale to assess positive and negative response to a statement along the lines: 1. Strongly disagree, 2. Disagree, 3. Neither agree nor disagree, 4. Agree, 5. Strongly agree. As soon as you expressed an opinion, they'd indicate that they have tested the issue and that such and such percent of interviewees agree or dissagree with you.

So when I first met Ernie Harburg, he immediately suggested that we should assess whether hypertension occurs in individuals with distinct personalities. The psychosomatic literature suggested that patients with hypertension might be tense and overly sensitive to environmental stresses. Furthermore, Harburg theorized that people who are internally angry but fail to express their rage might also be prone to developing hypertension. Conceptually Ernie's construct was on solid ground; patients with such personalities would be prone to frequent activations of the *defense reaction* which, among other things, would increase their blood pressure. Since the defense reaction is mediated by the autonomic nervous system, this line of research was right down my alley. But I was not quite sure whether personality could be realistically measured. What if the responses to the questionnaire were to reflect what people think about themselves rather than what they really are?

Son of the famous Yip Harburg who wrote unforgettable lyrics for many Broadway musicals, Ernie Harburg was a dynamic, intelligent, optimistic and fiercely independent individual.

"Oh, but we always take care of such issues. Trust me, the respondent will not have the foggiest idea what we are after. The sheer number of questions is overwhelming and we scramble them so that the individual will not remember how he responded to a previous similar item. For example, a question about anger might come after one about sex life followed by some nonsense about sporting events and so on. Furthermore the questionnaire that I intend to use has been repeatedly validated," said Ernie while laughing aloud.

"What do you mean by 'validated'?"

"It has been given to thousands of individuals and it holds together."

"So?"

"Well, we are not at all interested in answers to individual questions, rather we look at factors," said Ernest.

That was a new territory to me. Ernest instinctively understood the situation and proceeded to explain the basic principles behind the Cattell's 16 personality factor questionnaire. The questionnaire consisted of 178 questions dealing with the responder's behavior in specific circumstances, his general ideas about the world, his attitudes towards others, his preferred activities, his way of reasoning, as well as some questions with no apparent purpose. The answers to these questions tended to group around 16 distinct primary factors: At the time when we used the test, Dr Cattell and his coworkers were cautious in claiming what these factors might have described. Nowadays the authors of the test are less ambivalent and offer the following straight forward terms: Warmth, Reasoning (Concrete vs. Abstract), Emotional Stability, Dominance, Liveliness, Rule-Consciousness, Social Boldness, Sensitivity, Vigilance, Abstractedness (Grounded vs. Abstracted), Privateness (Forthright vs. Private), Apprehension, Openness to Change, Self-Reliance, Perfectionism, and Tension. These factors could then be further combined into global factors; Extraversion, Anxiety, Tough-Mindedness, Independence, and Self-Control.

Ernie Harburg's enthusiasm was infectious and by the end of the evening I was convinced. But Ernie was not done.

"If we pull this off we should not limit ourselves to psychometrics," he said. "A friend of mine, Noel McGinn, can design neat one-on-one behavioral experiments to test whether people actually behave the way they say they do."

Now, that was interesting! We decided to meet again.

It took just a few meetings to develop a research plan with Ernie Harburg and his colleagues from the Social Psychology Program. Almost magically everything lined up to facilitate our research. A new group of students was about to enroll in the University. All of them had to wait in long lines for a routine chest X rays in the University Health Service. The director of the Health Service was incredibly supportive. He let us recruit research subjects from the X ray line and gave us the use of two empty rooms in the Health Service. All students taking Psychology 101 could get credit for partaking in psychological testing and that certainly added to their cooperation. Dr. Hoobler also chipped in. He was interested in evaluating the usefulness of blood pressure self-monitoring at home. He purchased from his NIH grant a dozen blood pressure measurement devices with a stethoscope built into the inflatable cuff. His fears that students would not return the cuffs proved unfounded; in due time we obtained, without a hitch, home blood pressure readings in about one hundred students.

As soon as we got the keys, our Health Service headquarters became a beehive of activities. I was in charge of blood pressure measurements while Ernest and Noel McGinn got busy assembling instruments for personality measurements. Every evening we'd meet at the Health Service to discuss progress. A formal research proposal had to be developed for the approval by the Medical School Ethics Committee. Thereafter, we focused on log books, data forms, and on developing a system of cross-referencing various data sets obtained on the same individuals. Next, all data were cleaned and entered into coding sheets for computerized analyses. We pulled this all off in record time and within three months stood ready for field work. Students waiting for chest X rays in corridors of the Health Service proved remarkably cooperative. We pulled them off the line promising to return them back in time to complete the X ray, took their blood pressure, and entered the results into the log book. Eventually we collected blood pressure data on over 800 students.

Though we worked long hours, nobody complained. In retrospect, we were remarkably productive. In the short period of 14 months from the project's inception to my return to Zagreb, we collected data, analyzed them, submitted five papers, and all eventually got published. Two of these papers deserve comment, one for the lesson learned and the other for its interesting methodology and fascinating results.

For decades the University of Michigan Health Service routinely measured blood pressure and body weight of enrolling students. We located the records of medical, dentistry and engineering students from the 1939-1942 periods. Next, we got from the University of Michigan Alumni Association current addresses of former students residing within a 100 miles radius of Ann Arbor.[1] Eighty six percent of 240 contacted alumni responded and provided data about their current blood pressure levels and weight. To my disappointment, being overweight as a youth and staying overweight as an adult was a much stronger predictor of future hypertension than having a high blood pressure reading in college. In fact, students with the highest blood pressure readings were likely to have lower adult blood pressure levels. The converse was also true; those with the lowest college blood pressure levels had higher adult blood pressure values. This was a classic case of the *regression towards the mean* phenomenon; the likelihood that among outliers a repeated measurement will be closer to the average population value. If you want to know whether a student who scored high was lucky or is truly bright, give him another exam. By the same token, a dummy who failed the test might next time get a higher score by chance. Many golfers have had occasional good days but Tiger Woods is a true champion! The only way around the problem is to keep repeating measurements and average them out. I learned my lesson and from then on routinely took repeated measurements to characterize research subjects. Unfortunately, the regression to the mean is not universally appreciated. Not so long ago I visited a reputable research laboratory in Shanghai and was briefed about their work on biological effects of Tai Chi exercise. The exercises are designed to reestablish the balance between opposing yin and yang forces. The investigator informed me that his research strongly supports the usefulness of Tai Chi. A few weeks of exercise restored the balance in the body; plasma cortisol level increased in subjects who had low initial values and decreased among patients who originally had high readings! Unfortunately, his conclusions were based on one single reading at each time point. I suggested that more than one hormonal measurement would strengthen his conclusion. But when I started to explain the principle of regression towards the mean, the translator appeared overwhelmed. You do not need to know the language to figure out whether you are communicating with the other person. The longer I spoke the less focused were the eyes around me. It was time to move on!

In the other paper from that period we got fascinating results despite running a high risk of regression towards the mean.[2] To contrast personality characteristics, we recalled students whose systolic blood pressure in the line was either under 110 or over 140 mmHg. Eighty one of 100 contacted subjects responded and came in for testing with the Cattell's 16 personality factors questionnaire. Regardless whether their repeated blood pressure reading remained high or regressed, subjects with higher blood pressure proved consistently more submissive, sensitive, sociable and dependent than their low blood pressure counterparts. We could not confirm previous reports that persons with high blood pressure are excessively anxious. Admittedly, subjects with an occasional blood pressure elevation showed excessive anxiety, but those with sustained higher blood pressure (both the "in line" and during the personality testing) did not seem anxious. Similarly, subjects with occasional blood pressure elevation were more suspicious of others but this trait could not be confirmed among persons with sustained blood pressure elevation.

It was easy to construe how a submissive, outward oriented, not independent and suspicious individual might be constantly alert and therefore have higher blood pressure levels than his more relaxed counterparts. However, there are serious limitations to self-description personality questionnaires. To put it simply, the question is whether the responder knows himself well enough? Does his response describe himself or the person he aspires to be? Luckily, J. McLeod, one of the coauthors on the paper, was a behavioral scientist and in his doctoral thesis he developed a clever way to assess whether a person is submissive or dominant. First we asked the subjects to indicate on a seven point scale their attitudes towards six topics of interest such as capital punishment, juvenile delinquency, premarital sex, etc. Subjects were later matched into pairs with maximal disagreement on most of the six topics. About a month later, the pairs were recalled, seated with their partners, told they were matched by personality and would be able to work well together. The subjects were then handed copies of their own and the partner's opinions, given a fixed time for the debate, and urged to reach a compromise opinion, which they both might be able to support. After they became familiar with the procedure and before the discussion, each partner indicated on a scale how much they liked or disliked their prospective opponent. Three measurements of yielding were obtained by assessing changes in point scales: 1) Anticipatory yielding - how much they were willing to change before the debate. 2) Compromise yielding - the number of points actually moved as a result of the discussion. 3) Private yielding – how much the subject privately (unknown to the other partner) changed his opinion. Of 31 pairs with differing opinions, only 12 pairs had also high (>140) or low (<110) blood pressure. Regardless whether a comparison was of extreme of blood pressure readings or whether the entire group was divided by median blood pressure values the results were the same: People with higher blood pressure anticipated that they would not change their opinion, but actually lost the argument, altered their private opinion and, to boot, professed to like the person who proved them wrong. Now that was real submissiveness!

Eventually, the time had come to return to Yugoslavia. Any attempt to postpone the departure would be futile. In order to obtain the American exchange visitor visa, the Ford Foundation guaranteed that the Julius family would return to their country of origin after a two year stay in the United States. Two years and not a day later!

Deep down I was quite unhappy with my performance in the Ann Arbor University Hospital. I coauthored only one paper with Conway, a case report on the usefulness of methyldopa, the then new drug for the treatment of severe hypertension. Everybody kept telling me how important it was that the paper was published in the prestigious JAMA,[3] but this did not help very much. I was painfully aware of the score; I prepared four papers in my "secondary" research activity versus only one in my main job. Though I had ironed out most of the wrinkles in my relationship with Jim Conway, the time lost on unnecessary bickering had to have some consequences. So, I went with a heavy heart to say goodbye to him.

Without batting an eyelash, Conway dropped the bombshell:

"When you complete two years of purgatory in Yugoslavia come back, I will have a job waiting for you," he said.

I could not have expected a job offer in my wildest dreams. But James was a product of the British educational system; ready for give and take, objective, and not permitting emotions to affect his judgment. I was speechless. When I finally started to talk, sensing the danger of a long Slavic catharsis, Conway put a quick end to it.

"Of course, to have a career here, you will have to pass the foreign medical graduate exam which is quite tough. Do you think you can do that?"

"The biggest problem is not passing the exam but where the exam is given. One has to go to the American Consulate in Zagreb and this happens to be the best watched building in town. As soon as I enter the building everybody will suspect I am preparing to leave the country. But I will give it a try," I said.

Conway got us back on a normal business talk track.

"Good luck old boy, and let me know when you are ready." He then shook my hand and that was all there was to it.

The next day Dr. Hoobler, in his capacity as the Chief of the Hypertension Division, confirmed the invitation.

I was in seventh heaven. The next week while packing our belongings, I kept daydreaming and plotting scenarios for our return to the United States. It turned out that the road back to Ann Arbor would be even more tortuous then I envisioned.

**Bibliography**

1. Julius S, Harburg E, McGinn NF, Keyes J, Hoobler SW: Relation between casual blood pressure readings in youth and at age 40. J Chronic Dis 17:397-404, 1964.

2. Harburg E, Julius S, McGinn NF, McLeod J, Hoobler SW: Personality traits and behavioral patterns associated with systolic blood pressure levels in college males. J Chronic Dis 17:405-414, 1964.

3. Conway J, Zweifler AJ, Julius S: The place of methyldopa in the treatment of severe hypertension. JAMA 186:266-268, 1963.

# Round Trip to Zagreb
# 1962 – 1964

About three days after we returned to Zagreb, as soon as we unpacked and got over the jet lag, I made an appointment with Professor Arpad Hahn, my mentor and chief of medicine at the University hospital in Zagreb. Full of energy, Hahn was a short, rotund, cannon ball of a man. He had plans for everybody and everything in his Department. He'd arrive at the hospital at 6:30 AM having already completed the daily quick shave detour to the barbershop. First he would get the report from the head nurse. Next on the agenda was a session with the team of nightshift physicians. If they had admitted an interesting case or somebody who required immediate attention, Hahn would interrupt the report and rush off with the team to the bedside. He had a short fast stride and needed nearly two steps for what others handled with one hop. Because of his short step, Hahn's upper torso barely oscillated and it was quite a sight to watch the group moving through the corridors. The slumped bodies of taller, tired, nightshift physicians bobbed and weaved, and in the midst of them Hahn scooted like a huge mechanical toy. Occasionally and unpredictably Hahn would stop in his tracks and turn on his heels to take a quick peek at a ward's toilet, chemistry lab, or physicians' office room. He'd never say a word but if something was not in order, the person in charge was invited to the Chairman's office for a rather unpleasant conversation.

Once he reached the bedside, the small man grew into a giant. He loved to teach and was a fabulous diagnostician. First he'd start with the leading symptom such as pain, fever, vomiting, jaundice and the like. Next he posited differential diagnostic possibilities, that is to say he considered a number of diseases which could cause the symptom. Thereafter, he assembled clues to eliminate most of the initially considered diagnoses until he narrowed the field to one or two possibilities. Finally, he'd take a stance as to which was the most likely diagnosis and design a limited number of tests to confirm or refute his notion. However, from time to time Hahn would talk to the patient, complete a seemingly cursory physical exam and announce the diagnosis without comments. If you asked him to explain he'd shrug his shoulders and say that it "looked" like it or it "reminded" him of another case. His systematic mind processed and stored a tremendous amount of observations. The man had a fabulous memory. Once a patient on my ward stated he'd had a kidney stone on the right side.

"Yes," said Hahn "about fifteen years ago I took care of you in the Sisters of Mercy Hospital and you indeed had a stone but it was in the left kidney."

"Right side!" asserted the patient.

"Left," said Hahn in a calm matter of fact voice and left the room.

**"Rebro" University Hospital in Zagreb aerial view**

Two days later we got notes from the other hospital and sure enough, the stone was in the left kidney.

Together with facts, Hahn also remembered the elations of diagnostic successes and the disappointments of failures. That part of his experience, the emotional side, as well as the diagnostic "nose" he developed after long years of practice, he could not possibly transfer to others.

Hahn had set aside a whole hour to "debrief" me. He was in a good mood and attentive but didn't give me much of a chance to brag about what I had done and learned in America. Instead, he quickly steered the conversation towards medical and educational practices in the USA. Unfortunately, I learned next to nothing about the medical school curriculum and had only vague impressions about academic medicine. Hahn made it simple for me.

"So what is the biggest difference between the Department of Medicine in Zagreb and the one in Ann Arbor?"

"Well, they have an incredible number of full professors and a whole retinue of associate and assistant professors."

"Really! How many?"

"About twenty five full professors."

Hahn was incredulous, "Does that mean that the Department of Medicine has twenty five separate organizational units and twenty five different wards?"

He had organized the Department in Zagreb according to the Austrian and German models; one super-professor of medicine and a small number of full professors, each having his own Department and ward. At that time Hahn's department had sub-sections and wards of hematology, cardiology, and nephrology, each headed by a full professor. The gastroenterology ward directed by an associate professor was well on the way of becoming yet another sub-department. Hahn intended to also set up units of endocrinology and nuclear medicine. Having potentially six full professors was quite generous by European standards and Hahn could not in his wildest dreams visualize twenty five of them.

I instinctively felt that talking about the American model would not sit well with Hahn who had just completed a sweeping reorganization of his Department. As they say in Croatia, in the house of a hanged man you don't talk about the rope. But Hahn was curious by nature and would not give up. I had to answer his question.

"Yes there are 25 professors, give or take a few, in Ann Arbor but the department of Medicine has only thirteen divisions. This also includes divisions of infectious diseases and lung diseases," I said.

"Well, there are darn good reasons why the infectious and lung diseases are housed in two separate free standing hospitals in Zagreb," Hahn interrupted; "I sure would not want to have patients with tuberculosis, typhoid fever and an assortment of diarrheas in my department. Do you remember the pandemonium that broke out when somebody admitted a wrong patient to Professor Saltykow's room?"

I remembered it very well. Professor Saltykow, the famously cynical head of Pathology in the medical school was admitted to Hahn's department for management of diabetes. Out of respect for the old man, Hahn converted a physicians-on-duty room into Saltykow's private hospital room. One evening the senior night duty physician knocked on Saltykow's door and apologetically wondered whether the professor would mind if the night service were to admit a patient to the free bed in his room. The patient came from far away, was not mortally sick but had nowhere else to go and needed a thorough diagnostic work up. Next morning, as soon as another patient had been discharged, the intruder in to Saltykow's room would be moved to another ward. Saltykow agreed.

During the morning report, Hahn was told that the offending patient had already been moved to another room. Nevertheless, Hahn dashed off with his retinue to offer a personal apology to Saltykow. Old Saltykow met the group with the widest possible smile, a sure sign of troubles to come.

"Professor Hahn," he said, "you are in charge of the department of Medicine, aren't you?" Not quite knowing how to respond, Hahn said something to the effect that he tried his best to run a good department.

"Sure, but I just wondered whether you are running a leprosarium rather than a department of Medicine?" The patient admitted to Saltykow's room had leprosy!

Nobody would ever challenge Saltykow's diagnosis and right on the spot Hahn started issuing directives to isolate the patient, transfer him to the infectious diseases hospital and implement a program of disinfection. Leprosy is a communicable disease but one can get infected only after a prolonged and close contact with the patient. Hahn explained that disinfection measures were needed to reassure patients and the personnel rather than to prevent an epidemic. Hahn then spoke of the irrational fear of leprosy which went back to middle-ages when a low level of general hygiene facilitated epidemics. To prevent the spread of the disease, disfigured patients with signs of leprosy were forever isolated from society. Hahn then turned to Saltykow to explain that "for consistency's sake" his room would also have to be disinfected.

"Da, da," said Saltykow reverting for a second to his native Russian. "I much enjoyed your lecture and I understand the measures you instituted. Let me add that another reason for the fear of leprosy is the extremely long incubation period. For years after a contact with lepers, people could not be quite sure they were out of the woods. So, I too have good reasons to be afraid and would welcome partaking in all aspects of disinfection. Just think about it; I am eighty eight years old, have diabetes and in about 20 years I might develop leprosy!" The old cynic got in his last dig at Hahn's overreaction.

I was still smiling when Hahn interrupted the pause. “Joking aside, how do they deal with infectious diseases without having special wards?”

“They give the patient a single room, stick an infectious disease precautions sign on the door, don masks and gloves, wash hands and change white coats on the way in and out of the room, and that is about it.”

“I guess they have less infectious disease than we and under their circumstance that might suffice,” responded Hahn. “Rates of infectious diseases and tuberculosis are declining and one day we might find ourselves in a similar situation. I hope this happens on somebody else’s watch. I have my hands full as it is.”

The longer we spoke the more it became clear that Hahn wished to know what might be coming down the track but was not about to make any changes.

“Okay, he said, “I am a good enough mathematician to figure that thirteen Divisions of medicine and twenty five full professors means that some Divisions have a bunch of senior professors. One of them must be the boss or how else would that work?” I explained that each division had a chief of service who took care of administrative responsibilities and that other full professors were quite happy to do their research and not worry about finances, schedules, space distribution, and promotions of the faculty. To my best knowledge the system worked quite well. Most full professors had research grants which gave them considerable independence. They could hire their own personnel and, prior to forwarding a grant application to potential sponsors, the University provided a written guarantee that the applicant would have sufficient free time and laboratory space to complete the research. In other words, if his grant provided 50% of the salary then his clinical commitment was decreased by fifty percent.

“Yes, I know about their system of grants,” said Hahn, “This is a good arrangement for them but it would not work here. No way! Can you imagine some government office distributing moneys and determining who will have a lesser patient load in my department? You know how things are here; the government distributes the money up to the last penny according to its sense of social justice. If two Croats got grants the next one would have to go to a Serb, women may get grants because men had them and so on. If you add to this the inevitable political, religious and VIP meddling you can easily predict a broken system before it was ever implemented.” The VIP abbreviation Hahn used stood for "veza i poznanstvo" - the Croat acronym for "connections and acquaintances."

Hahn was getting quite excited. "And how would I prepare the budget if I do not know when and from where the money is coming?"

I should have simply responded with an "I don't know," and cut the discussion short. Stupidly I continued to explain details and in doing so, appeared to defend the American model. I spoke about the "overhead" portion of grants expecting Hahn would be pleased to know that his department might get a share of the grant money. He actually looked somewhat interested but then I chose the wrong words. "You see, the more professors you have, the more money comes to the department," I said.

Hahn was livid. "Sure, and each professor would remind me every day how much he is contributing to the department. They will all want to have a say in how the departmental funds are spent and, instead of being the decision maker, I'd become a consensus seeker. No, thank you! This would never work here. We are a hierarchical society, and without a clear direction full professors would be running around like street dogs, hungry, angry, barking, one at another. And day in and out I'd be busy resolving conflicts, imaginary or true ones. I might even be able to manage that but the invisible part, those countless behind-the-back plots, the shifting alliances, ambushes, and character assassinations would undermine the department. I intend to keep things as they are. There will be a strong Chairman and a few strong divisions with their own leaders. We are not ready for the American system my dear Julius."

The room fell silent. There was nothing I could do to break the tension. He had asked me about the American system, I explained how it works and Hahn erroneously concluded I was advocating for change. This happened more than four decades ago and only now, when I look at the present situation in Iraq, do I realize how smart Professor Hahn was. You cannot introduce change overnight; it takes time for a system of cooperation to develop.

Finally, Hahn started talking again. "Since you already passed the specialty board in internal medicine, you will be appointed as senior instructor. Your direct supervisor will be Dr. Mohacek in cardiology. As a first step towards a future academic career you must complete a dissertation for the degree of doctor of sciences in medicine. You will have the usual outpatient duties both in the cardiology clinic and also, when scheduled, in the night shift admitting clinic for general medicine. Eventually you should have your own clinic focusing on hypertension but I'd prefer that you work this out with your colleagues. If this comes to me as a proposal from Mohacek, I will approve it."

Hahn did not miss the opportunity to teach me how his hierarchical system works. We then discussed my doctoral dissertation. I explained that I'd like to test the relationship between personality factors and blood pressure levels to see whether I could reproduce in Zagreb the findings from our research in Ann Arbor. Hahn liked the idea. Finally he stood up to shake hands.

"Just a moment." He reached into one of the drawers and grabbed a handful of small banknotes. "Here, go and get yourself a decent haircut!"

Now that was more like it. The old joking, benevolent mentor and family friend!

"Too long for your taste?"

"Too disorderly," he said while pushing me in a friendly manner out of his office.

Quite happy that the difficult interview had ended on a positive note, I decided to stay out of Hahn's way as much as was possible. He gave me clear directions and it was up to me to implement them. But the stay in Ann Arbor had changed me more than I realized and I found it rather difficult to return to the old ways. Whereas Hahn remained a benevolent mentor, I became a bit of nuisance to him. I went to his office a few times to propose small improvements and each time I ran into a wall. Hahn sure knew how to put me into my place. I vividly remember the final incident which convinced me to forget innovations and try, the best I could, to fit into the old system. Most of the hospital physicians including myself had sloppy handwriting and reading other people's notes was nearly impossible. Our job was made easier if a patient had a referring letter or a discharge letter from a previous hospitalization. In return, we strived to do our part and dictate comprehensive discharge letters. Unfortunately, this was not a simple thing. We had no tape recorders and every physician had to make an appointment with one of the department's secretaries. As with everything else in Hahn's department, there was a strict pecking order for dictation: Hahn and other full professors dictated to the chief departmental administrator, the associate and assistant professors had the ear of her deputy, and all others had to vie for the time of the two remaining secretaries. These dictations were not scheduled in advance probably for good reasons; the secretaries had no idea what tomorrow would look like. The secretaries had a myriad of responsibilities.

Besides direct-typing of dictation, they handled telephones, acted as receptionists, bookkeepers, purchase order managers and organizers of all departmental schedules. However, they coveted the dictation sessions as the most interesting parts of their daily routine. On the day of dictation you'd call by phone or pop your nose into the department office to ask who could take dictation. The freedom to schedule their own time gave departmental secretaries a platform to curry favors or show dislike. Personally, I made my peace with the situation and, as they say in Dalmatia, I quickly learned "from whence the wind blows." At that time I was quite impressed with myself but I never walked into the secretarial office with my nose in the air. Eventually I established a good relationship with the entire typing pool and never had to wait too long. Nevertheless, I thought it appropriate to propose a solution to the problem and was quite optimistic that Professor Hahn would like the proposal. How stupid of me!! First, I should have remembered that Hahn rejected my ideas on three previous occasions. Second, I failed to understand that in order to act on a problem people must first be convinced that there is a problem.

Anyhow, I was excited by the important information I had just acquired and asked for an appointment with the Chief.

"Is this an emergency?" asked the chief departmental secretary.

"Not really."

"So why are you insisting to see him today?"

"I got some good news and I am excited."

"Aha, I am to tell Professor Hahn that Doctor Julius wants an appointment because he is excited," said the old fox. Ivanka, a slim gray haired woman in her early sixties, had seen it all before. "Come on," she smirked, "you have to give me some details. I must give him some idea what the appointment is about."

"I have learned that a store in town is about to receive a shipment of brand new typewriters!"

"What a weird thing to get excited about. I've been working all my life with typewriters and can't say that they ever excited me. I thought you had a baby coming. Fine, I will tell the boss you got excited about typewriters and want to see him right away. Let him figure it out. You can come at 5 PM and I will set 30 minutes for the meeting."

Leaving the room, I caught sight of her. Ivanka was mumbling "typewriters, typewriters" and shaking her head in disbelief.

"So what is up?" asked Professor Hahn. I had time to prepare for the meeting and gave a little speech about unreadable hospital charts and how this could be resolved if the physicians were to type notes and paste them into charts. Physicians could also type drafts of discharge letters and hand them to secretaries to type clean copies. The flexibility to type at their own schedule would eliminate waiting time and increase physicians' and secretaries' productivity. I indicated I had discussed this with many colleagues and they liked the idea. I closed my presentation with reference to my valuable information that a store in the city will soon get a large shipment of new Olivetti typewriters.

Back in 1962 the Yugoslav international commerce was changing hands from large government-owned export/import companies to a growing number of local enterprises. Under the old system big companies imported inadequate quantities of what they considered to be essential merchandise. Consequently, in economic terms, the demand was higher than the supply, everything was hard to get and the consumer bought whatever was available. The new system was designed by a group of communist reformers aimed at decentralizing the economy and introducing a modicum of competition between companies and regions. Ten years later this strategy paid off and the standard of living in Yugoslavia was better than in any other communist country. However, in 1962 amid the economic transition, nothing seemed to work. Instead of the previous universally meager supplies of everything, the stores in Zagreb periodically had either totally empty shelves or were awash with merchandise. You never knew what you would find in a store. Under such circumstances my discovery of the incoming load of typewriters was quite exciting. And to boot, the price was very good!

However, Professor Hahn was not impressed. He perfunctorily thanked me for the information and immediately started, layer by layer, to chisel away my argument.

"Stevica," he said using in a fatherly fashion the diminutive of my name, "two years in America and you have forgotten how things work here! Yes, your colleagues tell you they will be happy to type drafts but if I told them to do so, they'd rebel. The very next day somebody would let me know he didn't go to medical school to become a mere scribe. This is not America where an unemployed doctor might drive a taxi. The town is full of unemployed professionals twiddling their thumbs in coffee houses, complaining about bad luck, but none of them would in their wildest dreams accept a 'lower' position no matter how lucrative it might be. No comrade Julius, in this classless society we are more class conscious then we ever were!"

Hahn used the official party jargon to poke fun at the regime's egalitarian nonsense. As he continued the monologue, Hahn became quite animated. "I could probably take care of doctors but that would be only half of the battle. I'd have to deal also with the consequences of youthful yearnings by two aging Slovenians."

I started to laugh.

"Glad to see you got the idea, said the boss. He didn't speak obliquely out of a fear of listening devices. Saying things that only a few people comprehended was a favorite elitist sport, something done for the sheer fun of it. And understanding the convoluted reference to the communist regime was a proof of a listeners' intellectual maturity. Hahn was referring to Edvard Kardelj and Boris Kidric, two old time Slovenian communists and Tito's favorite theoreticians. Aided by a few Yugoslav academicians they drew a dividing line through Karl Marx's philosophical opus. The young Marx was good and the old one went a bit too far. Apparently before he wrote the "Communist Manifesto" which turned him from a philosopher into a politician and professional revolutionary, young Karl Marx paid attention to details of his proposed "classless society." According to him, a powerful centralized government was needed to steer the society towards communism. In a mature communist society every aspect of social injustice would be eliminated, the need for a centralized state would diminish, and eventually the state would wither away! Marx was too smart not to see how nebulous his concept of the withering state was. Based in his dialectical philosophy, Marx preached that social progress can only occur through a revolutionary "clash of opposite forces" and a "quantum leap" into a new social structure. How then could a communist society continue to evolve without a conflict? So Marx started to conceptualize what sort of balancing act would be needed to eventually do away with privileges of the state. Remembering young Marx's writings, Kardelj and company implemented "workers self-governance councils" in every enterprise, including the University Hospital in Zagreb. These elected workers' councils functioned independently of and parallel to the management structure of an enterprise. Managers were responsible for the day to day operation but every major policy or investment decisions had to be presented to the council. In the minds of Kardelj's group of social engineers, these councils would eventually counterbalance the excessive power of the state in Yugoslavia.

Hahn was a strong leader and a very smart man. He knew how to intimidate others and, workers councils notwithstanding, there was little doubt who was the boss in the Department of Medicine. But even the mighty Hahn had to abide by the new rules, and he was forced to regularly meet with representatives of nurses, janitors, office workers and hospital physicians. Needless to say, Hahn profoundly disliked this intrusion into his prerogatives.

"Stevo, you are asking for a change in the working routine of my office staff and I would have to discuss this with the workers council." Hahn said.

"You really would have to? This is such a small issue."

"Of course, but this is exactly the kind of small bone I'd permit them to chew on." Knowing he might have said more than he should Hahn stopped for a moment but soon continued to talk in a friendly, very deliberate, tone. "I am trying to make you understand the overall situation here. You are reform minded but you are wasting energy on things that are not your business," Hahn paused and finished the sentence in a slow - word by word - upward cadence "and such matters are outside the area of your expertise. Stick with what you know! I understand you had a fracas with Professor Radosevic and in such situations you are more likely to get my full support."

Hahn sure had a good intelligence service! I thought only Radosevic and I knew about our short quarrel and Radosevic had no reason to report it to Hahn. Apparently the walls of the Department of Medicine had ears! I started to explain my unpleasant exchange with the brilliant, cynical and very rigid chief of the nephrology service but Hahn stopped me in the middle of the sentence.

"We will come to that later. Let's finish what we started. I thought that from the conflict around your father's death and as a physician in Bosnia you had learned enough about our environment. Your proposal shows a remarkable lack of sensitivity. As I already said, some physicians will complain that typing notes is a menial job that 'high intellectuals' of their caliber should not engage in it." Hahn accompanied "high intellectuals" with the finger sign for a quotation mark. "But you totally neglected the other side of the class divide in our 'classless' society. Our secretaries complain about dictation but they love it. This is their chance to interact with physicians on more or less equal terms. They never protest against the activity per se. Sometimes they might whine about work overload but most frequently they complain about individuals. This or that doctor misses his dictation appointments, is rude, talks too fast and, believe it or not, picks his nose, has bad breath or emanates body odor. It all trickles to me through my secretary. I listen and rarely intervene because I know how much they like those dictation sessions. They'd never risk losing the privilege! And they do not expect me to act. They wish me to know what they think about certain doctors and that is all there is to it. If they complain about too much work, I thank them the next day for the extraordinary effort. It never fails."

The wise old devil!

Hahn continued talking. "So, if I were to tell them that as of tomorrow there would be no more dictation, all hell would break loose. So forget about that. But there may be some merit to having typewriters in physicians' ward offices. I will pursue this with Ivanka. Such things must be first cleared by the senior departmental secretary."

I arose to leave the room but Hahn was not yet done. "Let's now talk about your spat with Radosevic. You see, this was a clear medical issue and you had every right to bring it up. Did he really dismiss you out of hand?"

"It was worse than that," I responded. "In Ann Arbor we found *fibromuscular dysplasia* of renal arteries in many young women with hypertension. In eighty percent of them a surgical intervention brought the blood pressure back to normal level. I suggested to Radosevic that we start performing renal *arteriography* in hypertensive women suspected of having abnormal renal arteries. And you would not believe what he said!"

Hahn looked askance.

"He said 'Forget these American zebras. We do not have fibromascular dysplasia here.' If he dismissed me on some technical ground, I'd probably understand. He could have said that our surgeons do not know how to perform renal *revascularization* and that it makes no sense to look for something we can't fix. Or he could have said that our X ray equipment is not advanced enough to produce good images. But to assume that Croat women are somehow different from the rest of the world was too much for me. How could he know if he never looked for it?"

"So, you told him what you thought about it?"

"No, he turned his back and that was all that was to it."

"Look, even the best people can be wrong," said Hahn. "I am glad you didn't argue. Instead let's teach him a lesson. Go and talk to Zhuzha Saric. I have already spoken to her. She routinely passes cardiac catheters and doesn't think it would be difficult to place the catheter into renal arteries. Once you diagnose the first case, I will send the patient to

Switzerland for surgery and will talk with our chief of Surgery. I am sure when the ministry of health sees the first bill for the surgery they will find money to send a young surgeon abroad to learn the technique."

Hahn was amazing. To this day I do not know how he got enough information to talk with Dr. Saric before he spoke to me. Eventually everything evolved exactly as he predicted and we introduced renal angiography into our hospital.

I enthusiastically thanked Professor Hahn for giving me the opportunity.

"Stay just a moment longer," he said, "I want now to talk to you as friend of the family and not as a boss. You used to be easy going and cheerful but recently you became impatient and confrontational. Such a change in attitude is a sure sign of a person's internal problems. Something is bothering you and you must work that out. Yes, you have seen a better world and a better way of practicing medicine but you are returning to the old world. It makes no sense to be in constant conflict. You cannot possibly change how things are done here. Certainly not overnight. So, the question is whether you can adjust and start acting as you used to. Think about it, if I had magic wand and fixed everything you asked me to, would you feel better? If not, why not? It is up to you to understand yourself. In Freudian terms, are you 'projecting' your inner dissatisfaction like a soccer player who stares at his shoe after he miss-kicked a ball? And what are you dissatisfied with?"

"I see you fidgeting in the chair. Relax, I am not expecting an answer," Hahn continued. "If you decide you can't take it anymore you should leave. I will not be in your way. In fact I will help you but, please, take your time. I don't want you to do anything in haste. Here you have a very good career in front of you and over there – who knows?"

"Maybe there is a middle road to all of this," I said and proceeded to tell about Jim Conway's invitation to return to Ann Arbor. I explained that I intend to return home after a few years of additional education in America. Hahn would have none of it.

"Congratulations," he said, "I was concerned you'd end up practicing medicine in some small corner of America. That would not be fair to this country. Remember you had free education here. America is inundated with doctors and our state paid for your education to have doctors here not over there. But doing research and teaching in a leading American academic center is a very different proposition. We have very little to offer in that regard."

"But I intend to return to Zagreb." I said.

"No, you will not. And probably you shouldn't. You have caused me enough trouble as it is. I hate to think what you'd look like after a few more years of America. As I said, if you decide to leave I will help you in getting out of here. In the meantime work hard and try not to be a nuisance to everybody."

When I sought the appointment with Hahn, his secretary, Ivanka, showed no interest in typewriters. So a few days later, when Hahn told me she vetoed the idea, I was not surprised. Doctors are not expert typists, she said, and they would soon break the sensitive machines. She would then have to remove the old ones from the inventory, order new typewriters, enter them into the inventory only to hear they broke again. She just had no time to deal with the extra work! Hahn actually laughed when he relayed her words. Another lesson about human behavior and the departmental hierarchy for young doctor Julius!

I finally got the message and forced myself to forget about reforms. This proved much easier than I thought. After a while I was utterly cured. Not only did I stop asking for reforms, I failed to see the need for them. Some ethicist might deplore this sort of behavior but that is what most of us were trained to do in communist Yugoslavia. You could criticize, you could crack jokes about the regime but if you tried to change things, you got your fingers slapped. Or worse! As they say, only a fool hits the same wall twice - smart people find a way around it. And once you find the way you forget that the wall was ever there.

About once a week I was assigned to the night shift in the university hospital's general internal medicine clinic. Working late hours was not a problem for me. I am a night owl and never go to bed before midnight. On regular days the work load in the clinic was quite modest. A discharged patient might show up because he did not feel well, or a person that missed seeing his day time primary physician would walk up to the hospital and try to beat the system. But things were very different on admission days.

The city of Zagreb is nestled between the Zagreb Mountains to the north and the flood prone river Sava to the south. These physical barriers permitted growth only in an east–west direction and Zagreb evolved as a thin 12 miles long snake meandering under the mountains. The four city general hospitals were situated on foothills of the Zagreb Mountains to the north of the downtown. This reflected the medical ideas of the 19th century when infectious diseases and tuberculosis were a huge problem; the patients were generally kept in isolation and the fresh air was considered to have great healing powers. As the crow flies, the hospitals were relatively close but getting from one to another required passing through the congested downtown traffic. For a while the hospitals served well as regional centers to surrounding townships, but as Zagreb grew its hospitals became recognized as premier health care centers, and things started to change. The city proper had only about 300,000 inhabitants, but in the early 1960's Zagreb drew patients from a huge part of Croatia and Northern Bosnia. It was estimated that besides the city the hospitals in Zagreb served an additional region of at least one million people. Not belonging to Zagreb's regional health care system, these outsiders could go to whichever city hospital they chose. Some visitors were health care tourists, intent on getting a second expert opinion and visiting the city on the side. However, many were desperately ill and needed urgent attention. Most of the sick ones usually came by railway or bus and once they arrived had no idea what to do next. Not keen on having displays of human misery in their territory, the station masters would summon ambulances to take the miserable passengers on a hunt for help. To put some order in all of this and stop ambulances from crisscrossing the town, the four general hospitals in Zagreb arranged a rotation schedule; on any given day only one hospital would receive emergency cases. In principle this was a good solution but the chronic lack of free hospital beds created havoc. On the "receiving day," my hospital routinely offered only four beds for admissions. Four beds for a region of 1.3 million people!

On admission days our resident physicians handled the initial triage. Whenever it was possible they wrote emergency prescriptions, penned a short description of physical and laboratory findings and sent the patients home. But by around 10 PM we'd end up with ten to fifteen patients needing urgent hospitalization. And that was when my job started. As the second ranking physician in charge, I'd phone specialty hospitals in the town and try to "sell" them some of the patients. Zagreb had free standing hospitals for tuberculosis, infectious diseases, ear and nose diseases, dermatology, as well as obstetrics and gynecology. I'd call up physicians on duty to convince them that a patient had a problem in their area of expertise and should be admitted to their hospital. I became quite good at these negotiations and by midnight I'd send about one third of the patients to other hospitals. I also developed the "Oh, do I have case for doctor so and so!" technique to place a few internal medicine patients into a non receiving general hospital. This was a much more sophisticated job. First, you had to know the area of interest of some senior physicians in other hospitals. Next, you'd try to persuade the physicians on duty that doctor so and so would be pleased to add this patient to his collection of interesting cases. It rarely worked, but when it did, I got a round of applause from all my coworkers.

Around 11 PM, after the last train from Bosnia delivered the final batch of patients, the senior physician had to decide whom to admit. At that point the real life threatening emergencies had already been admitted. The remaining patients were not likely to die within hours but all of them were sick enough to require hospitalization. Yet only one or two could be admitted. Of necessity, these decisions were arbitrary but our senior physicians did the best they could do under the circumstances. A slight preference was given to people coming from afar and others were treated with sympathy and humanity. Patients living in the town were sent home with necessary medications and given appointments to be seen again during regular working hours. Most of them were eventually admitted.

There were two ways to cope with this pathetic lack of resources, become cynical or strive to be fair and keep a good sense of humor. Only a few of our senior doctors developed enough cynical insensitivity to isolate themselves from reality. Most were fair, did the best they could and helped the team with an upbeat attitude. One of the best was Dr. Vanja Mikulicic, a typical "upper town" Zagrebian. The term "upper town" describes Zagreb's initial geographic and social division into the aristocratic northern part and the southern bourgeoisie part. Old families from the foothills of the Zagreb Mountains used to be wealthy but in the 1960's the money had definitely rolled down from the upper town to the merchants and craftsmen of the city. However, the change in wealth had not altered the behavior of the two groups. The lower town people remained jolly, loud and agile whereas the upper town folks, many of whom had moved to downtown, remained reserved, well dressed and infallibly polite.

Dr. Mikulicic, in his early forties, cut an upright, slim and well proportioned figure. He had a great sense of humor and had a quick wit. When he let his magic work, people around him were short of breath from laughter, but Mikulicic remained unengaged and kept watching others with a benevolent smile. Neither the knee slapping type nor aloof, Mikulicic kept his composure not because he wanted to control himself but rather because that was the way he was. A consistently mild, undemonstrative and very fine man! And a good sportsman to boot!

When he played tennis he obviously enjoyed the game but left the screaming, grunting and swearing to others.

One late evening Mikulicic decided to admit an emaciated and profoundly dehydrated gypsy from Bosnia. The diagnosis was not quite clear but there was no doubt that the man was extremely weak. Rehydration and further laboratory work up were indicated.

"I guess he has nowhere to go," Mikulicic said. "Even if I were to give him some money I doubt Hotel Esplanada would take him. He is incredibly dirty. I mean dirt dirty! But our hotel will admit him. That is the way it is."

A typical Mikulicic insider joke - Esplanada was the best hotel in Zagreb and we were the best hospital in the town.

"Why don't you start the I.V. line here, hang a bottle of saline on the gurney and send him to the fourth ward. It will take us about an hour to get out of here. I am sure by then Sister L will clean him up a bit and we will be able to do a better physical exam."

"Sounds good, but what about lice?" I wondered. "Hahn will kill us if we mess up the ward."

"I told you we are better than Esplanada. They would not admit him at all but we will give him the best room in the house," responded Mikulicic. "I happen to know that the free bed in the fourth ward is in a two bed room. We will have to move the other patient to a larger room and our comrade gypsy will have an isolation room for himself. I will have to explain to the other patient that we are admitting somebody with a communicable condition. Luckily the other patient is a friend of mine and he will understand."

"You will kick out a buddy of yours for the sake of an old gypsy?"

"Sure, we will do what it takes. This guy is special, not only will he have a single bedroom but our cutest sister will take care of him," Mikulicic laughed.

In the past most of the nursing was done by nuns, but the communist regime decided to remove them from hospitals. Nevertheless, tradition prevailed and we used the term "sister" for all our nurses. Sister L was the best-looking but not necessarily the smartest nurse we had.

It took us about two hours before we found some time to examine the newly admitted patient on the fourth ward. When we came there the other patient had already been moved to the large room but the old gypsy was still on the gurney in the corridor. Dirty and untouched by human hands!

Just then Sister L came down the corridor. "Dr. Mikulicic," she said and flipped her eyelids to indicate innocent naiveté, "I know you will be mad with me but I just could not get myself to touch this dirty old gypsy."

"But he is a special case," said Mikulicic in his calm manner.

"What kind of a special case?"

"This is a very special and somewhat secret case. You see, Mr. Beganovic Mujo is in fact an Indian Maharaja traveling incognito, possibly on some important mission. He deserves special attention. Now that his cover is blown the Indian Embassy might call on us to check on him. This is not the kind of service foreign diplomats would expect to see."

L should have been tipped off by the patients' moniker. Mujo Beganovic was typical Bosnian Moslem name, but L fell for the joke in a big way. She gave the patient a thorough bath, combed his hair, dressed him in a brand new hospital gown and placed him in the nice room.

Mikulicic was quick! He came up with a doubly brilliant practical joke. Our patient was dark skinned, tended to mumble in his native Roma language and could be mistaken for an Indian. And Sister L was known to all of us as a dreamer. She knew she was pretty and felt she deserved a better life than the one that had been handed to her. She spoke of Hollywood and generally made no secret of her wish to leave the country. In her current dream the Maharaja was her ticket to a better world!

"Poor girl," Vanja said, "I wanted to crack a joke to defuse the unpleasant situation. Whatever you may think about L one thing is beyond doubt; she is a hard worker and a dedicated nurse. I did not want to bitch about her understandable lack of enthusiasm for a dirty job. I thought she'd get the joke after which I would say that he might not be a maharaja but ought to be treated as one. You know what I mean, the equality and humanity sham that we were taught in medical school and we passed on to innocent nurses in their school."

"The wolf ate the donkey." Vanja used the favorite Croat adage for an irreversible situation. "This being the case let's see how far it will get us. Let's go!" And with these words Vanja started to walk away.

"Where are we going?" I asked.

"To pay a visit to the concierge at the entry to the hospital. He has a city telephone line. From his little house we can call L through the hospital operator. It will look better."

On the way to the concierge, we passed by the closed door of the telephone operator.

"Wait a moment, the chess player is the duty operator," Vanja said, "and he is too smart to be fooled. I will have to level with him."

The telephone operator was a brilliant chess player with an incredible memory for voices and an uncanny ability for multi-tasking. If they had some free time, many night shift physicians would call him up to set a chess match. To my best knowledge nobody ever beat him, but all would clamor for yet another opportunity to play. It was a sheer pleasure to see him in action. He was well informed and while playing a brisk match he could chit-chat about many topics, comment on the newest gossip, and answer calls with ease, plug telephone lines into the right sockets, find the called person, and monitor the unanswered calls. The old fashioned operator's station resembled an upright piano with about six incoming lines and nearly one hundred plugs to various local telephones. Clearly he was smarter than all of us but his priority was to play chess and have a good job. He never pursued higher education nor did he try to move up the social scale. Being a telephone operator was just the right thing for him.

When we got to the gatekeeper's house, Mikulicic called up the operator, asked for the connection with L, and told the operator not to tell who was calling. "You can listen in if you so desire but you have heard nothing. Got that?" said Mikulicic. When he got L, Mikulicic pretended to be a member of the Indian consulate inquiring about the situation of "our comrade Maharaja Beganovic Mujo."

"If the typical Moslem name didn't do it, the title of 'comrade' should have alerted her that something was not right," explained Mikulicic. "But to no avail. She kept telling me that the Maharaja is in good hands, that she is personally responsible for his comfort and that he is as comfortable as he would be at home."

"Yes, he feels as good as at home-under his tent!" Mikulicic moved his head in disbelief. "We've done enough mischief for a night. Let's go to sleep."

The next morning during her early report the head nurse informed Professor Hahn that the Maharajah on the fourth ward was given a single bed room, and that according to her information he is doing reasonably well. Hahn ejected himself from the chair and headed for the door. "What? They admitted a dignitary to my clinic without telling me a word! I can't believe it," he screamed and rushed off towards the ward. It took him only one look to understand the situation. When we arrived to give him our report, Hahn right away launched an attack on Mikulicic.

"This was a very poor practical joke. You embarrassed my head nurse who bought the whole story and I embarrassed myself by overreacting. But worst of all is what you have done to Sister L. Poor soul, she is so excited and so pleased with her good deeds that I didn't have the heart to tell her the truth. You will have to fix that one, Mikulicic. And I hope in the future you will control your misguided sense of humor."

Instead of apologizing Mikulicic looked straight into Hahn's eyes and began to speak in a slow, calm and deliberate fashion. Sir," he said without raising his voice, "I cannot promise that. If we could not pass things off with an occasional joke we'd all die from stress. The situation with after-hour admissions is unbearable. Give us more beds and I guarantee we will behave with more decorum. I am not a comedian by nature, but I do what I have to keep my own and other people's sanity."

Hahn didn't tolerate even the slightest opposition and I expected him to explode. To my surprise he almost nonchalantly said: "Okay, but next time choose somebody who can take it." He never mentioned the incident again.

L had a hard time giving up her dream. I went to the ward to explain to her that it was a joke gone wrong but she first argued with me. Why would people from the Embassy call if he were not the real thing? After all it was a long distance and not an internal hospital telephone call. When I explained how we called from the gate keeper's office, she burst into bitter tears. Soon Dr. Mikulicic came to apologize. By and by we succeeded to calm her down. This was a miserable experience for all of us. People seem to have understood the situation. A good joke tended to have its own life, the story was embellished and new versions were circulated for years to come. This one died within a week.

There is a happy ending to the story. Eighteen years ago the results of our Tecumseh study attracted national attention. It started with a front page report by the venerable New York Times and soon my name appeared in many national newspapers. Telephone calls requesting interviews kept coming at a brisk pace. One day my secretary forwarded me a call from Texas with a terse comment. "One of those people with a heavy accent and I could not quite get her name," she said. It was L! She wanted to congratulate me and held no grudges. She had fulfilled her dream and got out of the country. It was not exactly Hollywood but she did find an American prince charming, was happily married and had a good life. We had a nice chat. I occasionally get congratulations and they are always appreciated. But this one was special.

As Mikulicic said; we needed humor to survive. The night shift on admission day was the worst but there was plenty of tension all around. I worked once a week in the cardiology clinic, and in addition to scarce free beds there were rotating shortages of just about everything. If the X ray films were available, the injectable contrast material to visualize internal organs was not. Some antibiotics would be available but not others. The one you used might disappear the next day and another would resurface. Some laboratory tests could not be done for lack of reagents and so it went. Life in general was not easy. Admittedly the communist regime became less stringent but was still far from benevolent.

Humor helped to cope with all of this and an uninformed visitor might have concluded that we were a carefree bunch. Jokes were flying left and right, laughter abounded and a sad face was rarely seen. The physician's lunch room, a smallish place with a nice view was our "comedy central." Between noon and 2 PM, physicians from all corners of the multi-specialty hospital would come to consume the free food, relax, gossip, and exchange jokes. In this small oasis we were free to poke fun at everything and everybody. Or so we thought.

One of the most fervent disseminator of jokes was a military physician. Let's call him Dr. X. Dr. X was a colonel, a former partizan, and a radiologist on "loan" from the military hospital. With his background, he must have been a party member but that did not stop him from unleashing a daily barrage of critiques of the regime and sharp jokes about individual communist functionaries. Some of his jokes were more bitter than funny. I recall the one about the prominent workers union leader, turned Croatian minister of social affairs. He and his wife were invited to the British Consulate reception. The Consul told the wife how much he enjoyed the recent concert of the Symphonic Orchestra of Zagreb and in particular the performance of the Beethoven's Fifth Symphony.

. "Do you like Beethoven?" he asked.

"Oh yes. He is a very nice guy. I frequently meet him in the street car number six on the way to the Republic Square."

Her response elicited embarrassed sighs and dead silence. Everybody became aware of her blunder. Obviously angry, her husband pulled her to the side.

"You stupid cow," he says, "don't you know that number six doesn't go to the Republic Square!"

Dr. X told many similarly hilarious stories. He became a lunch room favorite and everybody wished to sit at his table. However, one day he turned uncharacteristically quiet. A few days later he announced he would not be telling jokes anymore. I asked why he changed his mind.

"You would do the same if you were in my shoes," he said laconically.

After a pause, Dr. X eventually told us the story of his conversion to a "loyal citizen." The previous Saturday afternoon he had received an unusual telephone call from an intelligence officer. He said he wanted to speak with Dr. X but something else came up and he was wondering whether it would be okay if he were to call later. One does not joke with military intelligence and Dr. X told the officer not to hesitate to call again when he has more time. "Had I known what was coming, I would have acted differently but I gave him permission to call and there was no way out," Dr. X explained and continued to tell us the story. He spent a good deal of the afternoon guessing why the military intelligence wanted to talk with him. And why on the weekend? The longer he thought about it the more curious he became. The call came exactly at midnight. By then Dr. X was irritated and told the caller he'd prefer to go to bed.

"Too bad and I apologize," said the officer in an icy manner, "but we both are Army people and sometimes we have to cope with difficulties. May I remind you that I too am awake?"

"This is a telephone call. How do I know who you are? You might be a prankster," Dr. X protested. The stranger explained that in front of him he had Dr. X's personnel folder and then mentioned some details only a person in the possession of the dossier would know. The caller was particularly intrigued with Dr. X's high school education in the town of Knin in the province of Lika.

"You want to know what I did as a high school boy? You must be kidding."

"No, no - I am not kidding! Sometimes you need to get a full picture of a man," said the officer and continued the inquiry. He wanted to learn more about Dr. X's parents. Apparently they were poor farmers and farmers needed all able hands they could find to work the land. As a rule, farmers kept children on the farm and frequently opposed even elementary school education. How come then that Dr. X's parents decided to let him attend high school? Dr. X explained how his elementary school teacher, who thought he was an exceptional pupil, spoke to his parents and convinced them to send him to high school. It was not an easy decision for the parents nor was it easy for Dr. X. Day in and day out he spent three hours on the road walking to and from the school. He also worked on the farm as much as was possible. Most onerous was his permanent assignment to get up in the wee hours and milk the cows before he left for school.

The caller was quite impressed and wanted to get more details about Dr. X's difficult childhood. By then wide awake, Dr. X had a sudden revelation - his superior must have also proposed him for promotion. Only a few colonels get to advance to the rank of general and a thorough background check before the promotion was in order.

In telling us the story, Colonel X showed some of his old sense of humor. He laughed as he described how after his "epiphany" about promotion he began to embellish his tale. Clearly, the friendly caller wanted to enter into the biography more detail about his "proletarian" background and Dr. X helped the best he could.

The first night they did not get further than high school in Knin. At the end of the conversation the caller thanked Dr. X for the valuable information.

"Let's skip the Sunday and resume after working hours on Monday," he said. "I'd like to talk to you about your medical school and, if we get to it, a bit about your early times with the partizans. From then on your dossier is pretty complete and we should have no trouble wrapping everything up. Have a very good night."

Dr. X had a whole day to prepare and was eager to continue the weird conversation with the officer. Predictably, the Monday call came at midnight. The intelligence officer was very pleasant and gave Dr. X plenty of time to talk about his leftist activities in medical school, his graduation just before the Germans occupation and his decision to join the resistance movement.

"I forgot much of that old stuff and by the end of the interview I was quite impressed with my own story," Dr. X told us, "and I would have given myself a general's rank on the spot!"

At the end of the Monday chat, around 2 AM, the officer appeared sincerely pleased and promised to call next evening around 9 PM to complete the interview. Sure enough, the next evening the telephone rang at 9 PM sharp. This time the officer did all the talking.

"I enjoyed talking to you," he said, "I frequently have to call on all sorts of problematic individuals. But this time I had the privilege of chatting with a true patriot, a person of great integrity, a war time hero and a peace-time healer. But I am confused. What makes you tell those subversive jokes about your comrades? Nothing in you past warrants such a behavior."

Dr. X was stunned into a stupor and could not come up with a single word.

"Sorry to embarrass you," the officer said, "but I sincerely hope you will stop cutting the limb on which you sit. I've decided to keep this out of your dossier. It just does not fit there. But if you continue with the nonsense I will report you have changed sides in our struggle for a new society. Take it easy now and have a good night."

"I'd like to think that the man made me see the light. After all I am a communist, an officer, and why should I beat up on my own people?" said Dr. X as he came to the end of the story. "The officer seemed sympathetic but the words 'subversive' and 'side in the struggle' left little doubt what this was about. These guys are smoother than the ruffians who used to torture people. The technique changed, but there was plenty of threat and intimidation in that sophisticated conversation. You know what the folks say: 'A molting wolf might change his fur but never his nature.' And that's the truth."

Dr. X stopped coming to the lunch room and we missed him. But the general atmosphere in the lunch room didn't change. We heard a few less antigovernment jokes in part because the main source dried out. But if we heard a good one, we did not hesitate to spread it around. The secret service watched colonels but had no time for simple doctors. Or so we thought.

About a month later I realized I was wrong; the police had plenty of time for "simple doctors" and Dr. X was right when he said that a wolf changes his fur but not his nature.

Eventually I made peace with the overall situation in the hospital. In fact, Professor Hahn's dictum that I focus on medicine and forget organizational reforms proved helpful. There was plenty to do in the medical area. I began to work on my doctoral dissertation, we opened the new clinic for hypertension, renal arteriography became commonplace and the work in general internal medicine was rewarding.

Our faculty kept up with contemporary medicine despite substantial difficulties. Akin to the merchandize in stores, the availability of medical journals was sporadic and unpredictable. Somebody somewhere in some ministry kept allocating meager foreign currency funds to libraries. His budget must have expanded and contracted like an accordion. Some journals were never available, others did not show up for months, only to reappear as a bundle of missing issues. At times one, and other times another, medical library had more luck. This was hard enough but the real problem was with medical supplies. We knew enough about how and where to use pacemakers, peritoneal dialysis, and vascular grafts but the real trick was getting the necessary equipment. By chance I got involved in two medical premieres in Zagreb. In both instances it took energetic doctors and "important" patients to accomplish the task.

In the first instance, we admitted to cardiology a fifty year old patient with intermittent *atrio-venticular block* and classical *Stoke-Adams* attacks of fainting and convulsion. The patient felt quite healthy in between attacks. He was an employee of the ministry of internal affairs. The ministry had a huge number of departments; everything from traffic control, administration of jails, to secret police. The patient was modest and did not volunteer details about his work. We concluded he was a mid-level secret service man, a small cog in the machinery of the much feared ministry of interior affairs. But then telephones started to ring. Our patient was a former partizan and seemed to have thousands of concerned personal friends in all branches of the government. He needed a pacemaker and one of the callers had just the right connection. At that time most commercial roads led to Trieste, a large harbor on the border with Yugoslavia. By the end of the Second World War, Tito's partizans, including my brother Djuka, marched into Trieste claiming that this once ethnically Slavic town had been unfairly annexed to Italy after the First World War. The allies did not buy the story, and Yugoslav troops had to withdraw. However, Trieste was proclaimed an international harbor administered by, but not belonging to, Italy. As a part of the compromise Yugoslavia maintained a strong presence in

Trieste to protect the interests of the Slavic minority. Truth be told, the only thing to protect were graveyards with many Slavic last names. These reminders of Trieste's past had nothing to do with its contemporary ethnic composition; nobody spoke Croatian or Slovenian and only a few families felt they had Slavic roots.

In due time the pacemaker arrived and Dr. Pasini, a very capable surgeon, implanted the device without any problems. Thereafter, a combined cardiology and surgery team monitored the patient during the postoperative period. It was a smashing success, the patient had no new attacks and, in his own words, he felt "born again."

A similar scenario was replayed with the first peritoneal dialysis for treatment of advanced renal failure. This time it was not an obscure policemen but an outstanding man of letters. Mijo Mirkovic who wrote under the pen name Mate Balota was born in Istria, a Croat peninsula which after the First World War had been annexed to Italy. Theoretically an Italian citizen, Balota undertook to write in the local Slavic dialect. He was a prolific author and a political activist fighting to preserve the Croatian heritage of Istria. He studied in Germany, became a professor of economics in Zagreb and a member of the Yugoslav Academy of Science in Zagreb. He was admitted for advanced kidney failure and his family and friends rallied to help him. In Trieste they purchased catheters and numerous large bottles of fluids for dialysis. Dialysis is a method to remove impurities and maintain the chemical balance of the body. A large amount of fluid is infused into the abdominal cavity and after a while drained out. When repeated, this procedure greatly improves a patient's overall health. Our nephrologists led by Dr. Radonic knew how to do it and everything went well. The patient who was in a deep coma awoke, spoke to his son and started on the road to recovery. Unfortunately complications ensued and he eventually died.

I am writing this to underscore that the university hospital in Zagreb had excellent staff and practiced "cutting edge" medicine despite difficult circumstances. But there is another point to this account. The two "first" procedures were performed on privileged individuals but they opened the doors for other patients. In the 1960's, Yugoslavia was in transition from a totalitarian communist regime to a milder form of a one-party system. In the beginning, the communists openly called themselves "leaders of the masses" and ruthlessly implemented the Party line. In the sixties, they morphed into politicians. The former leaders did not exactly keep their ears to the ground to hear what the masses wanted, but a modicum of respect for public opinion did evolve. The word about therapeutic advances at the university hospital got out and soon pacemakers and solutions for peritoneal dialysis became regular merchandize available through local traders.

In the cases of cardiac pacemaker and peritoneal dialysis, I was only an interested bystander but I started making marks in my chosen field of hypertension. Despite Professor Radosevic's opinion to the contrary, quite a few Croatian women with hypertension had fibromuscular dysplasia of their renal arteries. Professor Hahn's plan worked, the first few patients were sent for *revascularization* in foreign countries and thereafter a surgeon obtained funds to learn the technique in the USA. Our hospital was on the way to be the first to offer the entire package of arteriography and surgery for patients with narrow renal arteries.

Dr. Mohacek agreed to let me have a small clinic for hypertension within the general cardiology clinic. The clinic grew and I became the lead physician of the combined cardiology and hypertension clinic. With this promotion came something I was not well prepared for - the right to admit patients to cardiology wards. Similar to the night-time admission clinic, the day-time cardiology clinic also suffered from a shortage of beds. Many patients needing hospitalization were turned away. Akin to other Slavs most Croats are prone to open expression of emotions and I cringed at talking to sobbing people. I had honorable intentions but more than once the bed went to the loudest bawler.

Presumably I was the boss but everybody got involved. My coworkers advocated for various individuals, not on medical but on social grounds. The patient was a friend, or he was known to be an exceptionally fine person, had a difficult family situation, was depressed and so it went. I suspect it was my fault because I too did not hide emotions. They saw me struggling, and my hesitation invited unsolicited input. One of the biggest meddlers was Franjo, the male nurse whom I had known since medical student days. He obviously had a great deal of empathy and vigorously advocated on behalf of his chosen patients. I was fed up with his behavior and planned to set him straight. This was a delicate task, the man was twice my age and we were rarely alone in the clinic. I thought I hesitated to confront Franjo because I was a kind and sensitive young man. Nothing of the sort! I was, in fact, a stupid greenhorn. One day when we had two free beds on the ward and I already had decided who would be admitted, the message came that one patient on the ward had a setback and couldn't be discharged. I decided which patient would have to wait an additional day for admission and proceeded to explain the situation to him. The man was obstinate and quite angry. "I already paid for the bed and shouldn't have to wait," he said.

"You did what?"

"I gave Franjo a blue envelope for you."

Oh, that shameful symbol of bribery - the blue envelope!! The general situation - meager physician's salaries and unavailable hospital beds - created an ideal climate for corruption. Knowing that some physicians accepted money for services, I made it clear to everybody who cared to listen that I would not take payoffs. Well, Franjo listened! And he was ingenious. All by himself he invented the Croatian equivalent of a Washington lobbyist. When I confronted him, Franjo first crafted the predictable elegy about his poor wife and children. Seeing that this did not work, he said he kept a tally and was willing - fair and square - to share the money. I got visibly angry and to pacify me Franjo proposed a 75 to 25% split in my favor! When he realized his words were falling on deaf ears, Franjo got very concerned and pulled the ace out of his sleeve. I ought to understand, he suggested, that when he solicits the money he takes all the risks and provides me a convenient shield. He just did not "get it" and thought I was driving a hard bargain to get the whole bundle!

What was I to do? Call the personnel department, rat on him, and destroy his livelihood? Put him to shame in front of the entire clinic crew, many of whom were likely to be equally corrupted? Try to explain to a cynic that I cared about my good name? I finally found the most practical solution and firmly told him that if he ever again asked for a favor the patient, no matter how sick, would not be admitted. It worked, he came to me only once again. "Doctor," he said, "this is my sister. So help me God!" And his sister was admitted.

Though I worked alone outside regular working hours and had no financial support, I made considerable progress with the thesis for the degree of Doctor of Sciences. In Ann Arbor we published a few papers about the association of personality types with elevated blood pressure in American university students. I wanted to see whether similar results could be found in Zagreb. This was by no means a foregone conclusion; the psychological questionnaire developed in the USA might not have been useful for assessing people speaking another language and living in very different circumstances. First I translated each of the 187 questions in the 16 personality factors questionnaire. The questions dealt with an individual's behavior under hypothetical situations. Besides cultural differences - the Americans play football and the Croats soccer- there were obvious societal differences between the two countries. A traffic cop in America is a less threatening figure than a stern militia man in former Yugoslavia. Americans used checks and frequently interacted with bank employees, Yugoslavia had a cash economy. No surprise then that I got stuck on a question describing an interaction with a bank teller. Should I instead ask about the ticket seller in the railway station? After a few attempts to change some of the questions I threw in the towel and translated everything verbatim. Let the chips fall where they may.

Nice words but to learn where the chips fell I had to find some optical scanners to read the response sheets. I anticipated it would be hard to locate a scanner in Zagreb but it took only one call to the School of Public Health. They were ready to help and their programmers agreed to complete a few first pass routine analyses of the data. Next I gave the questionnaire to 387 male medical students. Surprisingly the average values in Zagreb were not different from average values in the USA for 13 out of 16 personality factors. Furthermore all factors were distributed in a normal bell shaped fashion. This was very good news; the questionnaire worked well also in Zagreb.

Next, I took repeated blood pressure readings in 75 students on five different days and merged the data with answers to the questionnaire. The results were very reassuring. Akin to their American counterparts, the Croat students with higher blood pressure were submissive and sociable. The third factor, increased "sensitivity," which in the USA was associated with sustained BP elevation, was found only among subjects with wide blood pressure oscillations in Croatia. Realizing I had good material in my hands, I aggressively proceeded to complete all analyses, develop the structure of the manuscript, and write the thesis. I worked very hard and I found out something about myself; that I loved research and that I was good at organizing and executing studies. Working on the thesis gave me the opportunity to communicate with research-minded souls in Zagreb and I relished every moment of it. Doing research under those circumstances was a true passion and, frankly, most investigators felt that they were a better-than-average breed. Nobody is perfect and I too developed a bit of an elitist's superiority complex.

Doctorates in science were structured differently in the medical school in Zagreb than in most European countries. I had no thesis committee nor did I have a formal mentor. I submitted my thesis to the Dean's office for review by a standing committee. They approved the thesis and I had to give a 20 minute lecture in front of the entire medical school faculty. The lecture was more of an oratorical exam than a defense of the thesis; the purpose was to demonstrate my ability to craft a good and comprehensible lecture. After I gave the lecture, the faculty took a vote and conferred on me the title of "doctor of sciences pertaining to medical arts."

I brought a copy of the thesis to the USA and read it 42 years after it had been written. It had a good review of literature, clearly stated the hypothesis, described the research plan including statistical methodology, and presented the results in a systematic fashion. The conclusions were modest but the discussion was long and I widely overstated the importance of my modest findings. That was before I learned the American fact-based disciplined way of writing! There was nothing to be ashamed off in the thesis but I was quite saddened perusing it. The thesis was printed with a stencil copier on low quality paper. The typist used a typewriter with old fashioned unattractive letters and typed in an uneven fashion. Some letters perforated through the stencil and others left barely a trace. As a result the entire text was uneven, sloppy and hard to read. The printer must have been in a hurry. Some stencils were damaged, ink smudges and irregular ink lines crossing the text were visible on almost every page. What a shoddy way to present a decent piece of scientific work! I guess that was the best we could do under the circumstances. But the binding of the book is pretty and well preserved. The title: "Psychosomatic Characteristics of Students with Elevated Blood Pressure" printed in gold stands out against the dark grey background. I love that!

With time, the invitation to Ann Arbor seemed less important. I did not forget about it but I had no urgent reasons to leave the country. The clinical work went well and with the Doctor of Science title I completed requirements for the academic position of Docent. However, I was a bit too fast - some of my superiors had not yet completed their theses. So, the path to a professorial position was wide open but I had to wait for others to catch up. Professor Hahn was not about to break his hierarchical rules. I have a sneaky suspicion he used me as a tool to prod others to move on. Anyhow, on the whole, things were good.

Despite the lack of urgency, the Educational Council for Foreign Medical Graduates (ECFMG) exam, the first step towards practicing medicine in the USA, was looming. The exam was given at the American Consulate in Zagreb which was closely watched by Yugoslav authorities. I had to balance the inconvenience of telegraphing my intention to the police against the danger of becoming less proficient in the English language and American medical terminology. Eventually I decided to go for it. The language exam was a bit of joke and I easily convinced the Vice Consul I was able to distinguish "sheep" from "ship." In the next sitting, I had to complete the very tough medical part. The ECFMG was heavily weighed towards basic medical science and the longer you were out of medical school the more likely you were to forget the finer points of biochemistry, biology and physiology. And I took the ECFMG 11 years after the last medical school exam! This was difficult enough but even more perplexing was the multiple choice technique. All exams in my medical school were oral. I not only missed learning the technique, and there clearly is a technique to passing multiple choice exams, when it came to clinical medicine I could not fathom the basic concept of multiple choices. To me, the assumption that there is a right and a wrong answer to a clinical situation was unacceptable. But you do what you have to do and I passed the exam with flying colors.

So there I was with two tickets in my pocket, one for an academic career in Zagreb and the other for research work in Ann Arbor. Nothing will paralyze you more than two equally tempting opportunities. I kept vacillating which way to go and then, out of the blue, a telephone call set me straight.

The hospital operator forwarded the call to the cardiology clinic despite the rule not to disturb physicians during clinic hours. The operator apologized. "I would not otherwise bother you," she said, "but the caller insisted he must urgently talk to you. Furthermore, he is calling from Petrinjska Street." Petrinjska was a long street in the middle of the town, and the entire street could not have called me, but I understood the hint. The operator meant Petrinjska Street number three, the headquarters of the city police and when they call, everybody had better answer.

The caller introduced himself as an official of the Department of Internal Affairs. He told me he urgently needed to talk with me and that I should come to his office in an hour. "I am in the midst of a very busy clinic," I protested, "and cannot just walk out on my patients. It will take me at least three hours to complete cases I have already started."

He was not impressed: "If I do not see you here in an hour I will send a few militia men to pick you up. This might convince your patients that you had to leave. On the other hand your coworkers might not like to see you dragged away by police. I am trying to be discrete but you had better find a replacement and show up on time."

The man did not scream, he spoke in a frosty, monotonous voice and his slow delivery made the threat sound particularly ominous. What the heck was going on?

I first called a colleague to take over the clinic and then went to Chairman's office. As always, Professor Hahn was thoughtful and logical.

"Let's slow this thing down just a bit," he said. "Give me the man's name and I will find my way to him or, even better, to his boss. I know you haven't done anything wrong, but in this crazy world of ours that doesn't mean a thing. A neighbor, an angry patient could invent a story and off you go. However, whatever the issue may be, I know they don't have a strong case. In a perverse fashion, the fact that he threatens you is also a good sign. Police threaten when they hope to extract something. Had they thought you were guilty, instead of summoning you, they would give you a free ride in the Green Martin." Hahn was a bit old fashioned when he said "Green Martin," a term descriptive of police vans the Kingdom of Yugoslavia used to suppress riots and make mass arrests.

For a while Hahn appeared lost in thoughts but suddenly, as if he just awoke from a bad dream, he leapt from the chair and said with resolve, "I will let them have it. Come back in fifteen minutes."

When I returned Hahn was all smiles. "Your appointment has been postponed until 2 PM. You have four hours to collect your thoughts and I got what I wanted - show them they can't just barge in and interrupt the work of a hospital. They can do their arresting at night or early in the morning. Which is exactly what I told them!"

The doorman at Petrinjska number three knew I was coming, gave me the room number and when I knocked, the door opened instantly. In front of me stood a slender, dark haired, youngish person in civilian clothes. His face looked oddly familiar. He quietly pointed to the chair in front of the desk and turned to the stand in the corner of his rather small one window room. The rattle of the coffee making utensils interrupted the pantomime but the man did not yet utter a word. After a few minutes the water in small copper pots for Turkish coffee started to boil. “Sugar?” asked my inquisitor in the same quiet voice I heard on the telephone.

“Just one teaspoon, please.”

He gave me a small coffee cup, placed another one on his desk, sat in the chair and quietly stared at me. I presume the deliberately slow tempo and the very quiet atmosphere were supposed to make me tense. It worked, I was quite nervous but I decided to dissimulate. “You want me to ask questions but I won’t say a word. You wish to see me nervous and I will look relaxed,” said my inner voice while my public self explored the bottom of the coffee cup with great concentration.

“Would you like another one?”

“No, this was just right. More coffee would give me palpitations.”

“Oh, you doctors. Just like my brother, you use medical words to impress us mere mortals. I presume more coffee would make you uncomfortable. It relaxes me and I will make another cup for myself.”

At least we were talking. I pursued the hint that his brother was a physician. It turned out he was a respected pediatrician in my hospital. After that the conversion died off again. Finally the man decided to talk.

“You must wonder why I invited you.”

“Not at all; you wanted me to have some coffee,” I responded with the most angelic smile I could muster.

“I want to talk to you about your excessive association with foreign elements.”

He had just accused doctors of using strange language but his police jargon was not much better. Not “foreigners” but “foreign elements!” I admit to having a quirky sense of humor and the irony of the situation tickled me pink. I could barely stop myself from laughing. All of a sudden I felt relaxed as if it all was going to be a game. But it wasn’t.

When we left Ann Arbor, we invited friends to visit us in Zagreb. Many of them came and we were delighted to show them around. I was stunned to realize that my interrogator knew the name of each visitor, the dates of the visits and our whereabouts during each visit. I must hand it to them - we had no idea we were being watched.

As my investigator recited the names of visitors, I pointed out that all of them were friends, had legal visas, and I had no reasons not to host them. The young policeman was not at all impressed with my argument.

"You said all of them were your friends. Are you sure? Think about it," he urged. "What about a certain gentleman you met for lunch in the Esplanada Hotel?"

Freud was right when he suggested that people tend to suppress unpleasant experiences. It took some prodding before I recalled the lunch with the disagreeable American journalist whose name I have forever forgotten. He was a friend of a friend of my brother Djuka who at that time was a Yugoslav foreign correspondent in Latin America. The American mentioned the journalistic connection on the phone and we set a date for lunch. Things went badly from the word go. He rattled off a list of Croat Communist leaders and wanted to know which of them were considered more or less liberal. I could not oblige as I knew nothing about "who-is-who" in Croatian politics. A bit disappointed, he started to inquire about political opposition to communists and how he could get in touch with them. This was an inappropriate and dangerous line of questioning and I told him so. "Well, I disagree. To me these are very important questions and that is what I do for living," said the slightly tipsy journalist. After he downed another martini, the man started bragging about his connection to opposition figures in Czechoslovakia. "And when I talk to them I occasionally send a message to the regime. You see, I know my friends' homes are bugged and it is a pleasure to tell the regime what I think about communism." I could not believe my ears. What a cowardly swine! Playing hero from his safe journalistic position and wreaking havoc with other people's lives. I wondered how many people got arrested because of this mean agent-provocateur. I could not take it and this was the only time in my life that I walked away from the table in the middle of a meal.

After I gave him a short synopsis of the unpleasant lunch, the young police investigator asked the logical question: "So why didn't you report him to the police?"

"If he had asked me to put a ton of explosives under a government building, I probably would have told you. But this was a drunken weakling, an ineffectual braggart incapable of true action," I said. "Furthermore, he was leaving the very next day."

"You never know. You should have reported it and let us decide."

By and by the interview was coming to the end. There was not much more to talk about. The young police officer got up, brewed two new cups of Turkish coffee, put one in front of me and walked around the room with the other in his hand.

"Here is the story," he intoned in his slow methodical low key fashion, "I do not wish to see you in the company of foreigners anymore. Stop it before you get into real trouble. As a citizen you have the right of free association but I know my stuff and I advise you to drop all your contacts."

I protested, telling him I was the official translator for the Medical Society of Croatia and in that role I frequently saw various medical visitors. My interrogator was unmoved: "Resign, they will find somebody else to do the job." He then escorted me out of the office and as I was about to leave nonchalantly asked whether I was satisfied with the interview. The man must have been fresh out of the police school and he did everything by the book. First, project a sense of urgency, then threaten, and when the guy shows up create tension by not talking to him. During the interrogation be polite but not friendly. Fine, but asking me whether I liked the procedure was too much!

"Look, you certainly did the best you could to act in a civilized fashion. But you happen to have chosen an unpleasant profession and I could not possibly compliment you," I blurted out with some anger. The policeman produced a faint semblance of a smile and shook my hand.

Until that interview everything seemed just fine but the realization that I was watched and that the police wanted to limit my freedom changed everything. I was upset and deeply offended. Besides the emotional impact there was also a practical ramification to this development, I simply could not comply with my interrogator's request. I learned from the ups and downs of my life never to yield to a bully. If you cave in this will be the beginning, not the end, of your troubles. Had I acted as had been requested I would have admitted guilt and, as the secret police love to say, "you never know." You never know what they might request the next time. I'd become a marked individual, an easy-to-manipulate coward. The only way to deal with bullies is to fight back. You might lose but at least you will have their respect.

I mounted a counteroffensive with the help of a patient of mine who organized an interview with a top official in the Savska Street building of the Department of Internal Affairs of the Republic of Croatia. The official met me at the door and escorted me to his large well-appointed room. "Make yourself at home," he said, "I will have to finish a few things before I can talk with you." Next, a person entered through the side door with papers for signatures. While the official affixed his signatures, the bearer of the documents unashamedly trained his piercing eyes at me. I mean really piercing eyes! The man was looking right through me and I felt naked. A soon as he left another man entered through the same side door. He too kept sizing me up but his long look was less unfriendly. After he left the official quietly fiddled with his papers and then suddenly became animated and very friendly. I cannot prove it but I am sure that the two visitors were specialists in memorizing faces. Experts at recognizing faces are valuable to any police. The two men must have been placed at some important spots. After they concluded they hadn't seen me there, they probably sent some kind of an "okay" signal to the room. Maybe a little green light flashed in a remote corner of my host's overcrowded desk.

The well-mannered, top-level police official was apologetic. What happened to me was an unwarranted initiative of "our lower organs." Obviously the brain, apologizing for what his kidneys did! He insisted that the Zagreb police did not suspect me; they only tried to scare me and thereby decrease their work load. "You too, doctor, would love to decrease the number of your patients if you could," said the suave policeman. No doubt about it, but I nevertheless continued to complain. I argued that each of the visitors had a proper visa and if the police wished to decrease their work they should issue fewer visas.

"Oh, we certainly would limit the number of visas if we only could. But Ministries of Foreign Affairs, Culture, and Tourism want more exchange. Mind you, this is not specific for Yugoslavia. If you were to talk my counterpart in the FBI, he'd tell you he would love to close his borders." I guess he was right.

He then indicated that I had his fullest confidence and should feel free to exercise my right to associate with anybody I wished. "Was the guy in the Petrinjska Street rude to you?" he inquired.

"No, his manners were fine but his message was disheartening. He suggested that I behave as a dog. How could I possibly refuse to see people who were kind to me in the USA?

The high officer again stated I had his full trust and promised to order the Zagreb police to stop following me. At that point I got an absolutely brilliant idea. "I really appreciate your kindness," I said, "but I am not sure I trust your colleagues in the City Police. I will test their sincerity by applying for a long term exit visa for me and my family. If they trust me they will not deny me the visa."

"Right," said the official, "when I talk with them I will mention that you want to have the visa and I am sure they will grant it to you."

I am sure the high official from the Savska Street knew I had taken the ECFMG exam and that I contemplated leaving the country for good. But he honored his word and within a month we got our visas.

Next I went to the Consulate to seek the American visa. The Vice Consul suggested that my sponsors in Ann Arbor apply for an H1 visa. This might take some time but it would be well worth my while to wait for it. A few months later, when he handed me the visa, the Vice Consul congratulated me and indicated that, though it originally was developed for "cucumber pickers from Mexico," having an H1 visa was like having a piece of gold in my pocket. At that time I did not quite appreciate what the Vice Consul meant, but the visa indeed was a piece of gold in as much as it could easily be converted into permanent United States resident status.

Professor Hahn approved a two-year unpaid leave of absence for me. "Just in case you decide to return, the job will be waiting. But I doubt you will return," said Hahn.

We obtained tickets for the Italian liner Leonardo da Vinci sailing from Genova (Genoa) to New York, packed two large trunks and were set to leave when I took a cursory look at our passports. My wife's and stepdaughter's visas were fine but mine was not signed! It was late Friday afternoon and we were to leave the next morning. Being by nature a pessimistic realist, I immediately started to plan to cancel the voyage and arrange a new reservation date. Somebody suggested I ought to go to the passport office to see what could be done. I thought this was futile but to avoid looking callous to others I decided to go through the motions. It was late and the only person I could find was a cleaning lady. "The boss might be still in," she said, "he frequently works late hours and I have seen the light in his office. He is either working or left the lights on."

Well, the man was working! I explained the situation, showed him that the visa numbers in the three passports were in order and that by some chance mine was not signed. "You are quite lucky. Of all employees, you managed to run into the person that signs visas. I guess it was my fault and I am willing to redress the situation but we have a problem," said the hard working bureaucrat. "I have to sign this with a special ink which is locked in my secretary's desk. I am not about to break into her desk so we will have to do it with regular ink. If the guys on the border are sufficiently astute they will not let you leave the country." He was quite pleasant and indicated I have a better than fifty percent chance to pass without trouble. “If they stop you, tell them the story, maybe they will be reasonable. In the worst case, they will have to call me and I will confirm your story.”

“But we are crossing the border on the weekend. You will not be in the office and I would hate to see them disturbing you at home,” I lamented.

“They won’t,” said the official cordially,” they will call me Monday morning.”

While the man waited for the ink to dry all sorts of scenarios went through my mind; passing the border without trouble, negotiating my way out of the situation or being detained. The last would be an absolute disaster, as our trunks would reach the ship without us. But nothing more could be done. I thanked the nice man and left.

A burnt person blows on cold, they say in Croatia. Having half resolved one potential disaster, I started to obsess over all other aspects of the trip but everything seemed to go smoothly. We had our railway and boat tickets, my wife was in charge of the passports, our suitcases were in sight and the trunks were clearly marked, waiting to be loaded into the cargo section of the train. We came an hour early and I nervously paced around the railway station. I spotted the trunks in the station’s cargo room and went to chat with the personnel to make sure they would be loaded on time. “What is the problem,” said the senior cargo person. “You have to wait six hours before the train for Geneva arrives.”

“But we are going to Genova not Geneva,” I screamed.

The guy took a good look at the trunks. “I guess you are right, so we will send them to Genova. No problem,” he said phlegmatically. He has seen it all before.

Eventually we passed the border without trouble, the trunks arrived in time and we sailed to America without a hitch.

# Preparations (1964-1971)

If everything works out as I hope, this book will not be perused only by physicians but also by people who read my first book. "Neither Red Nor Dead" sold well to a wide readership interested in the Second World War, the guerilla resistance to German occupiers in Yugoslavia, post-war communist Yugoslavia, and how my nuclear family navigated though the obstacles around us. The present book describes my adventures in medical science and to understand it the reader must become somewhat familiar with the scientific background.

Part of the story of how I came to investigate blood pressure has already been told; it largely happened by chance and luck but the fact is that I ended up working in an important field. How do I know it is important? Well, first is the fact that when blood pressure is elevated over a long period of time, bad things happen. To generate increased pressure, the heart must work harder. The heart is a fantastic organ. Depending on the level of physical activity, for many years it pumps between two to six thousand gallons of blood a day. However, this reliable little machine has its limitations and if it must pump all that blood under a higher pressure it will enlarge and eventually fail. Our blood vessels are built to sustain a normal pressure and if the pressure is excessive they too start to suffer. The consequences of untreated hypertension are heart failure, kidney failure, stroke, heart attack, rupture of the aorta, and many other nasty things. The public health impact of elevated blood pressure is enormous. According to a World Health Organization report, the leading cause of worldwide mortality in 1996 was malnutrition (11.7% of all deaths) followed by tobacco use (6%) and by high blood pressure and its effects (5.8%). In the United States at the beginning of the 21st century, deaths from diseases of the heart and blood vessels outpaced deaths from all forms of cancer by 36%.

There is also other evidence for the overreaching importance of blood pressure. All fluids flow from a high to a lower pressure point. In our body, the left side of the heart develops the needed pressure to provide adequate blood flow to all organs. As the blood flows, some of the energy generated by the heart is lost due to the friction in resistance vessels. One would think that the longer the path the blood must take, the more blood vessels, the more energy would be spent. Therefore, large bodies which have a larger and longer network of blood vessels ought to develop higher blood pressure to provide the flow. Right?

Not at all! The blood pressure in all warm blooded animals (mammals) is the same, regardless of size. The normal blood pressures of a mouse and an elephant are very similar! The only known exception is the giraffe and for good reason: In the upright position, the head of the giraffe is much further from the heart than in any other mammal. To provide the flow against so much gravity, the long-necked creature needs a higher blood pressure and, in fact, the pressure of the giraffe is about twice as high as in humans or mice. But let us not discuss the exception to the rule. I gave you a partially wrong premise on purpose. In reality, the resistance to flow depends much more on the cross sectional area of the blood vessels than on its length; a short segment of a narrow vessel offers much more resistance than a very long segment of a wider vessel. The point here is that there is no a priori reason why the blood pressure of all animals ought to be the same. There must be something very fundamental to the blood pressure reading of 120/80 mm Hg. That uniformly narrow range of normal blood pressure across many species is also the optimal blood pressure; a higher blood pressure has been experimentally induced in rodents, sheep, pigs, monkeys and the consequences are similar to humans – more strokes, more heart failure and diminished kidney function, and more or earlier deaths.

Before I start complicating things, let me explain that it is not so hard to understand what blood pressure is and why it might be too high in some people. The pressure that the left side of the heart must develop to supply blood to the body depends on only two things; the amount of blood the heart must pump and the resistance to the blood flow in arteries. Thus, blood pressure (in millimeters of mercury) equals *cardiac output* (in liters per minute) times *vascular resistance* (in arbitrary units).

Cardiac output largely depends on the amount of fluids in the body and on the demand for blood. At rest, the demand for cardiac output depends on the *basal metabolic rate,* that is to say on how many calories the body needs to burn to sustain its basic functions. Some people burn more and some less fuel. Nervous tension and over activity of the thyroid gland increase the metabolic rate and, therefore, increase the resting cardiac output. When such needs as exercise and emotional excitement arise, the heart increases both its output and the blood pressure in order to meet the new demand.

Vascular resistance chiefly depends on the caliber of small blood vessels called *arterioles* and on the thickness (viscosity) of the blood. If the vessels are narrow or the blood is viscous, the vascular resistance increases and a higher pressure is required to push the blood forwards.

As simple as that! But, of course, this only explains "how" blood pressure can go up or down. The trick is in the "why." Why would some people have more fluid, narrower blood vessels or thicker blood? Presently we can identify the cause of the disease in only about 10% percents of all patients. That is the bad news; the good news is that because we know "how" the blood pressure increases we also know how to decrease it. And you guessed it right; blood pressure can be effectively decreased by drugs that get rid of excess sodium/fluid in the body, drugs that lower the cardiac output, and drugs that dilate arterioles. The even better news is that it has been proven, over and over again, that these drugs very effectively protect the patient from the ravages of hypertension. So, as things stand now, we can cure hypertension only in a few people but we can treat virtually everyone with hypertension.

Upon my return to Ann Arbor in 1965, I was assigned to work together with Tony Amery in James Conway's hemodynamic laboratory. Tony came to Ann Arbor from the Catholic University in Leuven, Belgium. He was a pleasant, quiet, reserved, and somewhat whimsical man in his mid thirties. He reminded me of Bosnians I had to deal with during my tour of duty in Gorazde; you could not extract a straight answer from the man. His personal life remained a mystery and when it came to science he'd cleverly turn around all my questions. He had encyclopedic knowledge of hypertension but it simply was not his style to give a direct answer. Akin to a clever psychiatrist, he'd ask what you thought about the topic and once you told him your preference he'd either agree or bit by bit poke holes into your argument until you realized you were wrong. He was so darn good at this somewhat demeaning and overly pedagogical technique that I never resented it. We worked well together and in the course of two years published three solid papers. Conway was quite generous and assigned one of us as a senior author in each of the papers. Tony returned to Leuven in Belgium where he became Professor of Medicine. He organized the seminal European study of hypertension in the elderly and developed the Leuven research group in hypertension. He died prematurely but the Leuven group he created (Fagart, Staessen et al) continues to advance the knowledge of hypertension.

Conway studied how aging affects responses to exercise in people with normal blood pressure. In his elegant design the exercise level was increased in a stepwise fashion, the blood pressure was measured beat by beat through an indwelling catheter in the *brachial artery,* and cardiac output was assessed when an experimental subject reached "steady state" (that is where the heart rate did not increase further after the individual exercised for four minutes at that level). To properly measure the cardiac output by the dye dilution technique, we advanced a catheter through the *cephalic vein* into the right *atrium* of the heart. Whether the catheter reached its destination was determined by the appearance of negative pressure values together with characteristic pulsatile pressure curves. Nowadays physicians can visualize the catheters directly and maneuver them to into the heart, but Conway's laboratory had no X ray equipment. Nevertheless, failure rates were very low and in most cases after pulling it back and pushing it forth, the thin catheter would slip into the atrium. At each steady state level of exercise, we also measured the amount of air breathed and the *oxygen consumption.* The oxygen consumption was useful to compare the degree to which the cardiovascular system was engaged with exercise. Not being too technical, let's just say that various people might be more or less efficient in exercising or they may have pedaled the bicycle at a different speed. The oxygen consumption reflects the amount of work they actually did and is therefore the best way of comparing various individuals.

Dr. Conway designated me as the senior author of the paper on the influence of age on response to exercise.[1] Healthy volunteers were divided in three age groups; young, middle-aged, and old. The elderly individuals had low resting cardiac output values but during exercise they were capable of increasing the output in a way similar to the other age groups. The respiratory capacity of the elderly group was also adequate. Thus, two major systems determining the capacity to exercise seemed to function normally with older subjects. Nevertheless, the elderly subjects quit exercising at lesser levels of work than the younger age groups. The remaining possibility to explain this limited exercise capacity in the elderly, namely that with aging the ability to relax arteries (vasodilate) may be diminished, also did not prove to be true. In fact, the vascular resistance in the elderly decreased <u>more</u> than in younger subjects. One possible explanation was that due to age-related muscle weakness the elderly had to use larger groups of muscles to achieve the same level of exercise as younger people.

When we wrote the paper, I suggested to Conway that old folks just might be smarter and less willing to exercise in response to our prodding. James, who otherwise had a very good sense of humor, was not amused. I think he read my mind. He was a true physiologist and what excited him failed to arouse me. He described a previously unknown fact that aging does not limit exercise capacity through reduced cardiac or pulmonary performance and that was good enough for him. It was a beautiful study in design and execution, the findings were new and contrary to previous assumptions but I did not get the point. So, they quit earlier because they use larger groups of muscles. "So what," I thought. "What did this mean and why were we doing it?" I was too green to see the implications of our research. If I were to write the paper today, I would have suggested that the exercise capacity of the elderly could be increased by exercise training designed to improve muscle coordination and muscle performance. But James was not given to speculation, he reported facts and it was up to others to draw conclusions.

The usefulness of our observation on aging became clearer in the next paper about the influence of hypertension on exercise performance.[2] In that paper, authored by Tony Amery, we compared hypertensive and normotensive individuals in different age groups. It turned out that there was something wrong with the regulation of vascular resistance in elderly hypertensives; they could decrease the vascular resistance in a parallel fashion to their elderly peers with normal blood pressure but the resistance remained higher than in comparable aged normal blood pressure volunteers. Now that was interesting; the residual increase of vascular resistance suggested that in elderly hypertensives the arterioles could not relax (*vasodilate*) to the same degree as in normal people of similar age. We did not pursue investigating the mechanism of the limited vasodilatation in hypertension. However, the observation was important and we could have not made it without Conway's previous effort to define the influence of aging on exercise capacity. What we had described at that time most likely reflected two different processes that act in parallel to decrease arteriolar dilatation; changes in physical properties of arterioles (the so called *vascular remodeling*) and damage to the inner lining (*endothelium*) of arterioles.

We also uncovered intriguing differences in the blood pressure response to exercise. Middle-aged and elderly hypertensives had larger blood pressure increases than normotensive individuals of similar age. However, in younger hypertensives the blood pressure increased less than in normotensive individuals. We concluded that excessive blood pressure rise during exercise was not a universal feature of hypertension, and that the steep blood pressure spike in elderly hypertensives was due to the rigidity of large arteries. Exercising greatly increases the blood flow (cardiac output). The pressure in a vessel is a product of its diameter and the amount of flow that passes through the vessel; if you pump more flow into a flexible vessel the size of that vessel enlarges and this minimizes the increase of pressure. If you pump the same amount of blood into a rigid system, the pressure would increase more. Since an excessive pressure increase to exercise was not seen in all hypertensives, we concluded that the blood pressure response to physical strain would not be a reliable predictor of hypertension. At that time various maneuvers to increase the blood pressure, so called "pressor" tests, were used to predict who would develop hypertension in the future. This practice was based on the firm belief that hypertensive patients respond to various stresses with larger blood pressure increases and that these repeated pressor episodes eventually "add up" to a permanently elevated blood pressure. The Amery paper was our first challenge to the theory of hyper reactivity in hypertension.

The second challenge came a few years later in an experiment executed by Rune Sannerstedt, the uncharacteristic Swede. When I think of a typical Swede, I see Bjorn Borg on the tennis court keeping his cool throughout the match, concentrating on the game, and appearing unexcited by the events around him. In his private life, he might have been funny and a few glasses of wine could probably loosen him up, but his public persona always emanated a serene calmness and stability. But Rune was as lively as they come. This blond, middle framed man in his early thirties showed emotions whether they were positive or negative. He was what I call an objective gesticulator; when appropriate he used body language but never overdid it in an Italian fashion. He came to Conway's lab from Goteborg where he completed his doctoral thesis on exercise in hypertension. His results in Goteborg were pretty much as Conway's; exercise did not elicit an inordinate increase of the blood pressure in young patients with hypertension. In Ann Arbor, he subjected patients with borderline hypertension to head up tilt and found that the heart rate, cardiac output, and blood pressure responses were similar to responses seen in normal volunteers.[3]

For the second time we could not find evidence for blood pressure hyperreactivity in hypertension. That hyperreactivity was absent in borderline hypertension was important. Borderline hypertension is a strong predictor of future permanent hypertension and if the hyperreactivity theory is correct then patients with marginal blood pressure elevation ought to be particularly prone to excessive pressure oscillations. Later this absence of hyperreactivity was very helpful in shaping my overall thinking about hypertension. However, before I move forward, let me draw an Aesopian moral from this part of the story. We think of science as an eminently cerebral and objective discipline where only facts count. By and large this is true. But, when it comes to scientific disputes one should not underestimate the impact of tradition, plausibility, conviction, and surprisingly, of semantics. In science plausibility, the appearance that something is logical and worthy of acceptance, can be ruinous.

Take the example of Max Pettenkofer, a highly respected professor in the medical school of Munich who was convinced that infectious diseases were caused by underground evil fermentation which was released into the air as a toxic "miasma." The bad air theory is still alive and in some instances, such as secondary smoke inhalation and heavy air pollution, it is absolutely correct. But it sure does not explain infectious diseases. When in the second half of the nineteenth century the evidence for the importance of bacterial infection started to increase, Pettenkofer refused to accept the news. He went so far as to send a letter to the future (1905) Nobel Prize winner, Robert Koch, requesting a sample of his germs causing cholera. Koch obliged and sent him a culture of living vibrio cholerae which Pettenkofer swallowed in front of a large audience. And guess what? Pettenkofer did not get cholera! The germ culture might have not survived the transport or maybe Pettenkofer took a large swig of bacteria killing schnapps prior to the experiment but the fact is that he did survive. This demonstrated Pettenkofer's strength of conviction but did not stop progress. Koch also proved, by using cholera as the agent, his classical postulate that 1) the suspected agent must be isolated from a sick individual, 2) that the isolated germ must be grown in the laboratory, 3) that the laboratory germ must cause the infection in an experimental animal, and 4) that a culture of the germ must be grown also from the artificially infected animal. Pettenkofer did not live to see Koch receive the Nobel Prize; in 1901 at 83 years of age he committed suicide.

The concept that blood pressure overreactivity could cause hypertension had a similar plausibility as Pettenkofer's theory of miasma. It was well known that such emotions as embarrassment, anger, fear, and sexual arousal elicit substantial blood pressure increases. It was therefore logical to assume that a tense life style might cause hypertension. Already in 1905 the Viennese physician Von Gaisbock described that a group of patients with high blood pressure who also had elevated number of red blood cells were "executives, bankers, or held other strenuous positions." Add to this an unintended semantic disaster, and you can easily see how the concept of excessive blood pressure variability as a precursor of hypertension could have taken root.

The semantic problem arose about five decades ago when effective drugs to lower the blood pressure became available. Before that, physicians knew how to measure the blood pressure but had no idea what to do with the results. However, the advent of effective anti-hypertensive drugs and the increasing knowledge that higher blood pressure levels are harmful to the patient called for reliable blood pressure measurements. To a physician's chagrin, in many a patient blood pressure was all over the map, one day very high and the next day normal. How could you then evaluate whether the drug you prescribed had an effect on a patient's blood pressure? Most of the early drugs had serious side effects and physicians hated to start treatment if it was not absolutely necessary. So drugs were prescribed only for patients who had "sustained" or "permanent" hypertension. Unfortunately, the term "labile" was chosen for patients whose blood pressure was not elevated at every clinic or office visit. The term was created to make a distinction between permanent blood pressure elevation ("stable") and occasional blood pressure elevation. Whereas "labile" is an appropriate antonym to "stable," the choice of "labile" was unfortunate; it unwittingly conveyed the notion of excessive blood pressure variability in such patients. And, as it turned out, there was absolutely no evidence for increased oscillations of the blood pressure in "labile" hypertension. In fact, Giuseppe Mancia's group in Milan repeatedly showed that the range of blood pressure variability is widest in patients with the highest average blood pressure levels. So, the evidence did not support the theory but the plausibility of the concept that emotional increases of blood pressure might lead to permanent hypertension continued to flourish. Investigators kept developing clever tests to demonstrate increased susceptibility to blood pressure oscillations in individual patients. The idea was that those with increased blood pressure response to a stress were heading towards a future permanent hypertension. The panoply of these tests was truly impressive. The cold pressor test required the patient to immerse his arm into a bucket of ice water and keep it there despite considerable pain. In the mental arithmetic test, the patient had to quickly subtract the number seven from a given large number and continue subtracting seven from the previous number, announcing the result. In the color confusion test, the word for the color, let's say blue, was written in another color and the patient had to read the color, not the word. In the isometric exercise test, the individual had to squeeze a device with his hand that measured the strength of his squeeze. Once the maximum strength was determined the patient was asked to keep squeezing the device at about 70% of his strongest performance. There is no doubt that such tests reliably increased the blood pressure but it remains doubtful whether any of them could predict future hypertension. The only test that comes close to being a predictor of hypertension is the blood pressure response to aerobic exercise. However, exercise does not provoke a hidden blood pressure hyperreactivity but, as we suggested in Amery's paper, blood pressure increases because patients heading towards hypertension already have stiffer large arteries.

The moral of the story is that science is not utterly pristine; scientists are frequently deceived by the plausibility of a concept and continue to believe that the idea is valid long after objective data suggest the contrary. And as a rule, akin to Professor Pettenkofer, the most arduous supporters of a failing concept are the last to change their opinions. The smart ones quickly realize that they are wrong. Pettenkofer's nemesis, Robert Koch, for a while believed he had found a way to cure tuberculosis by immunization. Later experiments proved that his "tuberculine" was well tolerated but utterly ineffective and Koch quietly moved in other directions.

I was dead set to work on borderline hypertension which we, at that time, defined as an occasional blood pressure elevation followed by normal blood pressure readings. For a while, Conway permitted me to study such persons in his laboratory but he rightly insisted that we do not call them "patients". We therefore used in published papers such noncommittal terms as "persons," "individuals," "subjects," or "volunteers." Nowadays borderline hypertensives are called "prehypertensives," and I am not quite sure whether this early labeling which implies that they are destined to develop a bad disease is really justified. But I am not blind to the advantage of using one rather than two words to describe a phenomenon and will use "borderline hypertension" and "prehypertension" interchangeably. Conway's insistence on a noncommittal term was not so much inspired by the need to protect these individuals from an ominous label as it reflected his doubt whether borderline hypertension was a true precursor of hypertension. Just as I failed to comprehend why Conway was so much interested in exercise, he could not see my point about investigating early phases of hypertension. As many of his peers in hypertension, Conway spent a great deal of time defining "true" hypertension. For him as a basic scientist, the first priority was to assure reproducibility of the phenomenon he was investigating.

We had lengthy debates and eventually I wrote a memorandum regarding why I wished to proceed with studies in borderline hypertension. The main point was that as hypertension advances the morphology and function of many organs changes which, in turn, cause secondary adjustments of various systems that control the blood pressure. In other words, if you find an abnormality in patients with advanced hypertension you don't know whether this is a cause or a consequence of hypertension. Studying hypertension at its very inception made a great deal of sense to me. In my memo I pointed out that borderline hypertension proved to be a reproducible condition. We used to recruit subjects with borderline hypertension by setting up blood pressure screening stations while students were waiting to register with the university health care center. Those having higher blood pressure were invited to come to our lab for further studies. It took on average about four months from the blood pressure measurement to the appointment in our laboratory. Nevertheless, during the second test none of the recalled students had clearly normal blood pressure values.

When we discussed my memo, Conway readily conceded that borderline hypertension might be a reproducible condition. But he took exception to what I thought was the strongest scientific argument - that the *hemodynamic* picture of borderline hypertension was different from sustained hypertension. In sustained hypertension the vascular resistance was increased; in borderline hypertension the cardiac output was high. I viewed this particular hemodynamic pattern, the so called *hyperkinetic state,* as a proof that as hypertension evolves the patient's circulation changes. Consequently, I argued that studies of early phases of hypertension are needed to understand what initiates the disease.

Conway was too good a debater to let me get away with such loose assumptions.

"I am not thoroughly convinced that borderline hypertension evolves into established hypertension, but for the sake of this discussion I will accept your argument."

Beware when somebody starts to patronize you! Bad news is ahead.

"But you are not suggesting that there is a hundred per-cent transition from borderline to established hypertension?" Conway inquired.

"Of course not, nothing in biology is a hundred percent."

"Okay, how much; a quarter, a half or five percent?"

"From what I read, it is more like fifty percent over a period of fifteen years."

Having set me up, Conway was now ready for the kill:

"What if I were to say that the ones with elevated cardiac output are a bunch of neurotics who will never develop hypertension? You see, anxiety increases the cardiac output but most hypertensives are not anxious. At least that is what you wrote together with Harburg; submissiveness and suppressed anger but not anxiety were associated with higher blood pressure values. For all I know, your anxious people with hyperkinetic circulation may be the 50% that never develop hypertension."

Conway knew his physiology and followed the literature. Nothing works as well as shoving the opponent's arguments down his own throat.

I was not ready to give up and quoted Levy's study of "transient hypertension" in the US Army. "Transient" meant that an elevated blood pressure reading was on subsequent exams followed by values below the 140/90 mmHg border of hypertension. That definition was very close to our definition of borderline hypertension. Whereas our findings were limited to young individuals, Levy and his coworkers also monitored people in their fourth and fifth decade of life. Over a period of five years, depending on age (higher in older subjects), the *incidence* of sustained hypertension was from two to five times higher in transient hypertension than in people with normal blood pressure. If a person had, in addition to transient hypertension, also a fast heart rate (*tachycardia),* his chances of developing sustained hypertension were from four to ten times higher.

Conway agreed that tachycardia was a good marker of the hyperkinetic state but invoked the perennial "one-swallow-does-not-a-Spring-make" argument. Yes, this may mean that the hyperkinetic state does evolve into sustained hypertension but one study was not good enough. More confirmation was needed. Never mind that the Levy study involved thousands of people and that it was meticulously executed; Conway was not convinced.

As I continue to describe my disagreement with James Conway, I do not want to sound harsh. He was a superb scientist, I learned a great deal from him, and he actually cared about me. He wanted to protect his younger colleague from investing time in a problematic enterprise. Furthermore, he had his own research plans and studies of borderline hypertension were not a part of it. We agreed to complete ongoing hemodynamic studies in borderline hypertension, but he made it quite clear that if I wanted to continue I had better get independent grant support and establish my own laboratory. I don't believe this was a threat but I also think he did not imagine I'd be able to generate support for my own line of research. With time I have mellowed but back then I was stubborn as an ox. The more James opposed me the stronger was my conviction to pursue borderline hypertension.

James Conway was not alone in his skepticism and the overall story of the hyperkinetic state is another example that hard facts do not always count in science. You'd think that prejudice is a social phenomenon; an unsubstantiated judgment or opinion about an individual or a group, but no matter how hard scientists try, science also suffers from preconceived judgments. And similarly to social prejudice, they have no idea why they think that way. The roots are forgotten and all that remains is a deep-seated narrow-mindedness. In "Neither Red Nor Dead," I described how Croats and Serbs have always been at each others throats not realizing that their hatred had been systematically fomented by Byzantium and Rome as these two aspects of Christianity diverged and aligned themselves with different political structures. If you asked a Croat whether he was fighting for western values, he might pay lip service to the idea but then he'd explain to you the main reason was that that Serbs are bad people and the Croats are the good ones. Well, it is not that dramatic when it comes to the hyperkinetic state but the disinclination to accept facts and the failure to recognize the root of the problem are equally irrational.

You will, dear reader, recognize that we are now in a zone only tangentially related to the main topic. But one tangent breeds another and let me now tell you the story about the Orthodox Archbishop on a visit to a remote parish.

"Father," he said to the local priest, "why didn't you ring the bells when I arrived?"

"Well, we've got thousands of reasons but first of all we don't have a bell," said the priest.

If I wish to tell a story about prejudice, I must first convince you that such a prejudice really existed. It did and to a degree, it is still alive today! After Levy's paper there was a steady trickle of new evidence about the importance of rapid heart rate as a predictor of future hypertension and as a risk factor for subsequent cardiovascular disease. The evidence came from reputable investigators and from well recognized studies. The Father of American cardiovascular epidemiology, Jeremiah Stamler, described the negative effect of fast heart rate in the Chicago Heart study, and the classical Framingham study, the Kaiser Permanente study, as well as the NHANES national surveys of health in the United States each had bad news to report about fast heart rate. We live in times of evidence-based medicine and prevention of cardiovascular diseases is a stated goal in many national guidelines. Nevertheless, these guidelines never mention the strong predictive power of tachycardia for future hypertension, sudden death and coronary heart disease. So why did that repeated hard evidence not find its way into the minds of guidelines producing opinion leaders? For a while I thought that the level of noise about the heart rate was too low to command attention, but presently I am inclined to attribute this oversight to the fact that opinion leaders tend to be opinionated. As I said earlier, semantics is important, and "opinionated" happens to be a tad more polite word than "prejudiced."

About ten years ago, I was dissuaded from the notion that tachycardia might have been omitted from guidelines by chance. At that time my good friend, Paolo Palatini, Professor of Medicine at the University of Padua, took a sabbatical leave in Ann Arbor. He had enough free time to search the literature and produce a well written and convincing paper about heart rate and hypertension.[4] We were amazed by the sheer volume of evidence and by the strong consistency of findings across different papers. In numerous large populations, heart rate correlated positively with blood pressure and in prospective studies tachycardia was a strong and statistically independent predictor of future hypertension. Furthermore, a fast heart rate forecasts future sudden death, coronary heart disease and also deaths from non cardiovascular causes. Our paper caused a minor boom as many investigators went back to their epidemiological files and uniformly confirmed the important predictive strength of tachycardia. Three years after Paolo's paper, the publication boom resulted in twelve new tachycardia-related articles. Paolo continued to publish reviews and original papers on tachycardia. You'd think this renewed interest would get the attention of guideline makers but nothing happened. Why, oh why?

I think the answer is in the history of hypertension. The practical need to obtain reproducible blood pressure readings forced early investigators to invent all sorts of measurement techniques. For example, on the theory that emotion caused unwanted blood pressure oscillations, Sir Horace Smirk in New Zealand isolated patients in a quiet room, threaded a long cuff-inflation tube through a hole in the wall to the adjacent room where the technician measured the blood pressure. Unfortunately, the byproduct of this approach was the utterly untested notion that patients not selected for treatment or research had a benign form of hypertension. And since emotions temporarily increase blood pressure and tachycardia is a sure sign of emotions, tachycardia must be benign. A classical case of self-enforcing circular reasoning; there is only a small step between "I decided not to treat this patient" to "Such patients do not need treatment."

But the notion that tachycardia is a benign sign of emotions perseveres. And just as the Croats and Serbs cannot explain why they dislike one another, the opinionated hypertension opinion leaders simply "know" that tachycardia is irrelevant. Damn the evidence! Don't confuse me with facts; my mind is made up!

It was very likely that hyperkinetic circulation was due to abnormal *autonomic nervous system* (ANS) control of the circulation in borderline hypertension. I do not know quite why, but since the first year in medical school I have been fascinated by the autonomic nervous system. As I progressed through medical studies, I got interested in many other topics but the ANS remained in the forefront. If you think of the human body as an orchestra with many players, the autonomic nervous system is its great conductor. It did not matter whether I was preparing for an exam in anatomy, pathology, physiology, internal medicine or any other clinical discipline, every teacher and every textbook had something to say about the autonomic nervous control.

When the famous Hans Selye came to Zagreb to give a lecture about his studies of stress, I became even surer that the autonomic nervous system was very important. Selye was a great orator and I was enthused not so much with details as with his way of thinking. He saw the outside world as a permanent challenge to the body and viewed physiologic processes as reactions to preserve vital functions of the body. Selye's mantra was *homeostasis,* the notion that if it is challenged, the body alters various functions in order to preserve its "internal milieu." This per se was not such great news; obviously the body has many mechanisms to protect itself. The news was his view that there is a purpose to everything in the body and that the investigator's duty was not only to describe but also to explain the purpose of a mechanism. This teleological view was in direct conflict with the empiricist's understanding of science as a process of gaining knowledge without any preconceived ideas. I suspected that everything Selye reported had first to pass the litmus test of usefulness and that the story was too smooth to be entirely true. Nevertheless, I was impressed with the complexity of the system and Selye's inclination to think about the larger context of things.

Selye's research was in endocrinology and at that time I had already started to lean toward a career in cardiology. Consequently, I listened with the typical attitude of an outsider. The overall scheme was fascinating, the details less interesting. However, my ears perked up when he mentioned the circulatory aspects of the alarm reaction. Now that was interesting! I started to read the works of Hilton, Hess, and Folkow about circulatory responses to danger and got forever hooked on the autonomic nervous system.

Having agreed with Conway that we disagreed, and that after we completed ongoing studies of borderline hypertension I'd be on my own, I started to plan my research future. In retrospect, I am amazed that I hadn't negotiated some middle of the road solution. By plunging headlong into an area which was of no interest to two of my immediate supervisors, I took a huge risk. Conway thought borderline hypertension was not worth investigating and, disappointed with his original work with ganglionic blockade, Sibley Hoobler was skeptical about the importance of the autonomic nervous system in hypertension. Nevertheless, Hoobler was the big boss and I had to talk with him. Unexpectedly the opportunity presented itself when Hoobler's car broke down and I drove him to work. After I outlined my intentions, Hoobler became very interested. We had reached the top floor of the parking garage and I found an open spot but Hoobler would not get out of the car. Right then and there he wanted to hear more. So we sat for a good half hour in the car discussing and gesticulating to the great amusement of people walking by. Hoobler liked the concept of investigating very early phases of hypertension. He offered me full support with one proviso, should the National Institute of Health approve his grant application to organize a cooperative study of prevention of secondary strokes in hypertension, I would have to help him with the project. That was fine with me.

Hoobler was a man of action. He made a few telephone calls and gave me the name of a man in the University Office of Research Administration.

"They have a nice chunk of seed money to help beginners' research careers," he said, "but they are not eager to part with it."

Hoobler turned silent and kept looking at his pipe. That was a good sign, he was concentrating. After a while he continued to talk, "The maximal amount they ever granted to an individual was eight thousand so you should ask for ten thousand dollars. These folks collect an additional forty percent overhead for each grant we bring in. After they pay our utility bills and provide cleaning crews, they are left with about thirty percent. That pays their salaries and then some. They are swimming in money and you should ask for more than they are willing to give. But they are not permitted to contribute to your salary and you must come up with a reasonable plan for additional expenses. If you are really bent on setting up a new laboratory, ask them to purchase some equipment. I can give you some space for your new operation."

Laboratory space is a cherished commodity in the academic world. Space is essential for execution of research but it is also a visible symbol of your status. By offering lab space, Hoobler proclaimed me an independent investigator before I had a grant or an academic appointment. But Hoobler was an experienced leader. Unbeknownst to me he had already started the process for my promotion to Assistant Professor. The space he gave me was a strong signal to the promotion committee that Hoobler thought I had good chances for academic success.

All of this was truly surprising and I wondered why he decided to support autonomic nervous system research, a field he personally abandoned a few years earlier. I was clever enough not to ask the question then or later. Apparently Hoobler was primarily interested in assembling a group of promising, young investigators and did not care what type of research they did as long as the project was potentially fundable and pertained to hypertension. He knew something we did not, that he planned to take early retirement and by expanding the faculty Hoobler wished to protect the future of his hypertension division. Hoobler strongly felt that clinical research must be supplemented by basic science investigations. Only a joint effort in both fields could create the right atmosphere for scientific productivity. Under his leadership, the hypertension division indeed became a beehive of interaction. We had regular "journal club" meetings to review and discuss new papers. These weekly evening sessions were hosted in homes of various senior professors in Ann Arbor. The hypertension division staff and David Bohr, Art Vander, Richard Malvin and Harvey Sparks from the Ann Arbor Department of Physiology always attended. Oscar Carretero from the Ford Hospital in Detroit frequently drove 40 miles each way to attend. When the topic was of interest, scientists from the Upjohn Company in Kalamazoo undertook a 200 miles round trip to join us. I enjoyed these sessions a great deal. It was a pleasure watching how uniformly brilliant brains but very diverse personalities analyzed a topic. All our basic scientists always provided an erudite background against which to analyze the paper. Was this really a new contribution, was the design appropriate, was the paper contradictory to previous work and if so who is more likely to be correct, is this line of research worth pursuing? They used similar analytic pathways but the ways they expressed themselves were very different. Some were quick and impatient, rapidly identifying the flaw and going right for the jugular to mercilessly pan the paper. Others were equally quick but willing to weigh the positive against negative sides of the paper. However, most preferred to first describe the findings and then, without being overly judgmental, open a discussion of pros and cons. While trying to appear neutral, some of them actually, slowly and methodically, built a one-way road to reach their predetermined conclusion. And since all had a good sense of humor the evening was never boring.

Besides building bridges to basic sciences, Hoobler was also a visionary. Long before anybody argued that the *metabolic syndrome* is a definable clinical entity, Hoobler decided the hypertension group in Ann Arbor must also develop expertise in atherosclerosis. He never articulated why but I am sure that he foresaw the need for an integrated approach to cardiovascular prevention. He first brought in Shafeek Sanbar, an MD who had just obtained his Ph.D. in lipid chemistry from the University of Oklahoma. Shafeek was very knowledgeable and whether the rest of us wanted it or not, he taught us the basics of pathophysiological and clinical aspects of *dyslipidemia*. However, he soon joined the U.S. forces in Vietnam and after the war returned to Oklahoma City to practice cardiology and law. Hoobler then recruited Dr. David Bassett from the University of Hawaii who became the head of the dyslipidemia program within the Division of Hypertension in Ann Arbor. David Bassett is one of the most meticulous, conscientious and honest people I have ever met. He worked hard and organized an exemplary lipid clinic, trained many lipid specialists and stayed in Ann Arbor about 30 years until his recent retirement.

At the beginning of the 1970's, Hoobler completed the recruitment of younger physicians. At that point the hypertension division had two senior physicians, Hoobler and Conway, and three junior doctors, Andrew Zweifler, David Bassett and myself. Among the juniors, Andrew Zweifler joined the division a few years before Bassett and I. Despite his seniority, Andy never pulled rank. Over a period of forty years the two of us supported each other, shared the clinical load and research resources, and worked with trainees and visitors. We never had a conflict, not because we agreed on everything but because of our mutual respect and readiness to consider alternative opinions. I was plain lucky to meet Andy; I would have never achieved what I did without his support. This for me was a most welcomed new type of friendship. In Croatia, we navigated through generally difficult and sometimes dangerous times, and we needed each other's support almost on a daily basis. Under such circumstances, details of private lives were exchanged as a matter of course and as a token of mutual confidence. Andy and I met under different circumstances. We didn't have the daily menu of miseries to bitch about, our steady private lives did not call for analysis and we did not have to visit on a regular basis. And yet our friendship is deep and we know we can rely on one another without any reservation.

Throughout four decades of our activities, the Ann Arbor hypertension group attracted a large number of trainees and visitors. Andy and his wife, Ruth, as well as my wife, Susan, and I, are sociable by nature and we were keen to help our trainees. Excess household furniture, dinnerware, television sets and appliances were circulated. There were frequent dinners, international recipes were exchanged, and once a year we had the famous pig roast. Contrary to our dietary preaching, every autumn we found a way to locate a suckling pig which was a small achievement in its own right. Unlike Europe where in many a butcher's window you can see these poor creatures with an apple stuck in their mouths, in Ann Arbor we had to network with farmers to get one. Next we'd improvise a pit, usually a hole in the ground on the farm of our chief laboratory technician Dick Rogers, and set the fire around 3 AM to complete the roast by mid-afternoon. Everybody brought food, beer and soft drinks, games were organized for children and we had a whale of a time. Dick Rogers eventually left the division and the ritual pig roast was replaced by huge barbecues in public parks and later behind our field clinic in the village of Tecumseh but the overall mood remained the same. Somebody up there must have liked us; to my best knowledge we were never rained out.

**Dick Rogers carves the pig**

The Ann Arbor tradition included also sporting competitions. For a while our softball team 'The Standard Deviates' kept winning tournaments, we had canoe races in the Huron River, and Andy Zweifler, who was an excellent tennis player, occasionally organized a "Hypertension Open" tournament.

Of course, it was not all milk and honey. From time to time our visitors had difficulties adjusting, some trainees were too ambitious, others too inert, occasionally there were personality clashes and, akin to difficulties I had with Conway, some trainees had different research interests than I. Nevertheless, I can say without any reservations that nearly all of them had a very good experience. Anytime we meet anywhere in the world, the former Ann Arborites love to reminisce about good old times in the hypertension division. Even those who seemed somewhat unhappy during their stay manage to pay regular return visits to the old place. Dr. Hoobler formally established the hypertension unit in 1946 and in the early 1950's the unit became an official division of the Department of Medicine. Military terminology not withstanding, in 1986 we celebrated forty years of the Ann Arbor Division of Hypertension and nearly fifty graduates returned to celebrate. Our fiftieth birthday attracted even a larger crowd. These celebrations facilitated interaction across different generations and regardless whether their terms in Ann Arbor overlapped or not, the former Ann Arborites continue to communicate among themselves. Whenever possible we meet for a dinner during larger professional meetings. I am aware of four such dinners and the last time in 2005, during the European Society of Hypertension meeting in Madrid, Susan and I dined with twelve former Fellows and visitors. We are proud of our extended Ann Arbor family and Susan deserves credit for maintaining contact with them.

It is not possible to describe each individual, but on the following page is a list of people who spent time in Ann Arbor during my stay in the hypertension division. Some were trainees, some started as fellows and became staff members or were recruited to the staff, and some came to spend a sabbatical in Ann Arbor. All of them contributed to our research productivity and helped sustain a vibrant academic environment in the division.

**Europe**
Accetto R
Armario P
Andersson O
Conway FJ
Cottier C
Garcia AP
Gundbrandsson T
Hansson L
Hitzenberger G
Ibsen H
Kiowski W
Kjeldsen SE
Leonetti G
Melis A
Nazzaro P
Osterziel K
Palatini P
Petrin J
Rubli H
Sanerstedt R
Smet G
Valentini M
Vriz O
Zemva A

**Latin America**
Lapco LT
Macal O
Sanchez R
Pastrana JT
Valesquez CJ

**Australasia**
Amerena JV
Esler MD
Fitzpatrick MA
Hynyor SN
Kashima T
Kingwell BA
Mejia A
Nicholls MG
Pascual A
Tanaka N

**USA**
Argyres SN
Bassett DR
Bisognano JD
Brook RD
Colfer HT
Correa RJ
Egan BM
Ellis CN
Ferguson JJ
Ferraro J
Gajendragkar SV
Gardiner WH
Gupta R
Hinderliter AL
Hollifield EW
Jamerson K
Johnson EH
Lazar JD
Majahalme S
Marcus R
Moon DY
Nesbitt S
Neubig R
Panis R
Putzeys MR
Quadir H
Randall OS
Rivas F
Rojas Vigo A
Romero JC
Sagastume E
Schmouder R
Schneider R
Schork NJ
Sekkarie MA
Shahab ST
Sakabai A
Sidiq M
Simon G
Takyyuddin M
van de Ven C
Weder AB
Wright JT
Yarows SA
Zweifler AJ

**Ann Arbor 40 Year International Reunion, September 12-13, 1986.**

Front row (left to right): *Hans Ibsen, Borje Johansson, David Bohr, Sibley Hoobler, Stevo Julius, Andrew Zweifler, Paul Cottier, Philippe Lauwers, Antoon Amery, Mystery Guest*
Second row (left to right): *James Conway, Hubert Loyke, Andrew Dontas, Murray Esler, J. Carlos Romero, Otelio Randall, Drew Fitzpatrick, Jackson Wright, Jr., Gerhart Hitzenberger, Edward Hollifield, Rok Accetto, Ernest Johnson*
Third row (left to right): *Ronald Chipps, Sandy Sanbar, Stephen Hunyor, Geza Simon, Gastone Leonetti, Spyros Argyres, Carlos Valazquez, Guy Smet, Ali Shakibai, Roberto Panis, Heinrich Rubli, Charles Ellis*
Fourth row (left to right): *Christopher Hough, Rune Sannerstedt, Karl Osterziel, Francisco Eng, Ramiro Sanchez, Wolfgang Kiowski, Robert Schneider, Alan Weder, Brent Egan, Sverre Erik Kjeldsen, Gary Nicholls, David Bassett*

As Hoobler suggested, I presented a petition for support to the Office of Research Administration at the University and they approved a grant of $10,000. In thinking about what to request, I first wished to learn what others in the United States were doing in the field of borderline hypertension. The $10,000 would not buy a lot of equipment for my new laboratory. Nevertheless, my budget request was split into one half for equipment and the other for travel. With the equipment funds we purchased a polygraph for recording of biological signals and a densitometer for measurement of cardiac output by the dye dilution technique. This was a good beginning, but nowhere near what I needed.

At that time, the news that cardiac output is increased in early phases of hypertension trickled in from Prague, where it was first described by Fejfar, Brod and Widimsky, to other countries including the United States. In the late 1960's, four American research groups confirmed the existence of this hyperkinetic state and I decided to visit them. All responded positively and I flew to Philadelphia, Pennsylvania, Syracuse, New York, Burlington, Vermont and drove to Cleveland, Ohio to get a feel about the field. Dr. Robert Eich's group in Syracuse and Carmelo Bello's group at the Temple University followed their hyperkinetic hypertensives for a few years to determine whether the phenomenon was reproducible but had no plans for future work in the field. However, the Cleveland group was very interested, and I established excellent working relationships with Ed Frohlich and a bit later with Robert Tarazzi. Ed Frohlich was then what he is now, a highly intelligent, energetic and productive investigator deeply interested in unraveling hemodynamic complexities of hypertension. He was particularly fascinated by the role of adrenergic receptors and described a subtype of the hyperkinetic state characterized by excessive sensitivity to stimulation of *beta adrenergic receptors*. These individuals did not necessarily have high blood pressure, exactly as one would expect if both the peripheral *beta two receptors* and the cardiac *beta one receptors* were hypersensitive. The cardiac beta 1 receptors would speed the heart but the beta 2 receptors would dilate the arterioles and the blood pressure would not rise. We were unable to reproduce Frohlich's findings in Ann Arbor. To the contrary, our subjects had suppressed cardiac responsiveness to infusion of the beta adrenergic *agonist* isoproterenol. Ed and I never had an argument about our opposing findings as we both understood that the results greatly depend on how study patients were selected. We chose our subjects on the basis of having elevated blood pressure during a screening session. Patients seen in Cleveland reported for examination because they had intolerable cardiac symptoms. Unfortunately, many investigators fail to appreciate how the "*selection bias*" can affect the results. A wider acceptance of this very important bit of truth could save many trees; tons of our professional Journals are filled with unnecessary debates and outright clashes. Even revered and towering figures occasionally fall into the selection bias trap.

The head of endocrinology in Ann Arbor, Jerome Conn, was first to describe a tumor in adrenal glands which secreted large amounts of *aldosterone* and thereby caused hypertension. In one of his earliest papers, he suggested that about twenty percent of patients with hypertension might have a hidden aldosterone secreting tumor. Indeed twenty percent of hypertensive patients seen in his department had the tumor. But Dr. Conn could not possibly avoid selection bias; patients suspected of having primary aldosteronism were preferentially referred to Ann Arbor from all over America. When the dust settled, it turned out that about ten percent of patients with hypertension had an aldosterone producing tumor.

Dr. Harriet Dustan recruited Milos Ulrych, a pupil of Jan Brod in Prague to Cleveland. Milos had a good dose of what I call "decompression syndrome."

All of us coming from a communist country had learned to be very careful. The regime was watching, one mistake could ruin your career and under those circumstances you had to be cautious. Because you knew that you were being watched you learned to watch others and to categorize everybody as either good or bad. In that polar world, not having information about informers could be harmful. Of necessity one developed a dual approach; to the bad guys you were fleetingly pleasant but careful not to spill the beans. With the good ones you openly analyzed the daily situation. And when I say "daily" I mean it; evaluating things around you became a way of life. Gossip became both a sport and a necessity; sport since you enjoyed juicy pieces of information and necessity because you had to be well informed. This slightly paranoiac way of life had its own rewards; the pleasant tension of the daily cat and mouse game with bad guys and the ability to draw a great deal of security from understanding, in strictly black and white terms, who is who and how to relate to every person around you. Nevertheless, we understood how demeaning and unproductive this life style was and we craved for freedom. But when we reached the West, everybody was polite, looked benevolent, and appeared evenhanded. It just wasn't true; all these folks couldn't possibly be as nice as they pretended. Hidden among pleasant faces there had to be somebody who didn't wish you well. Gossip is a taboo in America and it takes years before Americans confide to an immigrant what they think about other natives. Robbed of the regular diet of gossip, the brain of many an immigrant starts to play tricks.

During the second week into his visit to Ann Arbor, Djoka, a psychologist from Belgrade, smelled cigar odor in his room - a sure sign that the FBI was spying on him. My protestation that he was too small a fish for the secret services did not help. Convinced that the FBI would kill him, he sought refuge in our house. A bit later he decided that the FBI chose me to become his executioner. Nevertheless, he kept on coming to our house to chat with me, over a cup of coffee, about his scheduled demise. I had had enough and asked the logical question:

"If you think I will kill you, why don't you avoid me instead of torturing yourself and all of us on a daily basis?"

"Because I like you and know that you have no choice in the matter. You will do what they tell you but I am sure you will stretch it as long as you can. If I have to die, I could not have asked for a nicer assassin," replied his twisted brain.

I called his American sponsors and suggested they ought to return him to Belgrade. They purchased an airline ticket for him and a burly psychiatrist who spoke nothing but English.

"Coward, coward! No guts to do your own job," Djoka mumbled as I escorted him to the gate.

When they landed in Yugoslavia, Djoka rushed through passport control with the psychiatrist in hot pursuit. But the psychiatrist got stuck at the passport booth, the police officers insisted on checking his passport. The gate keepers spoke only a few words of English and unable to explain his plight to them, the psychiatrist became aggressive. Back in the mid 1960's you did not want to challenge the representatives of People's Republic of Yugoslavia. Nor did you have to speak their language to offend them. The passport people went into the slow motion routine. Scared to lose his patient, the psychiatrist tried to force his way past the guards. That was a big mistake; two policemen - one on each side- grabbed him. Greatly offended by this attempt to limit his personal freedom, the psychiatrist screamed at the top of his lungs. At that point Djoka regained some of his old sense of humor, walked up to the policemen and calmly explained that he, Djoka, was the lunatic and the ranting and raving piece of humanity they were wrestling with was his psychiatrist. Unfortunately, after a short remission the malady returned. Djoka imagined that his wife was after him and amid a loud confrontation in their tenth floor apartment he jumped to his death.

Years later on my sabbatical leave in Professor Folkow's lab in Goteborg, I ran into a guy from Prague who was quite disoriented by the calm, polite and very low-key Swedish demeanor. I attempted to help him but he grew dependent on me. His obsessive need for support started to affect our family life. He'd call at any time to complain about true or imaginary problems. A polite explanation that we were busy would not register; after a perfunctory apology he'd just continue the litany. I eventually asked him not to call and instead look me up during working hours in the department of physiology. But this weak attempt to limit my kindness to a specific time of the day backfired. If I was five minutes late, he'd call and ask for me. I finally explained the situation to Professor Folkow. Folkow, a giant in our field and a humanist par excellence, knew what to do. He first told the man that his hard work was much appreciated. Next he invited him for weekly evaluation sessions. Folkow combined these mostly laudatory but occasionally critical assessments with setting specific goals for the next week. That was all the man needed to regain balance.

Compared to the paranoiac psychologist from Belgrade and the anxious Czech in Sweden, Milos Ulrych from the Cleveland Clinic was a picture of health. Nevertheless, he too was affected by his past. He'd tell you what he was doing only to indicate that there was much more to the story than what he told you. I classified him as a charming nut and decided to cut the interview short. Free exchange of ideas is the basis of scientific progress and if someone does not want to join in the fun, that certainly is his privilege.

On the whole my survey of activities in borderline hypertension was not encouraging. Most Americans who studied the hemodynamics of borderline hypertension were satisfied to confirm the finding of elevated cardiac output. A few completed short term follow ups and only the Cleveland group planned further research in the field. Seeing that others did not share my enthusiasm, I tried to reignite their interest in the field. Dr Hoobler was very supportive and contacted an acquaintance in the life insurance industry. Eventually I got a small grant to organize an evening meeting during the American Heart Association Scientific Sessions. Ed Frohlich assisted me in setting the agenda but just a few invitees showed up. We thought we'd do better the next year. I sent very early notification but the second meeting attracted even a smaller group. Eventually I gave up the grandiose plan for a network of investigators in borderline hypertension.

❖❖❖

Not everything during my travels to various centers in the USA was in vain. In fact, my travel to Burlington, Vermont was a smashing success and had a major impact on my scientific career. Occasionally, younger colleagues come to my office to discuss their research careers. As if they had rehearsed together, most of them start with the same sentence: "Dr. J, what is hot in hypertension?" I usually explain they are asking the wrong question. New "hot" methods of investigation and new "hot" fields of research come and go. You must first review the contemporary and old literature to define what interests you. And you cannot be lukewarm about the chosen field. Only if you are a true enthusiast will you want to work long hours and endure the skepticism of the peer-review process. If a new method can answer your question, by all means utilize it. But you should not reduce yourself to the level of a sophisticated technician who knows how but lets others tell him what to measure. Finally, you ought to understand that even if you did your level best your ultimate success will depend on having some luck.

Strangely, my lucky streak started exactly on June 6, 1968 during my return from Burlington, Vermont to Ann Arbor via New York City. This was the day after the tragic attack on Robert Kennedy. Next to me on the small airplane to New York was a visibly shaken passenger.

"This would be a tragic loss but Bobby is still alive and I hope he will survive," said the man. There was a look of desperation in his eyes as he kept pointing at the word "alive" in the newspaper heading. I've seen this before, facing the unimaginable and unable to contemplate the impending loss most people hang on to the smallest straw of hope. As a physician I learned that extinguishing a hope, no matter how unrealistic it might be, is brutal and unnecessary. But this time around I surprised myself.

"I hope he doesn't make it," I said. "He was shot in the head from a very short distance and, though the assassin used a small caliber gun, chances of survival are minimal. And if he were to survive he'd be comatose for years. Or even worse; the bullet might transform this sportsman and highly intelligent leader into an uncoordinated mumbling imbecile."

"What are you," asked my fellow traveler, "a doctor or a gun expert?"

"Both, I am a doctor and during the Second World War I have seen a wide array of weapons and many head wounds."

"Aren't you a bit too young to have participated in WW II?"

"Definitely too young but I had no choice in the matter. Our family is Jewish, the Germans occupied Yugoslavia, I was kicked out of school and the entire family was in danger of detention in a concentration camp. So we joined the partizans resistance movement. At 14 years of age, I carried a gun and served as a courier between guerilla hospitals. Of all wounds, head injuries are the worst; the patient might linger for a while but the final outcome is invariably tragic."

Explaining that he too was Jewish, the man next to me continued to ask questions but his expressive eyes spoke volumes and I felt he didn't quite believe my war time story.

That morning everything seemed to go wrong. I had already broken the Good Samaritan rule and, instead of giving him some solace, I told him the brutal truth. Next, I broke my personal rule of not chatting about my past. People in Yugoslavia didn't crave to hear one more story about the good old bad times and my guerilla experience had even less appeal to an average American. But the bird was out of the cage and I tried to shift the conversation in a different direction.

"May I introduce myself? My name is Stevo Julius and I work in the University of Michigan hospital. Your accent tells me you are from the South. Am I right?"

"My name is Leo Fields. Yes, I am from the South. I live in Dallas, Texas and work in the Zale Company." The formal introduction trick seemed to work. The door to ask more was now open.

"Glad to meet you, Mr. Fields. What does the Zale Company do?"

"We produce and sell jewelry. Mr. Zale owns the largest jewelry business in the United States. I was on a business trip to Burlington. How about yourself? What took you to Vermont?"

I think Mr. Fields sensed I did not wish to talk about the war time experience and willingly gave me an opening. I jumped at the chance and spoke about the purpose my trip and my aspirations to do research in the field of borderline hypertension. Mr. Fields was a good listener but after a ten minutes long monologue I started to worry whether I had overdone it.

"No! To the contrary, I am quite interested, I appreciate medicine and would like to hear more about your plans," said Mr. Fields.

Just as I knew he had some doubts about my war story, I knew he was quite fascinated by the medical part of the conversation. Leo Fields was only a few years older than I and that created a pleasant atmosphere. By and by I opened up and, as I would have done with a friend in Yugoslavia, I came to lament the difficulties of starting a new line of research in a foreign country. After all, I was new, just a few people knew me, and though my boss supported me I wasn't sure things would work out. At that point, the conversation came to halt. Leo probably was not familiar with the Slavic urge to elicit sympathy by a bit of melodrama.

"This is a land of opportunities." said Leo, "If you persist, you will succeed."

This matter of fact statement woke me up. You do not bewail things and you do not show doubts in America. If you don't believe in yourself, nobody else will believe in you.

I regained my composure and asked Leo why he was interested in medicine.

"Medicine is my hobby. I believe in it and I support it. Recently I gave money to build the Pediatric Hospital in Dallas."

I was stunned. You do not travel everyday next to a man capable of building hospitals. Not knowing how to respond I uttered some general platitudes and fell silent.

"You know what," said Leo, "I think my uncle, Mr. Zale, might be willing to help. Why don't you join us for a family supper in his apartment in New York? Let's see whether you can rearrange your ticket at the airport. If that works out, we could drive straight up to the apartment."

"Thank you. It sounds like a great plan and I hope we can work it out," I said without falling back on usual qualifiers such as "I hope it is not inconvenient" or "Am I dressed well enough for the occasion?" I correctly assumed that the Zale family, which came from Eastern Europe, preserved an old Jewish custom from that part of the world. If you met or heard about another Jew in some sort of distress, you'd bring him for a relaxed meal with the family. There was always enough food in the kitchen for yet another person. Before World War II, my parents practiced that tradition and we had frequent dinner guests, mostly students but occasionally also emigrants on the way to another country and people separated from their families. I knew exactly what was coming. Leo would introduce me, I'd say a few words about myself and then we'd all sit down for a pleasant dinner.

Everything worked out and we were on our way to Mr. Zale's apartment. By now I don't remember the address but the Zale residence was in a relatively low building on one of the grand Manhattan avenues. A large group of people had already assembled in a huge dining room which, but for its size, resembled our dining room in Zagreb; a large table surrounded by sturdy upholstered chairs with tall backs, modest but true silver silverware, elegant porcelain dishes and a pleasant somewhat subdued early 20$^{th}$ century atmosphere. Leo introduced me as a new arrival to the States, a promising fellow who already had landed an academic job in Michigan, and a person in pursuit of his dream to start a new avenue of research. After that succinct introduction, it was easy for me to be thankful for the invitation, make clear that I understood the Jewish hospitality tradition and briefly explain what I'd like to do in research.

After dinner Leo asked his uncle to speak to me. Mr. Zale was a pleasant gentleman in his sixties who by appearance and demeanor could have easily been one of my Father's friends. Akin to gentlemen from Zagreb he wasn't pompous but had an aura of authority about him. The overall pleasant atmosphere helped me to relax and I made a focused presentation of my research interests. An excellent listener and a master in asking relevant questions, Mr. Zale must have concluded that I would be able to succeed on my own. You would not expect this from a rich businessman, but Mr. Zale had an extraordinary grasp of the system of research funding in the United States. He suggested that I submit a grant proposal to the Michigan chapter of the American Heart Association. These chapters, he said, were keen to help young investigators in their area. Later I could apply to a career development program of the American Heart Association and eventually, if I had generated sufficient preliminary data, I could apply for a more substantial support from the National Institute of Health.

Whereas this was a polite way of saying no, Mr. Zale was the first to explain how the system works and by and large I later followed his advice. In the whirlwind of having to change the airline ticket in order to accept the unexpected invitation for dinner, I had no time to daydream what might happen during the dinner. Not having any specific expectations, I wasn't particularly disappointed with the outcome of the interview.

"So you had a good dinner, met some interesting people and found out that even in America manna does not fall out of the sky," I said to myself. But Mr. Fields was quite upset. As he escorted me to the door he handed me his calling card.

"My uncle has better means than I and he could have given you solid support. But I would like to help you to the extent that I can. If you send me an official request and my contribution goes through regular

University channels I will send you $15,000 a year for two years. I hope this will help."

I was amazed and barely could find the right words to thank him. Mr. Fields lived up to his word and his contribution was absolutely crucial; I was able to complete the equipping of my new lab and had enough money left to hire a part-time technician.

I was wrong. In America, and only in America, manna can fall from the sky.

There were a few periods in my life when everything went wrong. For months after my father's tragic death, I'd wake up wondering what calamity would hit us next. But the chance meeting with Mr. Fields was a prelude to my first ever string of good news. For the next few years, everything seemed to line up beautifully in Ann Arbor. Dr. Hoobler got his grant for a cooperative study to investigate whether antihypertensive treatment can prevent the recurrence of strokes in people who had had a previous stroke. He was the principal investigator but the five year budget also funded the position of a national coordinator. Hoobler appointed me as the study coordinator and made it pretty clear that he would interact with the principal investigators in various study centers and with the scientific leadership of the study, but that all of the operational aspects were my problem. I was to lead the study center in Ann Arbor which initially supervised the recruitment of patients and later monitored every detail of the project. We reviewed reports for missing data and accuracy, transferred the data to coded sheets for entry into computers, and reminded the centers if the record of a scheduled return visit did not arrive in due time in Ann Arbor. I also responded to all study related medical queries and supervised the work of the study secretary and head nurse. Dr. Hoobler recruited a tall, imposing, no nonsense chief study nurse. Mrs. Bender, wife of a retired Army officer and a proud member of the Daughters of the American Revolution, who ran the study with a booming voice and unquestionable authority. The secretary was a young, precise, clever, witty and extremely good looking divorcee from Great Britain.

In my new position as the National Coordinator of the Stroke Study, I had salary support for the next five years. Though Dr. Robinson, the Chairman of Medicine, ran the department in low key fashion, there was a clear expectation that ultimately everyone had to generate his own salary support, either from clinical work or from funded research.

Backed by Hoobler's grant, I had a good five years of protected research time in front of me, a rare commodity in clinical academic medicine.

But the best of all news was that I finally succeeded to resolve a burning personal issue. For a number of reasons my first marriage had started to crumble. It started under a bad omen as my Father lost his life a few days after the marriage ceremony. All sorts of external forces added more strain to the marriage. The ensuing political difficulties, my exile to Bosnia, the meddling of our families, and the inability to get an apartment for independent family life all contributed to the tension. For medical reasons, we could not have children of our own. We had very different personalities and after 16 years it became painfully clear that neither of us could change. I tried to hang on but the sense of being hopelessly trapped eventually prevailed. I moved out of our house and in due course we obtained a divorce. There is no such thing as an "amicable" divorce but we settled all main issues in a civilized fashion.

I am sociable by nature and the empty freedom I had just obtained did not suit me well. Luckily at that point Susan Durrant, the secretary in the stroke study, entered my life. She was very attractive, lively, optimistic, straight-forward and had a great sense of humor. Chaperoned by another division secretary, we started to have lunches. I could barely wait for the next opportunity to relax in her company; she was such a contrast to the morose atmosphere of my previous marriage. And I fell in love. Admittedly I was an old fashioned formalist. To erase any trace of nepotism, I found her another job and Susan agreed to resign. Besides the formal reasons, we both wanted to test whether the relationship would survive the separation. And it did with flying colors! We got married, had children, had grandchildren and thirty five years later continue to be happy.

The stroke study also introduced me to Tony Schork, a brilliant statistician capable of explaining complex analytic issues in simple terms and a man of encyclopedic knowledge. Tony and I reviewed the literature on borderline hypertension and in 1971 we published a comprehensive paper.[5] That paper, more than anything else, convinced me that borderline hypertension was an important field and that investigating it might yield important results. Tony and I became the best of friends and after so many years we still continue to work together.

**Bibliography**

1. Julius S, Amery A, Whitlock SL, Conway J: Influence of age on the hemodynamic response to exercise. Circulation 36;222-230, 1967.

2. Amery A, Julius S, Whitlock LS, Conway J: Influence of hypertension on hemodynamic response to exercise. Circulation 36:231-237, 1967.

3. Sannerstedt R, Julius S, Conway J: Hemodynamic response to tilt and beta-adrenergic blockade in young patients with borderline hypertension. Circulation 42:1057-1064, 1970.

4. Palatini P, Julius S: Heart rate and the cardiovascular risk. J Hypertens 15:1-15, 1997.

5. Julius S, Schork MA: Borderline hypertension - a critical review. J Chronic Dis 23:723-754, 1971.

# A Lucky Streak
# 1972 – 1976

I thought working on the stroke study would be a tolerable but dull routine. I particularly dreaded attending meetings of various study committees. Back in Zagreb in a perverse show of pseudo-democracy, the Party would drag us to lengthy discussions of their predetermined conclusions. Since you could not change a darn thing, there was no point to the discussion. To cope with this dreadful waste of time, we learned to appear alert but completely block the input. You'd convert lengthy speeches into an irrelevant background noise for daydreaming. Only rarely, when the stakes were high, you'd expose the weaknesses of the proposal, stretch out the discussion and at the end deploy the ultimate defensive weapon, a suggestion to appoint a "study group." You couldn't derail the train but you might slow it down, and now-and-then force it to change track. I imagined the stroke study committee meetings would be equally unproductive and boring. I was wrong. If a decision has not been predetermined and the participants genuinely seek the best solution, committee meetings can be quite engaging. I have always been interested in people's behavior but the stroke study made me a true connoisseur. It was great to see different brains and personalities in action. Convinced that they were right, the self-confident and sometimes self-centered investigators led the charge only to be challenged by equally assertive people holding a contrary opinion. As sparks flew left and right, it was easy to see the strong and weak aspects of each argument. While appearing neutral, Dr. Hoobler waited for the opponents to run out of steam and then invited the cooler heads to comment. By and by the discussion would calm down. Eventually the best aspects of what initially appeared to be antagonistic proposals found their way into the final plan of action. This generally worked very well. However, persistent persons have a way of exhausting the audience and in rare cases, when a compromise was crafted solely to appease a hothead, the solution proved unworkable.

The most effective leader towards a compromise was Dr. Elbert Tuttle, a nephrologist from Atlanta, Georgia. It was a pleasure listening to him as he concisely, and in a typical unhurried Southern manner, summarized the issue. Polite and respectful, he'd use just the right adjective to disagree without offending anybody. Yet he was not wishy-washy and he knew well what he wanted to achieve. Rather than proposing a solution, he'd give you the facts in such a logical manner that you simply had to come to his conclusion. Having completed his comments, Tuttle quietly waited for others to finish the job. It never failed.

In retrospect, I realize how much I learned from the stroke study. Hoobler and Tuttle taught me two important rules of leadership; that you should give everybody a fair opportunity to speak and that leading with logic is much more effective than attempting to crack heads. The skills I acquired in the stroke study proved very useful decades later when I chaired a few clinical mega trials.

Dr. Hoobler was quite disappointed with the results of the stroke study. Lowering of the blood pressure reduced the incidence of heart failure and other cardiovascular events but failed to decrease the rate of recurring strokes. Unfortunately, Hoobler fell into the most dangerous trap in research; he was so sure his hypothesis was correct that he couldn't accept the negative. Nowadays you are requested to state your statistical analysis plan well in advance, but in the early 1970's you could pretty much do what you wanted. Hoobler kept asking Tony Schork to analyze the data in different ways. Tony diplomatically accepted additional work assignments but no matter how he sliced the cake, there was no evidence that blood pressure lowering decreased the incidence of new strokes in patients who survived a previous stroke. When I said he was diplomatic I meant that Tony Schork completed all requested analyses fully knowing that the results would not change. He gave Hoobler a bit of the same medicine Hoobler used on others; let them talk until they are exhausted. However, Tony was firm if Hoobler attempted to break some basic statistical rules. The conflict frequently revolved around the statistical "penalty" for repeated testing of the same population. All statistical methods are based on probability, a likelihood that an observation did not occur by chance. The more times you look, the more likely you are to find something by chance. The lowest accepted level of statistical significance ($p< 0.05$) means that in one of twenty tests you might find something that seems positive when, in fact, nothing is there. If you test the same population five times, there is approximately a one in four chance of declaring an accidental finding as significant. Statisticians use various corrections to atone for the effect of multiple comparisons. Each of these corrections raises the bar and makes it much harder to prove a finding is significant. Eventually

Hoobler had to concede that we successfully lowered the blood pressure but failed to decrease recurrence of new strokes.[1] About 30 years later, Dr. Stephen McMahon and his colleagues tested Hoobler's hypothesis in the PROGRESS study. Using different types of antihypertensive drugs and with a much larger *sample size* they showed that antihypertensive treatment can reduce recurrence of strokes. Unfortunately, Hoobler did not live long enough to see that his original hunch was correct.

About a year after my chance encounter with Mr. Fields and with his financial support, I assembled my own laboratory. Next, I submitted a protocol for hemodynamic studies of borderline hypertension to the institutional Ethics Board. Upon their approval, we began recruiting experimental subjects. However, my one room laboratory with a part time technician could not process many subjects and progress was slow. The lack of progress started to bother me but, unexpectedly, the lucky streak which started with the dinner in the Zale residence resurfaced.

A few months after the "grand opening" of my laboratory, James Conway came by to check out how things were going. I was delighted to show him around.

"Not a very luxurious outfit, I'd say!"

"Not too luxurious but certainly better than nothing. At least I am not crowding your laboratory," I said, "and as we agreed I will bring our output to your laboratory for weekly quality control sessions."

"That won't be necessary, my man."

I could not quite figure what that meant and looked askance.

Conway smiled: "I am leaving, old boy. Off to Washington to work with Ed Freis. Made me an offer I couldn't refuse. He either likes me or wants to get all that equipment. Because, you see, I bought the equipment with my grant money and I am taking it with me. I'll take everything, and I mean everything, from the tilt table down to the last chair!"

I was stunned and could not find the right words. Our relationship was correct but not cordial. I could not ask why he suddenly decided to leave. It would have been patently false if I fell back on the "we will miss you" routine. So I just stood there looking lost.

"Rejoice, you will inherit my large four room suite. I settled that with the Dean. He thinks he can stop me from removing my stuff but he is dead wrong. I got all the right papers and if needed I'll come in the middle of the night and take what is mine. So there!"

I still did not know what to say.

"Come on, you did not expect me to leave all that stuff to you. I got you the rooms and that ought to be enough."

"For goodness sake how could I have expected to inherit your equipment when a minute ago I didn't have the slightest idea you were leaving?" I protested.

"You are quick enough to put two and two together and you'd be surprised how quickly people get greedy. Give them a finger and they'll want both arms!"

"But you are not that way," continued Conway, "and as you said; what you have is better than nothing. It will be a good start and I wish you all the best."

He shook my hand and headed for the door.

"Call me if you need some help at night." I yelled as he walked down the corridor. Eventually I caught up with him and continued talking: "I am good at moving stuff. And it will be a pleasure. There is not greater fun in academic life than sabotaging a Dean."

Conway smiled. I was pleased to finally find the right words. I learned a great deal about research from Conway and I will always remember him as a somewhat rigid but objective, honorable and decent man. Meeting Conway was a new experience. Here was somebody who could take it and dish it out without ever getting personal. I must have been a big thorn in a certain part of his anatomy but he invited me back to the United States and to the extent that it was possible, he supported my career. How different from the temperamental bipolar friend-or-foe behavior I was accustomed to in Yugoslavia!

Back in the fall of 1969 when I moved into Conway's large research suite, he did not strip the laboratory as drastically as he threatened. He left behind a solid quantity of supplies as well as two large power stabilizers. These powerful stabilizers were symbols of Conway's meticulous approach to research. In Conway's lab we calibrated all measuring devices every day and there was a large day-to-day variability of baseline values. This bothered Conway and as the first step in resolving the issue he decided to monitor the power supply to the laboratory. He hit a bull's eye; the 24 hour voltage graph showed all sorts of short and long term peaks and valleys. Some surges might have reflected the usage of power hungry hospital machinery but the long term voltage undulations suggested that the power was not well regulated at its source. The installation of stabilizers resolved the calibration problems but on the "you-never-know" principle, Conway continued to mandate daily calibrations. Having those stabilizers was absolutely crucial for our future work and it was generous of Conway not to deny them to me.

Moving into Conway's laboratory promised to greatly increase my productivity but it also meant that I needed additional support for supplies and at least one full-time technician. I submitted a research proposal to the Michigan Heart Association and they funded it for two years. Using some of Hoobler's and my newly acquired resources, I had enough money to hire the first fellow to work exclusively on my research projects. And, again, my lucky streak continued. A few candidates applied and among them was Dr. Arturo Pascual, a naturalized American from the Philippines. He came to interview over the weekend and just as we were to unpack supplies for a nice picnic in nearby Dexter, his car collapsed in mortal agony. He had to return by Monday to his hospital located about 300 miles south of Ann Arbor and his town had no airport. My empathy for others and my personal sensitivity to deadlines did not help. I started to fret about Arturo's problem, became tense and my bad mood threatened to spoil the nice outing. But Arturo calmly placed a telephone call to his hospital and another to some relatives in Detroit, arranged everything and apparently unperturbed sat down to enjoy the event. I offered him the job on the spot and that was a good decision. He proved to be an excellent clinician, hard worker, and reliable organizer of research. During his two years in Ann Arbor, we published seven original papers in peer reviewed journals. Arturo left us to join the hypertension group in Detroit's Henry Ford hospital where he became renown for diagnostic skills and the excellent care of patients. After three decades in Detroit, Dr. Pascual took an early retirement and returned to practice medicine in the city of Cebu in his homeland. He had a strong sense of loyalty to the medical school in Cebu and great empathy for underprivileged people in his town. Arturo is a quiet man and I learned about his commitment to the old country by chance during a lecture tour in the Philippines. My wife Susan does not like long travels but on that occasion she did join me and we inserted into the program a few days for rest and recuperation. Arturo was delighted to assist us in arranging a "tourist" day in Cebu. It turned out we were guests of the Mayor of Cebu who arranged a marvelous tour and invited us for dinner. As we moved from one interesting place to another, when our guides mentioned Dr.Pascual's name all doors magically opened. Apparently Arturo had organized the Philippine community in Detroit and they made huge contributions of cash and medical equipment to the local hospital. Later, when he informed me he'd return to Cebu, I understood his decision. He was well rewarded for work in the United States and having no financial worries, Arturo wanted to practice medicine for the benefit of his community.

At about that time I got a "free" helper in the lab courtesy of a program developed by Dr. David Bohr in the Department of Physiology. He obtained a training grant in cardiovascular medicine from the National Institute of Health. This unique program reached out to undergraduate students long before they enrolled in medical school. Bohr led the program with great enthusiasm. He assembled a first rate faculty willing to work with students during the summer recess in the undergraduate school. He organized the didactic program, regular journal clubs, group discussions, and found individual mentors for each student. These mentors gave students the opportunity to actively work in various laboratories. Bohr selected very bright and active students for the program and this dynamic group of young people energized everybody around them. Instead of the slow down during the summer vacation period, all laboratories became beehives of excited activity. I was assigned to mentor Charles (Chuck) Ellis, a highly intelligent, energetic and sociable student from nearby Hillsdale, Michigan. He worked late hours, avidly read the literature, and within a few weeks fully understood the scope of the project. Since he did not have a medical license he couldn't perform invasive procedures, but he aggressively recruited normotensive volunteers for hemodynamic investigations. He also organized and largely executed a study of clinic and home blood pressure measurement among University of Michigan undergraduate students. During his short stay, he did enough work to deserve senior authorship on a paper about the venous tone in borderline hypertension and he was the senior coauthor on the paper about the utility of home blood pressure measurement. When he enrolled into Michigan Medical School, Chuck's biography brandished a paper published in the British Heart Journal and another in the highly respect Journal of the American Medical Association, a rare feat for an undergraduate student. I thought he'd have a stellar career in cardiovascular medicine but Charles ultimately chose skin over the heart and presently is professor and associate chair of the Department of

Dermatology at Michigan. Whereas our paths differ, we make a point of keeping in touch and continue to remain good friends.

My luck in getting good coworkers did not stop there, as I soon acquired my first foreign fellow. Lennart Hansson worked in the Sahlgrenska University Hospital in Goteborg, Sweden. His mentor, Professor Bertel Hood, a pioneer in Swedish clinical hypertension research, decided to send him to the United States for further training. It is a mystery how Hood got the idea to send Len to my lab. Whereas we published quite a number of high quality papers, I was by no means known internationally and I had not yet met Dr. Hood. The exciting news got even better when I realized Hansson would continue receiving salary from Sweden. What a deal! For a few days I was quite impressed with myself.

In his letter, Dr. Hood described Hansson as a capable and promising young scientist. This was an understatement. I have never before seen such a well-organized person. Hansson planned to come to Ann Arbor with his wife and three daughters and I took it upon myself to help them. We made reservations in a motel, prepared a list of suitable apartments, lined up a few decent used cars, prepared a package about Ann Arbor schools and assembled a folder containing useful information about Ann Arbor and various other issues such as obtaining a driver's license, social security number and a hospital employee orientation package. And of course, we went to meet him at the airport. Down the ramp came this tall, straight-as-an-arrow, heavy boned, but slim, blonde man with gigantic mustaches.

"Thanks for coming to meet us," he said, "but I did not plan on coming to Ann Arbor tonight. Anticipating this would be a long and tiresome trip for the entire family I've made reservation in an airport motel."

"Why the hell did I bother to make a reservation for him in Ann Arbor?" I quietly wondered. My European background precluded me from bringing up the other reservation in Ann Arbor. It'd be impolite to burden my guest with unnecessary choices.

Hansson introduced his family; his wife a slim, somewhat shy, elegant Swedish blonde and three equally blond daughters ranging from four to eight years in age.

"Welcome to America. By all means have a good rest and we will pick you up tomorrow," I said." In the meantime you can take a look at these lists of apartments and cars. We can get on with it when you recover from travel. The quicker you settle the basics, the easier it will be to enjoy the life here."

"That is very kind of you but I have already purchased a new car and hope to take the delivery in a few days. In the meantime I'd prefer to take a week of vacation with the family. We can then start apartment hunting."

"But don't you need a driver's license before you can use a car?"

"I have an international license. It ought to be OK for a few weeks. Thereafter, I will take a driver's exam and obtain the regular license."

"I see you are concerned about our welfare and that is appreciated,' continued Hansson, "but please don't worry. We are quite accustomed to taking vacations in other countries and we will have fun. I do not expect any problems but should I need to call you, we have your telephone numbers."

I was taken aback by his "do not call me, I'll call you" response and by what appeared as a standoffish attitude. After a few months I learned that if you got his confidence Len was an intrinsically open and sincerely warm person. I wish I could do it his way. Being a bit reserved could have saved me from many disappointments.

After a week, Len came to the office. He had already obtained the driver's license, got the social security number, established an account with the local bank and applied for the medical license. And somewhat sheepishly he admitted to having already rented an unfurnished apartment which he outfitted with leased furniture. This would be an unusual feat in the present era of internet but in the early 1970's, Len's performance was phenomenal. I have no idea how he got the needed information without a Google search!

Len continued to surprise us. At the end of the first year when he had to file a tax return, Len went to the University of Michigan personnel department to show them a document he brought from Sweden. According to such and such a paragraph in the United States-Sweden tax agreement, he did not have to pay American taxes! I regret not to have gone with him in order to see the jaws of our bureaucrats drop wide open.

As soon as he started to work with us, Len did put me to a serious test. At that time Europeans used beta blockers for treatment of blood pressure but in the United States such drugs were approved only for angina. Not having personal experience, we were not quite convinced that beta blockers could lower blood pressure. Len teamed up with Gerhard Hitzenberger to convince us to the contrary. Gerhard came from Vienna to work with Dr. Hoobler and with our clinical pharmacology group. He was a knowledgeable, thoughtful scientist with the typical pleasant (gemutlich) Viennese disposition. He has recently retired from his position as full professor of medicine at the legendary University of Vienna. Both he and Len showed us very convincing data that propranolol lowers blood pressure but both disagreed with Brian Prichard's notion that if it were given in stratospherically high doses propranolol would lower pressure in everybody. In fact, Len wanted to study patients on moderate doses of propranolol to discern why some people fail to respond to the drug. A few weeks after his lecture about propranolol, Len dropped on my desk an excellently composed and well-argued research proposal. This created a twofold problem for me. Whereas I investigated borderline hypertension, Len wanted to study patients with established hypertension. Furthermore, propranolol was not yet approved for treatment of hypertension. I knew Len's parallel program would slow down my own work. Unwittingly Len turned the table on me; I was now in Conway's position having to decide whether to let a younger colleague interfere with my core program. Some people are very good at explaining to themselves why they must do something they thought they would never do. I am not one of them. The choice was perfectly clear; would I behave as Conway or not? I simply had to let Len do his thing, otherwise I'd be looking at a bastard in the mirror every morning.

However, the fact that propranolol was not yet approved for hypertension created a problem. While I kept looking for the right strategy to circumvent the obstacle, an ambiguous letter from the Food and Drug Administration arrived on my desk. Facing a dilemma the FDA took refuge in a Delphic oracle. They wanted all physicians to know that if propranolol were not safe for human use it would have not been approved for angina. So far so good, but then the letter reminded all physicians that the drug had not been approved for hypertension and if they were sued for using it to lower blood pressure "the FDA could not predict" what the court would say. What a cop-out; the letter protected the FDA's collective glutei but had no guts to forbid the use of the drug for yet unapproved indications. I took solace from the first part of the letter, Len prepared an application to what was then called the Committee for Use of Human Subjects in Experimental Purposes presently known as the Institutional Review Board, and they approved the project.

As with any other protocol, I very carefully reviewed all details of Len's proposal and asked him to justify every step. He was not defensive and accepted all reasonable suggestions. The work he completed in Ann Arbor became part of his Ph.D. thesis in Goteborg. Len took the first set of hemodynamic measurements after a four week course of placebo. He then obtained baseline values and compared them to the response to intravenous injection of propranolol. Patients were then treated with oral propranolol for four weeks and the hemodynamic measurements were repeated.[2] Intravenous propranolol decreased cardiac output but vascular resistance increased and blood pressure remained unchanged. Oral propranolol lowered blood pressure and compared to the initial response the vascular resistance decreased. However, cardiac output remained suppressed well below baseline values. Clearly chronic beta adrenergic blockade had different effects than the acute blockade and with chronic oral propranolol "something" induced the vascular resistance to decrease. When Len assessed the difference between responders and non-responders, none of the measures he took predicted who would respond to the treatment. The degree of beta adrenergic blockade assessed with isoproterenolol infusion was equal in responders and non-responders, plasma renin levels were equally suppressed in both groups and measurement of baroreceptor sensitivity did not provide an explanation for the difference in blood pressure response.

Len's search to determine how beta blockers decrease blood pressure triggered further research in our hemodynamic laboratory. Most interested in the topic was Geza Simon, Hungarian born and educated in the US, a charming, clever and complex man. Geza came from Michigan State University in Lansing where he acquired a solid background in animal models of hypertension. On the surface Geza was an "equal opportunity" cynic; he belittled his own and other people's work and tended to put everybody, himself included, into their right place. His quick mind and a dry wit were disarming and nobody took offense. If I enthusiastically expounded on the importance of our work, he was likely to remind me that we all are just trying to make a living and that there is nothing so special about what we do. But he worked hard, knew his science, and socially was a really nice guy. It was fun to be around him. You never knew when the barb would come. Here is a typical example: In the 1980's during the International Society of

Hypertension Meeting in Kyoto, I organized a dinner for the Ann Arbor hypertension group in a traditional Japanese restaurant. Since everybody's stay in Ann Arbor may not have overlapped, I introduced the guests as they arrived. Sverre Erick Kjeldsen from Norway got an important Award from the Society and belatedly came to the dinner directly from the ceremony. We were all relaxed, had taken our coats off and, of course, everyone was sitting on the floor. From our low vantage point the fully clad and naturally tall Norwegian looked even taller and more formal.

I introduced him and mentioned with pride his important award.

"What award ?" asked Geza. "Was it for the best dressed young man?"

That was vintage Simon; witty, deprecating and uncalled for. But Geza could not help himself; that is the way he was.

Geza studied timolol, a new potent beta blocker. First he reported that the baroreceptor sensitivity did not increase during the treatment. Thus, timolol did not lower blood pressure by increasing the sensitivity of baroreceptors.[3] In the next paper, he infused beta adrenergic agonist isoproterenol to test the sensitivity of cardiac beta adrenergic receptors.[4] The hypothesis was that people with higher receptor sensitivity might better respond to the drug, but no such relationship was found. Furthermore, neither the degree of beta blockade nor the decrease in plasma renin activity predicted the blood pressure response. In the last experiment, Geza injected atropine and phentolamine at rest and after chronic treatment with timolol.[5] The parasympathetic tone (assessed by the parasympathetic blocker atropine) did not change but the alpha adrenergic tone (assessed by alpha receptor blocker phentolamine) increased during the treatment. This suggested that, in the absence of vasodilating beta receptor activity, the vasoconstrictive alpha adrenergic activity may prevail. In this sophisticated paper Geza pointed out a possible cause of resistance to the treatment but did not explain how beta blockers lower blood pressure.

Geza left us to take a position at the University of Minnesota in Minneapolis where he continued with research and practice in hypertension and advanced to the rank of Professor. We keep in touch and I was not surprised that upon early retirement he decided to spend considerable time in Hungary. Geza had strong feelings about his native land and wished to contribute to better and more available health care in

Budapest. Though he pretended to be a cynic, deep down Geza was not a misanthrope.

Harry Colfer, a staff cardiologist who in 1982 spent a research year with us, took the last stab at the beta blocker enigma.[6] He investigated two possibilities 1) that, because they slow the heart rate and increase the stroke volume, the beta blocker might cause a greater stretch of cardiopulmonary receptors which might decrease the sympathetic tone, and 2) that beta blockers lower the blood pressure by triggering *autoregulation*. He found no relationship between the cardiac stretch and the blood pressure response and before I report his finding on the second point I must explain what autoregulation is. All blood vessels in local tissues have a capacity to maintain a steady flow; if the flow is too low they dilate and when the flow is high they constrict. Colfer investigated whether beta blockers might initially excessively lower the cardiac output and thereby trigger a vasodilating response. And indeed the initial nadir of the drop of the cardiac output predicted the future blood pressure decrease. Harry's study was complex to execute and called for further research. Unfortunately, Harry, who loved sailing, left the University to practice cardiology in Petoskey near Lake Michigan and we did not continue investigating beta blockers.

These were sophisticated studies but what brings the pressure down in some patients receiving chronic beta adrenergic blockade has not been explained to this day. Let me say this more categorically, forty years ago we started using beta blockers for treatment of hypertension, we continue to use them and still have no idea how they work!

Before he left, Len Hansson got also involved in a study of the antihypertensive agent clonidine. Sibley Hoobler was one of the first American investigators of this new antihypertensive agent, and after the FDA approved it, our clinic in Ann Arbor routinely used clonidine for treatment of hypertension. We became aware of a sudden increase of blood pressure in a few patients in whom the drug was temporarily stopped for various reasons. When during our regular patient conference I presented yet another such case, Len Hansson remembered that a blood pressure overshoot had been reported in the Swedish literature. Stephen Hunyor who came from Australia to work with Dr. Hoobler was also present at the conference. He too had experience with clonidine and published two papers on its mechanism of action. It was known that clonidine works on the brain to decrease the sympathetic tone. During the discussion we wondered whether, akin to other drugs which act on the brain (alcohol, morphine, cocaine, barbiturates), clonidine might cause withdrawal symptoms. Len, Stephen and I prevailed on Hoobler to join us in an investigation of the blood pressure rebound after cessation of clonidine. After our Ethics Committee approved the protocol, we recalled five long-term users of clonidine for studies in the Clinical Research Center of the University Hospital. There we gave them placebo instead of clonidine, measured blood pressure, drew blood for determination of *catecholamines,* and collected 24 hour urine samples before and after discontinuation of the drug.

We expected to see some increase of blood pressure but we got much more than we bargained for.[7] A few hours after switching to placebo, all patients complained of restlessness and those who had already gone to sleep woke up. In a while all developed tremor, severe headache, and some complained of nausea and stomach pain. Whereas patients on clonidine usually had dry mouths, most of them started to over-salivate on placebo. Seeing patients shifting in the bed, unable to control their restless legs, drooling, shaking and in obvious distress was bad enough but the blood pressure readings were absolutely stratospheric. Twelve hours after withdrawal of clonidine the average blood pressure increased from 149/103 millimeters of mercury (mmHg) to an average maximum of 216/161 mmHg! In parallel the urinary catecholamines increased by 400%. We first injected intravenous propranolol which almost immediately stopped the tremor but had no effect on the blood pressure. Addition of the alpha blocker phentolamine brought the systolic as well as the diastolic blood pressure down by an average of 47 mmHg.

Thus, discontinuation of clonidine caused a veritable storm of excessive sympathetic drive to muscles (tremor) and blood vessels (high blood pressure) combined with an increased parasympathetetic tone (salivation, abdominal pain). Unexpectedly, our publication about this autonomic storm caused a major semantic storm. People in the Boehringer Ingelheim Company were taken aback with our choice of words in the title of the paper. They objected to the “Blood Pressure Crisis” phrase and leaned on Hoobler to call it “Blood Pressure Rebound.” But we were young and unwilling to bend. It was a crisis, darn it! Anybody who had seen how patients suffered, and had frantically worked through the night to control blood pressure, would call it a crisis. So it remained “Crisis!” I am sure this didn’t make us popular with the industry but they should have been thankful to us. In the patient “P.C.”, a 29 years old woman with stage 1 hypertension, the blood pressure rose from a resting hospital reading of 120/80 to a peak of 200/135 mmHg! What if she had a surgery and, according to routine practice for preoperative preparation, the anesthesiologist discontinued treatment? Somewhere amid the surgical procedure the blood pressure would skyrocket, the surgeon would have to abandon the operation and in the worst case the patient could have died.

I am proud of this piece of clinical research. In fact, the "Crisis" paper is the only instance in my entire career where our work might have saved some lives. But the industry was right to object to the multiple publication of the same work. When Stephen Hunyor, our sociable, tennis playing, energetic and hard working colleague returned to continue his successful academic career in Australia he published a paper similar to Hansson's article published in Sweden in a local medical Journal. I am sure Stephen was right to alert Australian colleagues but a bit of better coordination between him and Hansson including a mutual acknowledgment of the other paper would have helped. Things happen!

During the two years of his stay in Ann Arbor, Len Hansson and I developed a strong and long-lasting friendship. Len had a stellar career; he was Professor of Medicine in Goteborg and later in Uppsala. He directed a large number of mega trials to examine the effect of antihypertensive treatment on cardiovascular outcomes. Len Hansson was a tireless advocate for better and more aggressive treatment of hypertension. His work was soon recognized and he became president of the International Society of Hypertension and later of the European Society of Hypertension. All of this did not go to his head. He remained loyal to his colleagues in Ann Arbor. Any time we appeared together on a program he'd point out with pride that I was his mentor.

Len had a dry, teasing sense of humor and he loved practical jokes. However, he practiced it only with friends and continued to maintain some distance from people he did not know well. When he realized he had cancer of the prostate, Len preferred not to talk about it with others and took it in stride with dignity. Over the seven years of his battle with cancer, we'd from time to time talk about his health. Len was never bitter nor did he ever ask for pity. He died prematurely and kept his sense of humor until the end. I had seen him in good form about two months before he called me up.

"Stevo, I have some good and some bad news for you," he said. "What would you like to hear first?"

"Give me the good stuff first."

"Gerd and I just completed a tour of the Mediterranean and the big ship stopped in Dubrovnik. You were right; Dubrovnik is a piece of paradise. We danced on the deck with the beautiful walled city in the background. It was fantastic."

"I would not lie to you, there is nothing like Dubrovnik and I am delighted for both of you. Let me now have the bad news."

"I don't think I will make it this time around," he said in a measured, matter-of-fact manner. "The chemotherapy destroyed my bone marrow and I am very weak. Nothing seems to help. Doctors in the hospital are quite pessimistic and I was discharged to home care." His voice was getting weaker. "I had better lie down."

"Can I call you tomorrow to hear a bit more about it?"

"Yes, that would be fine. Bye, bye."

The next day he could barely talk. A few days later Len died. Many of his old friends and admirers showed up for a dignified commemorative service in his church in Stockholm. I wrote the following In Memoriam for the journal Hypertension in the United States:

*"On November 8, 2002, at 62 years of age, Lennart Hansson lost a long battle with cancer. He died in the quiet dignity of his Stockholm home surrounded and supported by loved ones—his wife Gerd, his daughters, and stepdaughters. Len suffered his predicament with discipline and determination. He kept a full traveling schedule up to 3 weeks before his demise. Nobody listening to his recent lectures could have suspected that he had any personal cares or that he was coping with pain. And in the last week before his final hospitalization, Len took a cruise through the Mediterranean, willing himself to spend some quality time with his beloved wife.*

*His behavior in the last few weeks was typical of Lennart Hansson's entire life. He always could decide what was important, set his priorities, and focus his incredible energy on achieving specific goals. Luckily for all of us, the study of hypertension was the love of his scientific life. He left behind him an impressive opus that will continue to guide research into the treatment of hypertension for many years to come. For him, the treatment of hypertension was both an academic discipline and a practical problem. And as his curriculum vitae shows, he systematically prepared himself to master both sides of the issue.*

*Lennart Hansson was born on June 16, 1940 in Landskrona, Sweden. He graduated from Goteborg University Medical School in*

*Goteborg, Sweden, where from 1968 to 1978, he was first a resident and later an instructor in medicine. In Goteborg, Len trained under Bertil Hood and Lars Werko. Hood was one of the early European pioneers in the epidemiology, treatment, and clinical aspects of hypertension. Werko was an expert in the hemodynamics and clinical pharmacology of hypertension. Recognizing Hansson's talent, his mentors encouraged him to seek further education in United States. He came to the University of Michigan in Ann Arbor in 1971 for a 2-year term as a resident and research associate in the Division of Hypertension. Upon his return to Goteborg, Len completed his doctoral thesis on ß-adrenergic blockade in hypertension. The thesis included a study of the hemodynamic effects of propranolol, which he completed in Ann Arbor. In 1978 he became a docent (associate professor) of medicine in Goteborg and moved from Sahlgrenska to the Ostra University Hospital where he organized a hypertension treatment and teaching center. A decade ago he became Professor of Medicine at the University of Uppsala in Sweden.*

*Knowing that his academic work would be fruitful only if its results were applied in clinical practice, Dr Hansson invested considerable energy in the education of the public and professionals in the treatment of hypertension. He was one of the first chairmen of the Swedish League Against Hypertension, and he represented Sweden in the Nordic Working Group on Hypertension. In 1980, he became secretary and, in 1984, president of the International Society of Hypertension. He was instrumental in writing two ISH-WHO guidelines on the treatment of hypertension. More recently he was president of the European Society of Hypertension.*

*His huge bibliography is remarkably consistent. He maintained a steady interest in the hemodynamics and natural history of hypertension, in the questions of how the drugs work, who benefits most from treatment, and how aggressively one should treat blood pressure. In the last 2 decades, Len applied his knowledge and his universally recognized leadership skills to organize large scale treatment trials of outcomes in hypertension. The results were the landmark Hypertension Optimal Treatment Study (HOT) and the Swedish Trial in Old Patients with Hypertension (STOP-1), proving that aggressive treatment of blood pressure yields better results and that lowering the blood pressure is useful even in very old patients. His mega-trials comparing outcomes with various drugs, STOP-2, Nordic Diltiazem Study (NORDIL), and Captopril Prevention Project (CAPPP), confirmed that blood pressure lowering by any means is useful, but that different drugs might have different effects on specific morbidity and mortality in hypertension. His latest project, the Study on Cognition and Prognosis in Elderly Hypertensives (SCOPE), opened a new field by investigating whether treatment of high blood pressure in elderly patients might be useful also for preserving cognitive function. Full of energy and a zest for life, Len had many interests beyond his work in hypertension. He was a wine connoisseur, a keen golfer, an expert in American jazz, a collector of modern art, and an avid reader of novels. In political discussions nobody could shake his love of America. He was also eternally loyal to the University of Michigan; the best gift you could give him was an "M" jersey, and he made sure that each of his grandchildren got one. Len had a keen eye for his own and other people's weak spots, and he used his marvelous dry sense of humor to put everybody in his rightful place.*

*Len was open and had no fear of telling the truth. People who sought advice from him got just that—honest assessment, be it negative or positive.*

*Our profession has lost a great and unique contributor. Those who knew him well have lost a loyal, dependable, and precious friend."*

The European Society of Hypertension established an Annual Lennart Hansson lectureship. I was honored to give the inaugural presentation. You learn how much you liked somebody by how frequently you think of him. Len is right there next to my late brother and memories of one of them regularly pop up. I miss them both.

We lost Len in 2002 but back in the early '70's my lucky streak continued.

Nineteen seventy four was a particularly eventful year. The Michigan chapter of the American Heart Association renewed my grant again, we accumulated sufficient data to send a grant proposal to the National Institute of Health, and in eight years I advanced through the ranks from instructor to full professor of medicine at Michigan. This was pretty darn fast; the University routinely kept people in the untenured position of assistant professor for seven years and then either offered them tenure as associate professors or politely told them to pack up. I quickly zoomed through the process and never had to charm anybody, plot, or belong to any fraternal, religious or political group that would promote my interests. It really went on merit; I was productive and that was all that counted. Nothing like that would ever happen in the old country. It is amazing how unaware I was of the basic fairness of the system. Until 1971, when I got the tenured associate professorship, I truly did not know I was doing well. To understand my quandary, remember that I came to Ann Arbor from the conspiratorial "us and them" atmosphere of communist Yugoslavia where having daily input about what others thought or said about you was a basic survival tool. Over there, the group of "us" sought information about the threatening "them" and the inner support within the group was very important. In fact, we became a mutual admiration club. If I gave a good presentation at grand rounds, friends would hug me on the spot and as the word spread around I'd get a steady stream of laudatory, and frequently overblown, input. Not so in America! Everybody was polite and subdued; the most feedback that I'd get was a flat and unemotional "you did well" statement. Back in Zagreb "you did well" meant I just squeezed through! Had I done really well, my friends would have used all sorts of bombastic adjectives.

During the first two years in Ann Arbor I was technically still on an unpaid leave of absence from Zagreb and needed to decide whether to return. Confused by the lack of input, I sought an appointment with the chairman of medicine, Professor Robinson. That wasn't a very good idea. Dr. Robinson happened to be a low key, almost shy, person with an idiosyncratic nervous throat-clearing cough.

"Dr. Robinson," I said, "people from the University Hospital in Zagreb are urging me to return. I now have a visa which permits me to stay in the USA. If I had a reasonable chance to pursue an academic career here, I'd prefer to stay in Ann Arbor. What would you advise me to do?"

Dr. Robinson was not accustomed to such straight forward questions and he immediately got possessed by a salvo of coughs; at least six distinct attempts to clear his throat. Bad sign, I thought.

"Well, I was just looking at the situation in the Veterans Administration Hospital," said Dr. Robinson, but another urge to clear the throat prevented him to finish the sentence right away. "I think there is a good chance to get a salaried position from them and there is enough of hypertension there to channel the salary to a specialist like you."

More cough!

"So, you are telling me you don't have a salary source right now but hope to get one in the future," I told myself while waiting for the rest of the conversation. "Who can plan a career on such a vague promise? And why the VA and not the main University hospital?"

After a long pause Dr. Robinson resumed talking. “As far as the academic career is concerned that, of course, depends on your future performance. From what I have heard so far, you are doing well.”

Again that bland “you are doing well”!

“But I cannot advise you whether to stay in America or not,” said Robinson. “That is a personal decision and I am sure you will take all other factors into consideration. Good luck!”

End of interview!

In my judgment, all I got was a lukewarm “maybe”, a routine “you are doing well” and a noncommittal ‘you are on your own buddy’ statement. Not too much to hang your hat on. Crestfallen, I went to talk with Dr. Hoobler.

“That was fine,” said Hoobler, "and it is all you need. He told you he is working on getting a permanent salary position for you. It does not matter where the salary comes from. All the VA positions are merged into the big salary pot of the department of medicine. At most, you’d have to do work one day a week in the VA clinic. The important thing is that Robinson decided to get you a permanent salary, which is a precondition for academic tenure.”

Realizing I wasn’t yet convinced, Hoobler produced another puff from his pipe.

“OK, you occasionally need a translation,” said the boss. “In essence, Dr. Robinson told you that if you do not kill a patient or embezzle money you have a straight forward career in front of you. He told you everything you needed to know. What the heck did you expect; was he supposed to jump off his chair and hug you?”

In fact, that is exactly what I expected. Anyhow, Hoobler’s translation was very useful.

As I already said, 1974 was a very good year for me but it also held a big surprise for all of us in the hypertension division. Out of the blue Dr. Hoobler decided to step down as the head of the division. He grew tired from a series of setbacks in private life and sought to divest himself from administrative responsibilities. In the previous decade he went through two divorces and was well on the way to his third marriage. We knew a bit about it and insensitive as only young people can be, we cracked jokes; “If his first marriage lasted three decades, and the second finished in seven years, I hate to think how short the third will be,” was the standard comic line. It was not comical at all and I am glad to report that the third marriage lasted until his death. We should have known better. There is nothing funny about other people’s personal problems and we should have appreciated that despite personal difficulties he remained an effective leader of the hypertension division.

Hoobler's division was an organizational anomaly. Other academic departments of medicine either did not focus on hypertension or delegated research and clinical practice in hypertension to interested individuals in divisions of cardiology, endocrinology, nephrology or general medicine. Consequently, Hoobler had to constantly defend the free-standing status of his division in Ann Arbor. His strongest arguments for independence were the research productivity of the division, its standing as an international training center in hypertension and the divisions expertise which attracted patients from the entire State of Michigan and from many neighboring states. Hoobler felt his present staff had a sufficient standing to protect the division from takeover attempts. Nevertheless, he informed us about his decision to pass on the baton in an indirect and somewhat ambivalent manner. He invited Andy Zweifler and me to separate meetings. I can only describe what he told me but I suspect he said something similar to Andy.

"I met with Dr. Robinson to inform him I will step down as division chief. He knew of my plans in general terms and was not surprised. We then discussed the overall situation and he thinks the division is doing well and its organizational standing is justified by present circumstances. However, should the circumstances change he might have to change his assessment."

Hoobler stopped for a while, played with his pipe and waited for the message to sink in. But I didn't get it and had nothing to say.

"Julius, when you want, you can be passive though deep down you are dynamic and aggressive," Hoobler continued, "I like the aggressive part but the passive one gets on my nerves. You played that game with our morning meetings and I did not enjoy the process."

Hoobler referred to weekly division meetings which he scheduled at 7:30 AM. I am a late night owl and sluggish early in the morning. Despite an honest effort to adjust, I'd usually arrive late and too sleepy to participate in the meeting, I'd just sit and vegetate in my chair. For a while Hoobler bitterly complained, but he eventually accepted my bio-chronological limitations and moved the meeting to a more acceptable time.

"Okay," Hoobler said, "let me explain what Robinson said. He'd be willing to appoint a younger guy as division chief but will watch him with eagle's eyes. If that person is able to keep the division at its current levels of productivity or even improve it, that will be fine.

But if the division underperforms, he'd rather dismantle the group than give another person a chance to straighten it out. So whoever gets the job of division chief will work under a difficult onus. And if the division gets offered as an entrée on the takeover menu, you will have no say in deciding who eats you."

I pretty much understood the gist of the situation even before Hoobler's drastic explanation but did not know how to respond.

"Well I certainly will do my best to help Andy keeping the division afloat. We are good friends and we won't have any problems working together."

"Man, sometimes you are remarkably obtuse. I am considering proposing you for the position and Dr. Robinson will accept my recommendation. He told me so. I can choose my successor."

I was astounded. Where I came from, the advancement depended on your seniority, personal connections and political standing. I came to Ann Arbor from a far away country, had no personal connections, no political affiliation and Andy had seniority in the department. So I assumed Hoobler spoke to me to explore how I'd react to Andy's appointment but I never considered being a candidate. This was one of those "only in America" moments and it came on suddenly without premonition.

"I am stunned. I never expected such an offer," I said.

"And you didn't get one. I told you I am considering proposing you and 'considering' is the operative word. And I used the same word with Andy. He too is being considered."

"I need some time to digest this," I replied, "to either take myself out of the consideration or, alternatively, provide you with good arguments why you ought to lean my way. You can then decide."

"No, I won't," replied Hoobler, "the two of you will have to work this one out. Andy also took some thinking time and I set an appointment tomorrow to talk with both of you. Let see how it goes."

I knew exactly what Hoobler would do at that meeting - I had seen his method in the stroke study. He'd listen to our position statements and then slowly lead the discussion until we both agreed who'd be a better leader. Our friendship might suffer but that was not Hoobler's problem. The clever, manipulative old fox!

As it happened, Hoobler didn't get to see us arguing. That evening Andy Zweifler called to tell me he called Dr. Hoobler to inform him he was not interested in the position of division chief.

Andy is a remarkable person. The prospect of promotion moved this introspective man to reassess his priorities. Andy had an abiding interest in social issues and felt the new job would leave him no time for his other agenda. We had a long talk and decided Andy should focus on clinical services, medical school teaching and the relationship with our referring physicians, whereas I would pursue research and promote the national and international standing of the division. This seemingly nebulous assignment as ministers of internal and foreign affairs served us very well. Over the next three decades, we harmoniously promoted the development of the hypertension division. I owe a great deal to Andy – without his understanding and support I would have been much less successful. And I should add that I had an unfair advantage. The experience in the post World War II communist Yugoslavia made me a cynical and inactive armchair liberal. Deeply disappointed with the politics of social justice and doubting that things can change, I took refuge in science.

**Andrew Zweifler**

Andy spent a year in the Meharry College of Medicine in Tennessee helping to modernize the teaching of black physicians, he persistently advocates for a national health care program, and in Ann Arbor he and his wife Ruth were involved in a number of practical and time requiring local projects to protect minority rights. Andy retired five years ago but he continues to see patients in a fee-free clinic in a nearby community and he actively raises funds for this important activity.

David Bassett, the head of our lipidology section, also spent a great deal of time on social issues. He was a conscientious objector to wars and together with his Quaker friends engaged in a prolonged struggle to legitimize their stance. He personally chose a Gandhi-like passive resistance to get the government's attention. Every year, he would deduct from his taxes a proportion equal to the Pentagon's percentage of the national budget. He deposited the money into a separate bank account from which the Treasury could claim the money if they guaranteed that the funds would not be spent for military purposes. I am sure that his "come and get it" stance attracted the government's attention and I have no doubt David knew how risky his defiance was. Anytime along the line, he could have been accused of the crime of tax evasion, but David was ready to face the consequences. Eventually, others joined him and that movement exists to this day. The group lobbied for congressional approval of their position. I do not know where that stands today, but to my best knowledge nobody got arrested.

Brent Egan, one of my most impressive trainees, also had to sort out his diverse interests. After a stay at the University of Milwaukee, he moved to Charleston, South Carolina where he is a professor of medicine and pharmacology. After outstanding work on various aspects of the *metabolic syndrome,* Brent decided to focus on improving the community care of hypertension in the "stroke belt" in the Southeast United States. Recently, he coauthored a proposal approved by the National Institutes of Health and funded by a $2.7 million grant. The project will implement a preventative health care program in impoverished, largely African-American, communities of South Carolina. Brent also dedicates a large proportion of his time to personal religious development and to providing pastoral care in Charleston.

My very narrow and a bit selfish focus was good for a straight line academic career but I greatly admire people who pursued different paths. We all tried to do similar things, leave a legacy of hard work on issues we decided to tackle.

Though I am cynical about social issues, I did not lose interest in questions of justice and equality in my immediate microcosm and when it came to the welfare of our colleagues, Andy Zweifler and I worked in unison. Just about a year after I came back to Ann Arbor, Dr. Hoobler brought Dr. Carlos Romero from Mendoza in Argentina into his laboratory. In Argentina, Carlos worked with Juan Carlos Fasciolo, a professor of physiology and a renowned investigator of the renin angiotensin system. Carlos was an eternally optimistic individual, and so vivacious that as soon as you were in his company, no matter how you felt before, you'd liven up. Everything about him was dynamic; his way of speaking, his loud laughter, and above all his amazingly restless feet. Regardless of the physical position, whether he stood up or was sitting, Carlos's feet tapped at a breathtaking speed. When you first met him his apparent restlessness could put you off but you soon realized it wasn't a sign of inner nervous tension; Carlos was a focused and patient investigator capable of quietly working in the lab for hours on end. However, in social occasions he'd tap madly but nobody minded this particular expression of eagerness in an outward oriented and interested person. And his hilarious self deprecatory descriptions of daily adventures in the new world were irresistible. As all foreigners, Carlos went through a difficult adjustment period but he handled it with a roaring laughter. His very first weekend in Ann Arbor Carlos wanted to see an American style football game and following the crowd he ambled towards the big stadium. He heard about famous American hot dogs and to get himself into the right mood Carlos stopped at the first vendor he could see.

"Can I have a hot dog?" he asked.

"Sure you can. Here it is."

It tasted rather good. Soon Carlos asked for another one and was immediately served.

"Thanks, that was very good but I must go now. How much was that?"

"Nothing!"

"Why, I ordered and I should pay," protested Carlos.

"Because I am having a tailgate party with my family," responded the 'vendor'.

At that point Carlos would embellish the story with details of how stunned the host must have felt when a stranger approached him and how he, Carlos, having already gotten a free ticket from a colleague thought he must write to his leftist friends in Argentina. Contrary to what they told him, in the ruthless capitalist America everything was free.

Carlos produced another funny anecdote almost every day and, needless to say, everybody loved him. Andy Zweifler hired Carlos's wife, Sylvia, to work in the laboratory where Zweifler investigated thrombo genesis after vascular injury. Sylvia was preparing to qualify as a psychiatrist in America and did not mind spending some time in the laboratory. However, though he was a medical doctor, Carlos had long ago decided to pursue research and not practice medicine. We soon realized Carlos was a knowledgeable and extraordinarily talented scientist. Unfortunately, the department of medicine in Ann Arbor had a standing policy of hiring only practicing physicians. Within that system, Carlos could neither have an academic career nor was he in a good position to develop an independent line of research in Hoobler's laboratory. Admittedly Hoobler struck some kind of a deal with Dr.

Robinson which allowed him to keep Carlos for a longer period of time, but in our judgment that extraordinary position outside the regular medical school advancement track was not suited for a scientist of Romero's caliber. Eventually both Andy and I independently urged Hoobler to do the right thing for this excellent man and he concurred. He arranged an interview at the Mayo Clinic which Carlos passed with flying colors. Carlos Romero is a professor of physiology at the Mayo Clinic where he has a distinguished scientific career.

A few months after conversations with Hoobler and Zweifler, I officially became the chief of the hypertension division, a position I held for thirty years. At about the same time, I got another piece of good news. A patient of mine worked in the Alumni Office of the University where he cultivated relationships between the University and its alumni. In the course of his duties, he interacted with Mr. Frederick Huetwell in Detroit, a University of Michigan alumnus who had inherited a great deal of money. A serious physical handicap tied Mr. Huetwell to a wheelchair and he preferred a modest and restricted existence. His old alma mater was the only passion of his life and Mr. Huetwell allocated money every year to various causes at the University of Michigan. In 1974, he decided to donate some money for research in hypertension and rather than working through the Deans office, my patient steered the Huetwell grant directly to the division of hypertension. The contribution was relatively small, between five and six thousand dollars per year, but it gave me the flexibility to initiate research in areas not covered by other grants. Each year I gave Mr. Huetwell a concise report on how we spent his money. I'd regularly invite him to visit the laboratory but the reclusive Mr. Huetwell never took me up. After four years he stopped sending money and I thought that was all that was to it; easy come, easy go. Mr. Huetwell died and in 1995 the University of Michigan received 16.8 million dollars from his estate. Mr. Huetwell was precise in his philanthropy. He directed contributions to a number of specific projects at the University of Michigan ranging from building an On Campus visitor's center to various student fellowships. Included in his directive were four professorships in the medical school and a professorship in history. In 1995, the regents of the University of Michigan appointed me as the first F.L. Huetwell Professor of Hypertension. The interest on a capital fund of two million dollars generates the Professor's salary whereas the income from another endowment of half a million dollars provides laboratory support. I am proud of the Huetwell Professorship which gave me an extraordinary freedom to concentrate on research at the end of my active career.

Over the years I often wondered what Mr. Huetwell might look like. I met him only once during a conference and his appearance matched the mental image I had formed of him; he was a small and shy man with a grossly deformed spine which limited his ability to breathe and bound him to the wheelchair. He intended to support medical research and the department of Medicine gave him an overview of ongoing studies. Mr. Huetwell barely said a word but his lively eyes scanned the room and you could see he was assessing every speaker. The man knew what he was doing. However, towards the end of the conference he was visibly uncomfortable. I kept my presentation as short as I could and resisted to elaborate on how helpful was his earlier support to the hypertension division. And there was plenty I could have said about our progress.

**Bibliography**

1. Hypertension-Stroke Cooperative Study Group. Effect of antihypertensive treatment on stroke recurrence. JAMA 229:409-418, 1974.

2. Hansson L, Zweifler AJ, Julius S, Hunyor SN: Hemodynamic effects of acute and prolonged β-adrenergic blockade in essential hypertension. Acta Med Scand 196:27-34, 1974.

3. Simon G, Kiowski W, Julius S: Effect of beta adrenoceptor antagonists on baroreceptor reflex sensitivity in hypertension. Clin Pharmacol Ther 22:293-298, 1977.

4. Simon G, Kiowski W, Julius S: Antihypertensive and ß-adrenoceptor antagonist action of timolol. Clin Pharmacol Ther 23:152-157, 1978.

5. Simon G, Kiowski W, Julius S: Effect of systemic autonomic inhibition on the hemodynamic response to antihypertensive therapy with timolol. Int J Clin Pharmacol Biopharm 17:507-510, 1979.

6. Colfer HT, Cottier C, Sanchez R, Julius S: Role of cardiac factors in the initial hypotensive action by beta-adrenoreceptor blocking agents. Hypertension 6:145-151, 1984.

7. Hansson L, Hunyor SN, Julius S, Hoobler SW. Blood pressure crisis following withdrawal of clonidine (Catapres, Catapresan), with special reference to arterial and urinary catecholamine levels, and suggestions for acute management. Am Heart J 85: 605-610, 1973.

# Master Plan and the Panorama 1969 - 1982

Upon acquiring Conway's laboratory, I developed a graphic plan of action and in November 1969, I proudly hung the framed board of the plan in the laboratory. The picture is mushy and unreadable. At present, the only purpose of the illustration is to prove that such a plan actually existed.

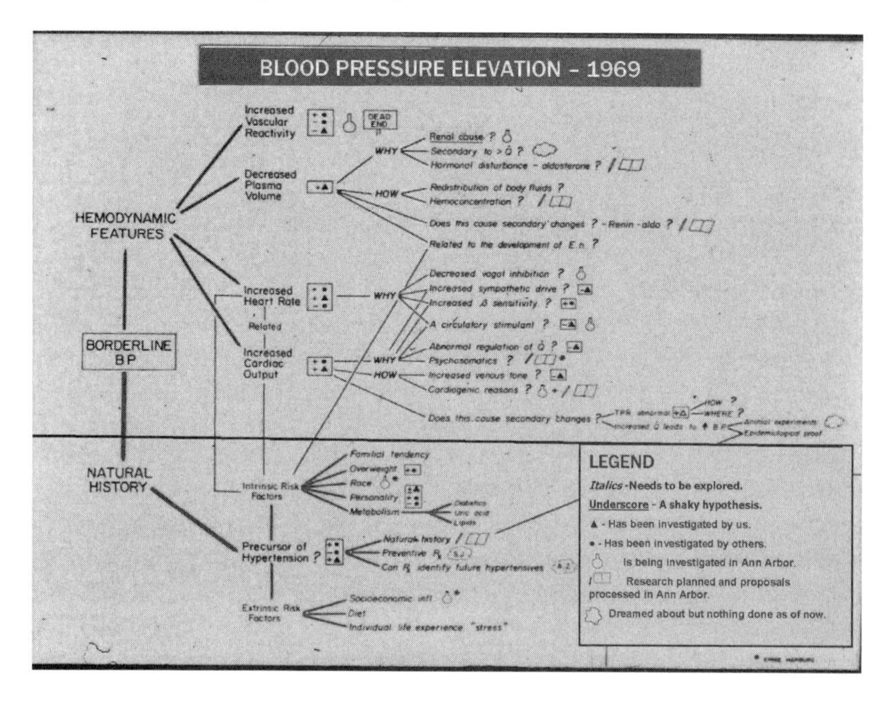


The original graph disappeared and the only remnant is a faded old-fashioned transparency slide. In the last decade as we all yielded to the insidious pressures of electronic projection, the pale slide of the original plan almost became the victim of the new trend. You'd think I have a sense of history, and maybe I have, but my legendary sloppiness gets in the way of the Platonic appreciation of the past. My office is disorderly and chronically overcrowded and I occasionally get possessed by the urge to throw things out. In one of these attacks, I marked for extinction a large number of albums with celluloid slides.

In the last moment, I decided to take a look and found the "Hypertension 1969" slide under a heap of obsolete slides. Later in the book I will redraw the master plan in more legible fashion and refer back to it.

The master plan described hemodynamic investigations we had already completed and envisaged an ambitious plan of further studies. In parallel, we anticipated that sometime in the future we would have to get "out of the laboratory" into the "real world." Most of our volunteers were University of Michigan students. Those with prehypertension had higher sympathetic activity than students with normal blood pressure levels. We preferred to describe this difference as "increased sympathetic activity in borderline hypertension" but this was by no means proven. It was quite possible that the difference in sympathetic activity reflected the way we recruited the students. Our testing was lengthy and somewhat threatening; we used needles, catheters, and injected dyes or various chemical agents. We did not coerce people to cooperate, in fact we paid a mere pittance for volunteer services. However, the motivation to cooperate might have been different in subjects with borderline hypertension than among volunteers with normal blood pressure readings (*control subjects*). Subjects with borderline hypertension might have volunteered because they were anxious about their health. On the other hand, our control subjects might have been unusually phlegmatic and self assured individuals. What if the discrepancy was due to a decreased sympathetic tone in our overly "cool" normotensive volunteers? If you compare a tall and a short man, you right away know the direction of the difference; over six feet is tall and less than five feet is short. Period! But we had no idea what is the expected average sympathetic tone. The only way to resolve the problem was to do a larger study in a representative sample of the general population and determine the reference point for average values. Such a study would take us from the field of pathophysiology, an area we were comfortable with, to the esoteric turf of epidemiology. But we knew what was coming and we kept systematically preparing for a future "real life" study.

The disappearance of the poster-size original is a story in itself. Conway's lab was on the ground floor of the free-standing Kresge Research Building. We knew we had a cockroach problem in the old Kresge building but there appeared to be a social contract between them and us; they stayed away during the day and we did not much ask what they did at night. We complied with all the right preventative measures; we threw out luncheon bags, hid the food in refrigerators, and kept a vigorous level of general hygiene. But the darn creatures can survive on practically nothing. Occasionally people from the facilities maintenance came by and sprayed insecticides with no visible consequences. The cockroaches must have had a ritual to inter their own; I had never seen a dead member of their tribe. At best, our sprayers had a tactical victory on the way to a thorough strategic defeat. After such a spraying, the cockroaches were a bit less evident but they'd soon reappear in force. Why I thought our laboratory would be an exception is not clear. Yes, we kept it very clean and disinfected all the equipment but there was no inherent reason why the bugs should not have used the easy highway of pipes and cables going from one floor to the other. With time, the population explosion among our co-tenants became unbearable. I dutifully called a person in the Dean's office that I jokingly called the "Dean for cockroaches" and he promised all sorts of help, but things did not get any better. In fact, cockroaches became more daring and in the upper floors, where we had our offices, one could occasionally spot them during the day. It was predictable that some of them would also show up in the lab. Emerged in my research, I lost judgment and kept using any possible Freudian mechanism to deny the reality. Luckily at that point I had two excellent and straight-forward co-workers, Hans Ibsen from Denmark and Brent Egan from Michigan. Each of them came separately to talk to me and both urged me to close the laboratory. I agreed. It was a hard decision. I knew we'd never again be as productive as we used to be. Eventually, I donated some of the equipment to the Clinical Research Unit in the main hospital but just as any other investigator, we had to sign up for its use. We also had to compete with others for time slots in the research unit. Though we made the right decision, I doubt it made a great deal of difference. The old hospital was also cockroach infested but the clan living there was smart enough not to show up during the day. The stupid cockroaches in the Kresge research building dropped right on the sterile white sheet covering the subject's chest! This was too much to take and we put ourselves out of business.

The decision to close the invasive laboratory meant also giving up the research suite in the Kresge building and we took the master plan poster off the wall. Up in our 6$^{th}$ floor offices all the walls were occupied and I stashed the poster somewhere behind the obsolete equipment in the utility room. A few years later, the university started to build a brand new hospital right behind the old one. The old building was slated for demolition and the hospital leadership made elaborate plans to move everybody and everything into the new building in two days. As I was a division chief in the department of medicine, the hospital director came to inform me about their plans for moving the division into the new facility. Physicians usually look upon the hospital administrative leadership as a tolerable nuisance and I was not an exception. However, this time around I was impressed with the precision and thoughtfulness of their plan and when the time came, the move was indeed completed as planned and without a hitch.

As is customary, the director asked me whether I had any questions.

"What will you do to prevent the cockroaches from moving into the new hospital?" I blurted out.

The poor man was stunned. He delivered a good routine talk and had answers to all possible questions but that was a new one. "Well," he said after a very long pause, "the new hospital is impeccably designed. The cockroaches usually live around the food. The kitchen in the new hospital is in the basement and fully automated. The food is placed into insulated containers, loaded on little robot carts and the robots scoot on tracks to elevators in various sections of the hospital. There the robots open the elevator door and dial the desired floor for delivery. It is cute and worth seeing. If you would like, I will arrange a demonstration for you."

"Thank you ever so much. That must be quite a scene to see," I said, while mentally imagining the little creatures conveniently hitching a ride to various wards.

I knew I'd overdone it and politely demurred. I had cockroaches on my mind and he didn't. It would have been better if I had not obsessed about the bugs and paid more attention to the move. During the move to the new hospital, the poster with the master plan disappeared forever!

Between 1969 and the closure of the hemodynamic laboratory in 1987, we found a wide spectrum of pathophysiological abnormalities in borderline hypertension. It was likely that an autonomic nervous system imbalance was the common denominator for all those abnormalities, and during the ensuing eight years we scrupulously executed our "master plan" of research. The main strategy was reasonably simple: if the sympathetic over activity caused the observed abnormalities then a blockade of sympathetic receptors ought to abolish each of these abnormalities. Since fast heart rate and elevated cardiac output were the most prominent findings, we first studied whether this "subtraction by blockade" strategy might abolish the hyperkinetic state in borderline hypertension. The starting point was an observation we published in Rune Sannerstedt's paper.[1] (For some reason that paper got stuck in the review process and though its findings were known to all of us, its publication was greatly delayed.) The intravenous injection of propranolol, which blocked beta one and beta two receptors, decreased but did not normalize the cardiac output in borderline hypertension. Furthermore, the blockade had a minimal effect on heart rate which remained significantly elevated. In the abstract of Rune's paper[1] we concluded that the "*Elevation of the heart rate in borderline hypertensives is not mediated through the beta adrenergic system and may result from decreased parasympathetic inhibition or form a different intrinsic myocardial pacing*."

Let me remind you that the heart rate is controlled by both branches of the autonomic nervous system; the sympathetic speeds the cardiac pacemaker and the parasympathetic slows it down. Theoretically, it was possible that the generator of the rhythm in the heart (pacemaker) operated at a different "intrinsic" speed in these individuals. To sort this out, I designed a study in which we first injected large doses of propranolol to block the sympathetic beta receptors followed by equally potent blocking doses of *atropine*, a blocker of cardiac parasympathetic receptors.[2] If this stepwise blockade did not normalize the heart rate, we would have to assume that people with borderline hypertension inherited over-zealous pacemakers. The results were exciting.

As expected, people with borderline hypertension had elevated resting heart rate and cardiac output (shown as *cardiac index*). The complete blockade of cardiac autonomic nervous system receptors totally erased the difference in cardiac output and heart rate between the two groups. Note that I now talk about the autonomic nervous system and not only about sympathetic over-activity. Both components of the autonomic nervous system were involved. Beta blockade decreased but did not normalize the heart rate and cardiac output. To abolish the hyperkinetic state we had to also block the parasympathetic receptors. A closer look at the graph enables you to assess the relative contributions of the sympathetic and parasympathetic nervous system for the

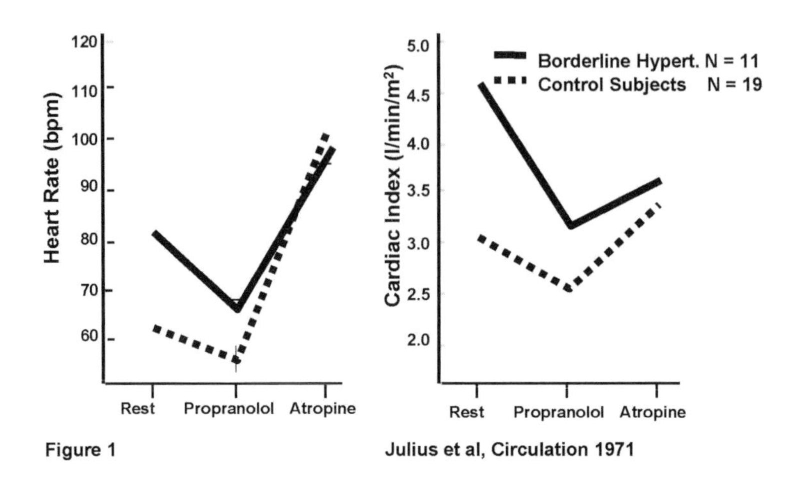


**Figure 1** **Julius et al, Circulation 1971**

maintenance of resting cardiac output and heart rate in both study groups. The sympathetic blockade with propranolol caused a larger decrease of heart rate and cardiac output in the borderline group than in the control subjects. Thus in the resting state, subjects with borderline hypertension had increased cardiac sympathetic drive. Conversely, the parasympathetic blockade with atropine elicited a lesser increase of the heart rate and cardiac output in borderline hypertension. Consequently, the borderline hypertension group had a lesser parasympathetic inhibition than control subjects.

The sympathetic and parasympathetic branches of the autonomic nervous system are organized in an "opposed but interrelated reciprocal fashion;" when the tone in one branch increases, the tone in the other branch decreases. The link between the two branches takes place in the well localized cardiovascular control center in the brain stem (*medulla oblongata*) where two anatomically separate areas (a sympathetic and a parasympathetic neuronal nucleus) are functionally integrated. Thus, our findings of a reciprocal abnormality of the sympathetic and parasympathetic tone in borderline hypertension strongly suggested that the hyperkinetic state is a complex autonomic nervous abnormality which emanates from the brain stem.

Nevertheless, another interpretation of our findings was also feasible. The response of an organ to the autonomic nervous tone can also depend on the sensitivity of its receptors. The Cleveland group had shown that some people with faster heart rate have supersensitive beta adrenergic receptors. To resolve this issue, we decided to also test the sensitivity of beta adrenergic receptors in borderline hypertension. Whereas I viewed this investigation as a "house cleaning" effort to tie up loose ends in our research, I am delighted that we did so.

In this project I cooperated with Ottelio Randall from the division of Cardiology. A fully trained staff cardiologist, OT loved research and decided to join our group. He was welcomed and we enjoyed the company of this sharp and mature scientist. He was a good mathematician and his brain operated on a level which, with my poor background in algebra, I could hardly comprehend. He was interested in arterial compliance, an area in which I had no expertise. However, we cooperated very well in hemodynamic investigations and, by and by, OT became more of a member of our rather than of the cardiology family. Dr. Randall rose to the rank of Associate Professor but eventually decided to leave Michigan. Presently he is professor of Medicine and leader of the clinical research program at the Howard Medical School in Washington.

To investigate the sensitivity of beta adrenergic receptors, we injected the synthetic beta adrenergic agonist isoproterenol.[3] After the injection, the heart rate and cardiac output increased less in borderline hypertension than in the normotensive group. This decreased sensitivity of beta receptors was particularly pronounced in the subgroup of borderline hypertensives with normal cardiac output. We reported these findings to reinforce our previous claim that the autonomic nervous abnormality in borderline hypertension originated from the brain. Since the cardiac beta receptors under-responded to stimulation, our original findings could mean nothing else but that the actual tone in cardiac sympathetic nerves was increased. This was a good enough observation at that time, and we moved on with other aspects of our research. However, an investigator should never forget his old findings, be they negative or positive. If you worked systematically and each piece of your research was solid, some old data might assume a new meaning. With experience, our research horizon widens and a piece of old data all of a sudden neatly fits into the jigsaw puzzle. In my case ten years after the initial observation, the decreased sensitivity of beta adrenoreceptors helped me to explain the hemodynamic transition from a high cardiac output to a high vascular resistance state in the course of hypertension. And about thirty years later the decreased sensitivity of beta adrenergic receptors partially explained the strong connection between hypertension and obesity.

Having taken care of this aside let me now return to the topic of the neurogenic increase of the heart rate and cardiac output in borderline hypertension. The dramatic decrease of the cardiac output after the autonomic blockade in borderline hypertension suggested that in addition to increased heart rate the overall strength of the myocardium might have also been increased. Sympathetic stimulation is known to have a potent inotropic effect on myocardial contraction, but the cardiac inotropy also depends on the Starling mechanism; an increase venous filling of the heart increases the strength of the cardiac ejection. Luckily the measurement of the cardiac output with the dye dilution technique also permits assessing the volume of the blood in the heart and lungs (called either "*cardiopulmonary*" or "*central*" blood volume). Though these volumes were "back calculated" from dye-dilution curves, they barely correlated with the cardiac output measurements. That was good news as it documented the independence of the calculated cardiopulmonary blood volume from the measured cardiac output. If a person has a larger cardiopulmonary blood volume, it is assumed that he also has an increase venous return to the heart. This in turn, elicits a larger stroke volume and a stronger myocardial ejection power (through the Starling mechanism). And in both groups the cardiopulmonary blood volume did indeed strongly and positively correlate with the stroke volume.[4]

As Figure 2 shows, subjects with borderline hypertension responded to the same degree of venous fillings (cardiopulmonary blood volume, horizontal axis) by a much higher stroke volume (vertical axis).

In other words, their cardiac force was stronger and they ejected more blood into the aorta. A subgroup of volunteers in this experiment also participated in the cardiac blockade studies. At rest, subjects with borderline hypertension had significantly elevated stroke volumes but after the blockade with propranolol and atropine both groups had similar stroke volumes. This definitely proved that the increased cardiac ejection strength in borderline hypertension was due to over-activity of the autonomic nervous system. It is fun to look at your old publications and catch mistakes. In that paper, we claimed that the plasma volume in both groups was similar. We did not prevaricate, technically this was correct, but any time you deal with volume measurements you have to reference the data in some way to body size.

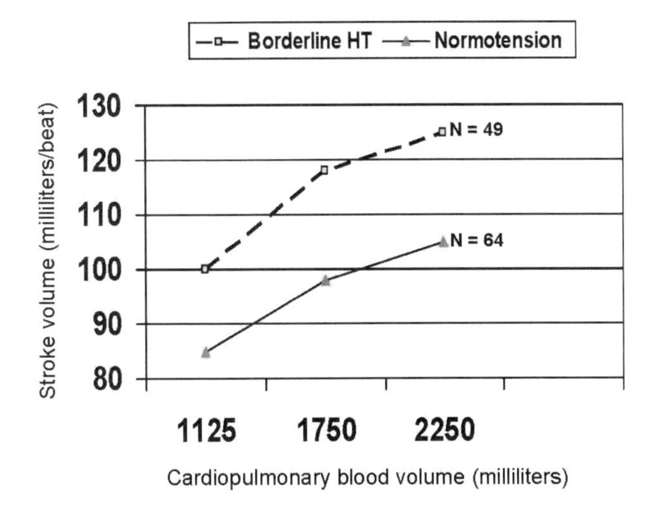


**Figure 2** **Adapted from Ellis, Julius: Brit Heart Journal 1973**

Larger people have larger cardiac outputs and bigger blood volumes. Everybody seems to agree that the cardiac output ought to be "normalized" (made comparable between people of various sizes) by dividing it by the complex and probably inaccurate measurement of body surface (cardiac index). Unfortunately, there are no agreed upon standards to account for the relationship between body size and blood or plasma volume. My dear friend, Robert Tarazi, from the Cleveland Clinic argued that the plasma and blood volumes ought to be divided by a patient's height on the somewhat questionable theory that the capacitance vessels are arranged longitudinally so a longitudinal measure ought to count. I countered that when you become overweight, you "grow" centripetally while your length won't change; dividing the volume by weight seemed reasonable to me. Anyhow, had we accounted for body weight our results would have been very different; our borderline hypertension group was always about 12 kilograms heavier. Expressed per kilogram of body weight, their plasma volume would be significantly lower. So why is that important? Well, if the plasma volume is decreased and cardiopulmonary volume is normal, that means that there was a redistribution of the blood from the peripheral to the central capacitance space. This, in turn, suggests that the veins on the low pressure side of the circulation are also affected by hypertension. If so, you must wonder about the mechanism of this veno-constriction and, you guessed it right, I think the sympathetic overdrive affects also the veins.

Unfortunately, I am not yet done with the cardiac output story and we have to return to that topic. Since the cardiac output is one of the components of the blood pressure (again, blood pressure = cardiac output times vascular resistance) early investigators suggested that the high cardiac output was responsible for the blood pressure elevation in borderline hypertension. Already in 1968, Conway and I argued that something must be wrong with the vascular resistance in early phases of hypertension.[5] That point was that the meaning of a vascular resistance can be understood only in its relationship to the level of cardiac output. We followed up this notion on a larger group of individuals in 1971.[6] In Figure 3, you will note that the two bars of average vascular resistance on the left side are exactly the same in both groups. If you concluded that the vascular resistance in borderline hypertension was normal, a look at the right side of the graph ought to give you some food for

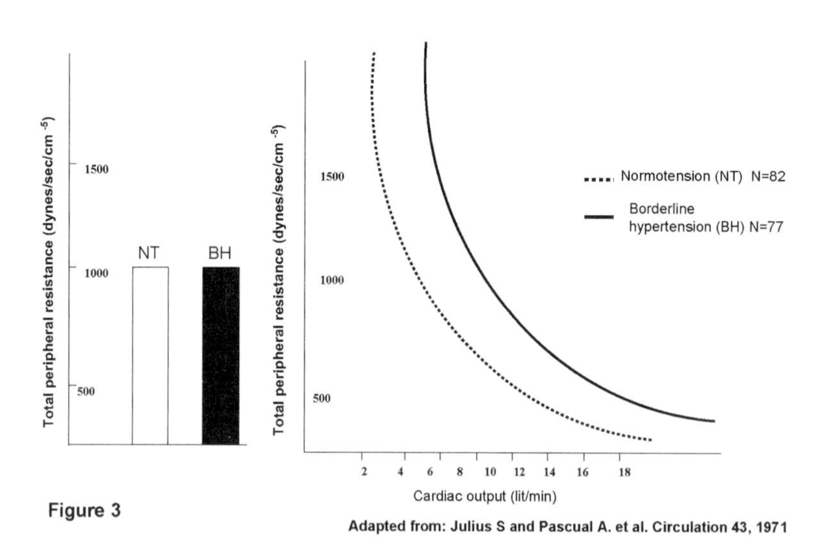


Figure 3

Adapted from: Julius S and Pascual A. et al. Circulation 43, 1971

thought. Both parabolae on the right side show a negative relationship; the higher the output, the lower the resistance. It is easy to see that at each level of the cardiac output the curve is shifted towards higher resistance values in borderline hypertension. Remember, the resistance of a vessel depends on the size of its lumen; a constricted artery offers more resistance than a dilated one. In other words, the right side of the graph shows that the borderline hypertension group did not sufficiently vasodilate to accommodate the increased cardiac output.

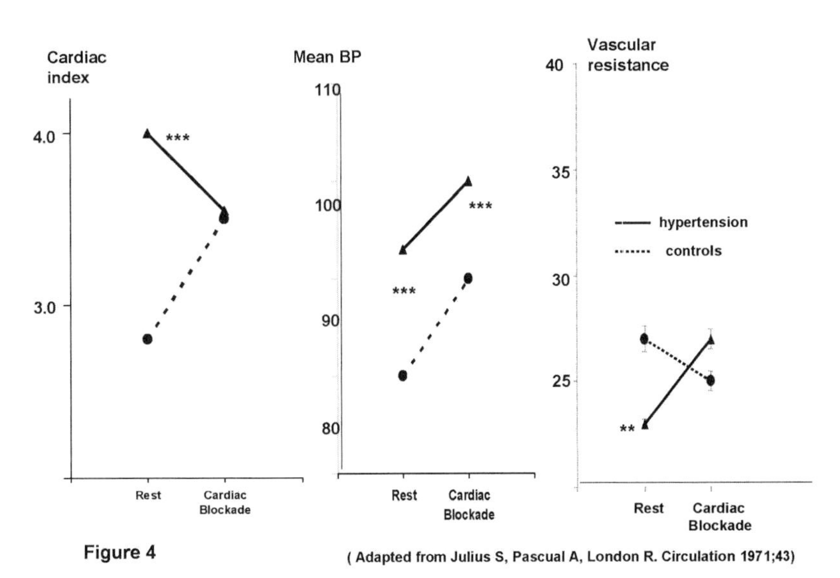


Figure 4 ( Adapted from Julius S, Pascual A, London R. Circulation 1971;43)

As Figure 4 shows, the same year (in 1971) we actually demonstrated the existence of this abnormality of vascular resistance in borderline hypertension.[2] If the high cardiac output causes the blood pressure elevation, then the cardiac blockade, which abolished the high cardiac output (left panel), should also have abolished the increase of blood pressure in borderline hypertension. But the blood pressure remained stubbornly elevated!

In fact, as the middle panel shows, after the cardiac blockade the blood pressure in borderline hypertension increased in a similar fashion as in control subjects. However, the hemodynamic underpinning of the rise in blood pressure was very different in the two groups; the resistance in the control group decreased whereas in the borderline hypertension group the vascular resistance surged upwards (right panel). The reader should excuse me for utilizing a presently much abused work, but "surge" is just the right way to describe the observation. Look at it; a few minutes after the autonomic blockade, the vascular resistance in borderline hypertension increased by about 40% over its resting value. This simply had to be a reflex response. Nothing else could so quickly shift from one mechanism of blood pressure elevation to another.

In that paper we also investigated the hemodynamic response to the sitting position, to mild exercise, and to expansion of blood volume by infusion of dextran. I consider this one of our best papers and will quote what we said at that time. "*In the recumbent position, patients with borderline hypertension have increased cardiac output and "normal" peripheral resistance. Under all other experimental conditions, the peripheral resistance in patients with borderline hypertension was elevated. Increased resistance was accompanied by a decrease of the cardiac output. Nevertheless, whether cardiac output was high or low and resistance normal or elevated, patients with borderline hypertension maintained mild elevations of the blood pressure.*"

No matter what we did, something kept the blood pressure consistently at a higher level in borderline hypertension. Furthermore, the resting hemodynamics of high cardiac output and "normal" vascular resistance in borderline hypertension was the exception; under all other conditions the vascular resistance was elevated. It was also amazing how different the stimuli were that we used to alter the resting hemodynamics. There is nothing intrinsically similar between blood volume expansion and sitting or between sitting and mild exercise. Nevertheless, each of these maneuvers uncovered an abnormality of vascular resistance in borderline hypertension.

Many readers might wonder why we bothered to study the blood pressure response to these diverse experimental maneuvers. Well, we were after the "hyperreactivity" crowd – the scientists who maintained that borderline hypertension is a state of increased blood pressure responsiveness to environmental stimuli. We had a hunch that hyperreactivity was not the problem and this series of experiments convinced us not to pursue the hyperreactivity line. Here are our conclusions from that paper: "*Among patients with borderline hypertension, there is no evidence of a hyperreactive pressure response to different stimuli. The blood pressure is maintained on a higher level, but acute pressure changes are regulated in a fashion closely resembling that in normotensive subjects.*"

The conversion from a high cardiac output to a high vascular resistance after blockade with propranolol and atropine in borderline hypertension was very much on my mind when Murray Esler, a witty and cool Australian, arrived in Ann Arbor. Murray worked in Melbourne with Professor Paul Korner, a superb circulatory physiologist, and Professor Paul Nestel, an expert in lipids and metabolic issues.

Murray was a hard working, easy going, pleasant and knowledgeable independent investigator. We had similar research interests and after he looked at the data, Murray concurred that the quick hemodynamic change (five minutes after completion of the cardiac blockade) was more likely a reflex adjustment than a hormonal reaction. Since cardiac receptors were already blocked, the reflex vasoconstriction must have been mediated by unblocked arteriolar alpha adrenergic receptors. We agreed that if we were to add a blockade of alpha adrenergic receptors after cardiac blockade this ought to abolish the surge of vasoconstriction, and reduce the blood pressure in borderline hypertension. Conceptually this was a good idea but we did not know whether this triple blockade (propranolol, atropine and alpha blocker) would be safe and well tolerated. We took some solace from Dr. Hoobler's early results with "chemical denervation" of the autonomic nervous system to predict who might respond to surgical sympathectomy. In Hoobler's experiments, the blood pressure did not fall below normal levels as long as the patients remained in the recumbent position. However, if they stood up, the blood pressure tumbled to dangerously low levels.

I also had some clinical experience with ganglionic blockers. In hypertensive emergencies, you'd put the patient in a tilt-bed and the degree of tilt determined how low the blood pressure fell. We expected that our planned triple blockade would be safe if we kept the study volunteers in the recumbent position until the effect of the receptor blockade dissipated. Murray went on to read about alpha blockers and of the two compounds available for human use, the short-lasting *phentolamine* better fitted our purpose. Intra venous injections of phentolamine were used for diagnosis and treatment of *pheochromocytoma*, a rare tumor capable of causing bouts of extremely high blood pressure readings. In essence, a pheochromocytoma is a tumor of sympathetic nerve cells that secrete huge amounts of adrenaline and/or *norepinephrine*. Based on Murray's reading, we anticipated that all effects of phentolamine would clear two hours after the injection. That was fine but nobody knew what the appropriate dose was. The drug had never before been used in our specific circumstance in normotensive people and after a previous blockade with propranolol and atropine. Eventually Murray decided we should use the maximum recommended dose for treatment of pheochromocytoma. And then

Murray threw in the bomb; he'd volunteer to test the effect of the triple receptor blocker cocktail on himself!

When I used the propranolol plus atropine injection, a certain Dr. Jose had already published a paper on the effect of aging on the cardiac pacemaker. He claimed the procedure was safe and gave the doses of propranolol and atropine needed to achieve a complete blockade of cardiac receptors. I showed Jose's data to our institutional Ethics Committee and they approved the procedure. What were we supposed to do with a brand new procedure? Obviously, there had to be a first experimental subject but I was very reluctant to inject the stuff in my distinguished colleague from Australia. Seeing that I hesitated, Esler pointed out that the Ethics Committee would never approve the procedure without some safety data. This meant that we had to test the procedure before we could ask for permission. But they wouldn't give us the permission to seek a volunteer for the first test. To break out of this vicious circle, we had to do the first tests on ourselves.

Esler's argument was perfectly logical but I continued to fret. However, Murray exuded a uniquely reassuring self confidence. If he said "it will be OK," you just knew that it would be OK! After a while he calmed me down and we started to talk about details. We'd do it discretely without an audience, I'd inject the stuff, we would place a "just in case" line for saline infusion. A few days later Murray sailed through the procedure without any problems, we reported the results to the Ethics Committee, reminded them of our excellent safety record and in due course they approved the procedure.

Murray's ability to relax everyone around him became legendary. I can actually provide biological data about the "Esler effect." Our laboratory procedure always included a catheter in the brachial artery, a catheter in the right atrium to the heart, and occasional withdrawal of arterial and venous blood. After everything was in place, we'd wait 15 minutes before taking the first "resting" measurement. But was this really a resting state? Just try for a moment to put yourself into the position of our experimental subjects. You lie on a table, while white coat silhouettes circulate around you quietly calling out the time in a NASA-like fashion and raring to inject this or that chemical into your system. Yes, they whisper to you reassuring words, ask how you feel, but are you truly relaxed? Nobody knows, but I do know that our volunteers relaxed more when Murray Esler was around. His measurements of resting cardiac output were on average about 15% lower than mine!

Murray Esler decided to enroll into his study patients with mild established hypertension (State 1 hypertension by current terminology) rather than subjects with borderline hypertension. This, in part, reflected his conviction that the results in "established" hypertension would have a greater impact but it was also simpler for him to recruit patients from the hypertension clinic than to organize yet another blood pressure screening of students at the University of Michigan campus. This was a departure from my own work in borderline hypertension but I had to agree. If borderline hypertension was a precursor of future established hypertension then the findings similar to what we found in borderline hypertension ought to be present also in established hypertension. In retrospect, I am glad Esler recruited patients with a more advanced form of hypertension. Back in the mid 1970's not too many people were convinced that borderline hypertension eventually evolved into established hypertension. Murray decided to incorporate into the total autonomic blockade protocol also measurements of epinephrine, norepinephrine and plasma renin activity. I thought this was superfluous; after all, the results of the blockade would give us quite a clear picture about the role of the autonomic nervous system in hypertension. How would those additional measurements help us? I was totally wrong. When we analyzed the data, Esler's additional neuro-hormonal profiling of patients proved very helpful.[7] As it turned out, the complete autonomic nervous receptor blockade decreased the blood pressure only in patients with high plasma renin and norepinephrine levels. (Figure 5) And in them, as predicted, the blockade decreased the blood pressure through a substantial fall in the vascular resistance.

Esler's paper was published in the prestigious New England Journal of Medicine and became a classic in the field. At that time, most experts in hypertension accepted Arthur Guyton's dictum that the autonomic nervous system could not play a role in the evolution of human hypertension. Dr. Guyton, a remarkable man whose textbook of physiology influenced many generations of physicians, developed a nephrocentric model of blood pressure regulation. In his computerized model of circulation, supported by his experimental data, Guyton suggested that the nervous system has a small "gain" in the overall regulation of the blood pressure. However, Esler's work proved the opposite. At least in one subgroup of stage 1 hypertension, those with high renin and norepinephrine values, the blockade of the autonomic nervous system brought the blood pressure into the normal range; ergo that group had a neurogenic form of hypertension. As simple as that, as Ross Perot would say.

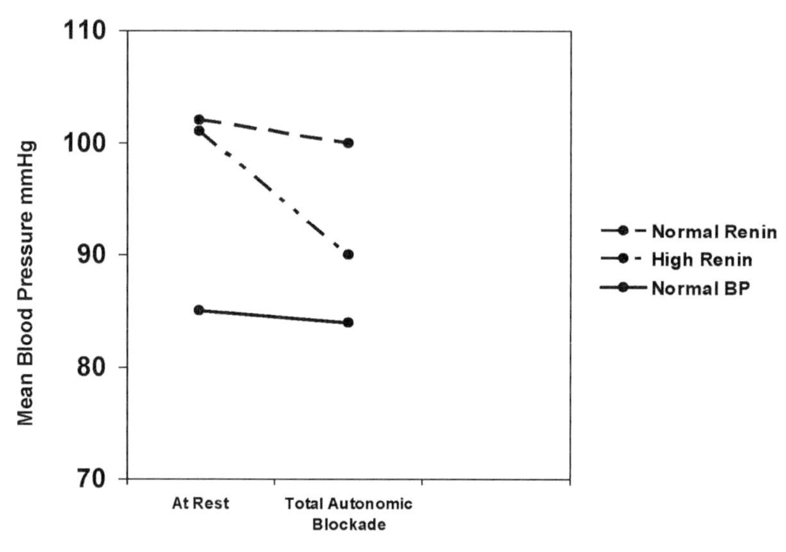


Figure 5

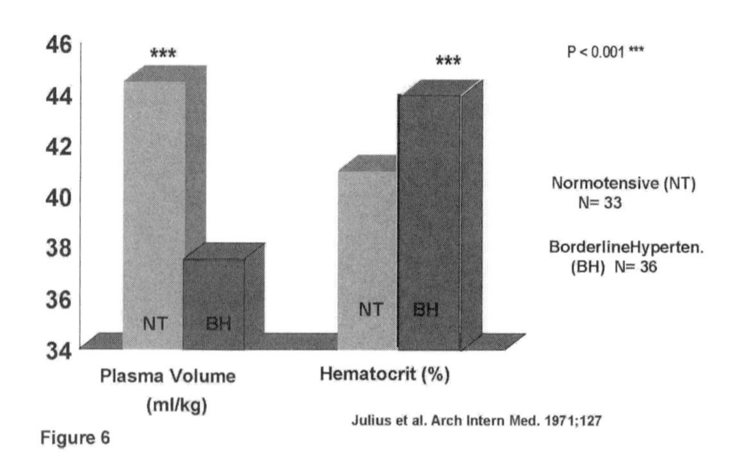


Figure 6

Unfortunately, we are not done yet. Besides numerous hemodynamic abnormalities I've already described, we also found that the borderline hypertension group[8] had decreased plasma volume and an excessive amount of red blood cells (*hematocrit*). (Figure 6) Similar findings were described in epidemiological studies and in studies of patients with established hypertension but no one expected such abnormalities to occur also in people with a minimal blood pressure elevation. The finding was yet another confirmation of our notion that borderline hypertension ought to be taken seriously. Contrary to the belief that a "little" blood pressure elevation was an innocuous, almost "cosmetic" condition, abnormalities in prehypertension involved multiple physiological systems.

We wondered why an increased amount of red blood cells, a hematological problem, would be associated with the hemodynamic state of prehypertension. Since high hematocrit values were associated with decreased plasma volume, we suspected this was not a "true" increase in the red cell mass but rather a state of hemoconcentration. In other words, akin to acute dehydration, a normal amount of red blood cell may appear high because there is less fluid (plasma) in the blood. In dehydration, be this by loss of fluids or by fluid deprivation, the fluid content of the entire body is decreased. However, Robert Tarazi in Cleveland has shown that in hypertension the total amount of fluid is normal and the decrease is limited only to the intravascular fluid content. Presumably more fluid was filtered out from capillaries into the extra vascular fluid space.

So, what could cause a permanent "leak" of fluid from the intravascular space into the extra vascular fluid compartment in hypertension? It was known that after the blood passes through high resistance region of the arteriole, the pressure in the post-arteriolar portion of the circulation starts to decrease precipitously. Physiologists always assumed that when the blood reaches to the smallest capillaries the pressure is close to zero. They so assumed because no one had a reliable method to measure capillary pressure. But if that hypothesis was incorrect and the capillary pressure in hypertension was just a few millimeters higher than in normotensive individuals, a state of permanent decrease of plasma volume would ensue.

I cannot prove or disprove the hypothesis of a higher capillary pressure in hypertension but in one of our experiments we unexpectedly elicited a precipitous decrease of plasma volume which could be explained only by a sudden increase in capillary pressure.[9] In most laboratory experiments, we measured plasma volume with Evans blue, a dye which binds to plasma proteins and remains in the blood for a long period of time. Depending on whether the plasma concentration of the same injected dose of Evans blue was higher or lower, we could calculate the intra-vascular fluid (plasma) volume. During the original experimental protocol of cardiac autonomic blockade, we noticed that soon after the propranolol injection the plasma physically appeared to be much more blue than in the resting state. That was a good enough reason to focus on more precise measurements of blood volume during the experiment. We could do this by three methods: the Evans blue concentration, the concentration of proteins in the blood, and by red blood cell concentration. Seven minutes after the propranolol injection we detected a substantial decrease of plasma volume with each method. (Figure 7)

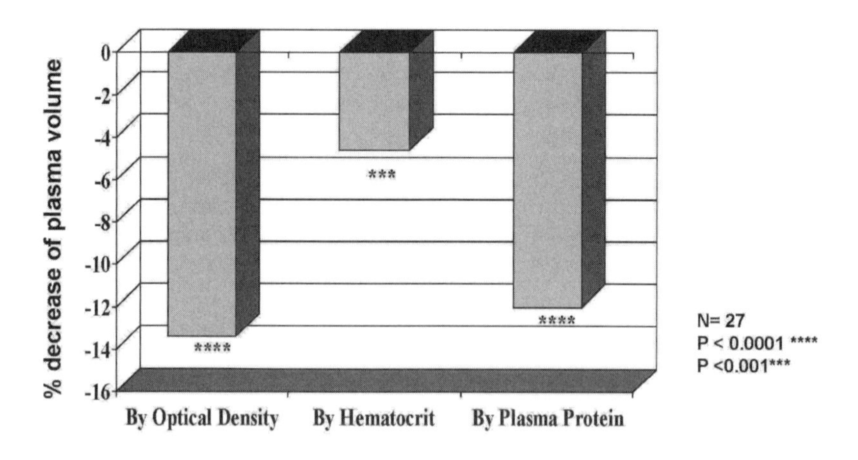


Figure 7 Julius et al, Proc of Soc Exp Bio Med 1972;140

To understand our conclusions let me remind you that blood vessels contain two types of vascular sympathetic nervous (adrenergic) receptors; the beta two receptors which dilate the blood vessels and alpha receptors which constrict the vasculature. After blockade with propranolol, the dilating (beta adrenergic) effect of sympathetic stimulation is removed. Our technical word for this phenomenon is “unopposed” alpha adrenergic tone; after beta blockade only the alpha receptors respond to sympathetic stimulation and they do their thing by causing arteriolar vasoconstriction. In that protocol we also injected atropine after propranolol to block parasympathetic autonomic nervous receptors. The plasma volume remained decreased. Propranolol and atropine altered hemodynamic parameters in different directions but it did not matter whether the cardiac output, central venous pressure, intra arterial pressure, or vascular resistance increased or decreased, these changes had no effect on the plasma volume. The likely explanation was that excessive stimulation of unopposed alpha receptors in small post capillary veins impeded the drainage from capillaries, increased the capillary pressure after blockade with propranolol and thereby filtered more plasma from capillaries into the extra vascular space. This conclusion was supported by previous findings of Jay Cohn who calculated a capillary pressure index after infusion of sympathetic agonists and found that the index negatively correlated with plasma volume. Of course, in my monomania about the autonomic nervous system, most exciting was the finding that excessive sympathetic stimulation can cause both a decrease in plasma volume and an increase in hematocrit!

It is remarkable how much we resolved during the period described in this chapter. On the next page, I redrew the "master plan" to show our progress. Underlined are the questions posited in 1969, questions to which we provided answers by 1982.

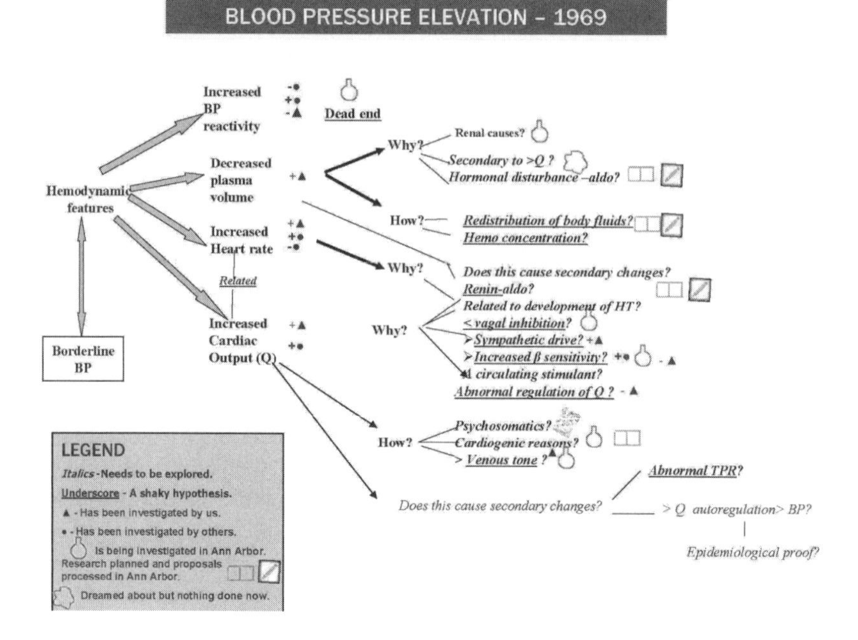


A. There is no evidence for blood pressure hyperreactivity in borderline hypertension – the blood pressure is set at higher level but is regulated in a normal fashion.

B. Plasma renin is elevated in a subgroup of patients.

C. In subjects with hyperkinetic borderline hypertension, the sympathetic stimulation of the heart is increased whereas the parasympathetic (vagal) inhibition is decreased.

D. Rather than being increased, the sensitivity of beta adrenergic receptors in borderline hypertension is decreased.

E. Venous tone is increased in borderline hypertension.

F. The vascular resistance (TPR) is increased in borderline and in stage 1 hypertension. A subgroup of patients characterized by high renin and norepinephrine values has a "neurogenic" hypertension. In this group an alpha adrenergic blockade decreases the vascular resistance and abolishes the increase of the blood pressure.

G. The plasma volume is decreased in borderline hypertension. Conversely, the hematocrit is increased. This constellation is due to a redistribution of plasma from the vascular into the extra vascular fluid compartment. Sympathetic over activity can increase the capillary pressure and thereby cause this shift of fluid in borderline hypertension.

This has been a long chapter. Instead of searching for additional closing words, let me show a picture of people involved in some of the earliest studies.

**The picture was taken in 1987 during the celebration of 40 years of the Hypertension Division in Ann Arbor. In the front row from left to right are:** *Arturo Pascual, Charles Ellis, Stephen Hunyor and Ramiro Sanchez.* **Standing in the back are**: *Wolfgang Kiowski, Murray Esler, Lennart Hansson and Ottelio Randall.*

More about Kiowski and Sanchez in the next chapter.

## Bibliography

1. Sannerstedt R, Julius S: Systemic haemodynamics in borderline arterial hypertension: Responses to static exercise before and under the influence of propranolol. Cardiovasc Res 6:398-403, 1972.

2. Julius S, Pascual AV, London R: Role of parasympathetic inhibition in the hyperkinetic type of borderline hypertension. Circulation 44:413-418, 1971.

3. Julius S, Randall OS, Esler MD, Kashima T, Ellis CN, Bennett J: Altered cardiac responsiveness and regulation in the normal cardiac output type of borderline hypertension. Circ Res 36-37 (Suppl. I):I-199-I-207, 1975.

4. Ellis CN, Julius S: Role of central blood volume in hyperkinetic borderline hypertension. Br Heart J 35:450-455, 1973.

5. Julius S, Conway J: Hemodynamic studies in patients with borderline blood pressure elevation. Circulation 38:282-288, 1968.

6. Julius S, Pascual A, Sannerstedt R, Mitchell C: Relationship between cardiac output and peripheral resistance in borderline hypertension. Circulation 43:382-390, 1971.

7. Esler M, Julius S, Zweifler A, Randall O, Harburg E, Gardiner H, DeQuattro V. Mild high-renin essential hypertension: Neurogenic human hypertension? N Engl J Med 296:405-411, 1977.

8. Julius S, Pascual A, Reilly K, London R: Abnormalities of plasma volume in borderline hypertension. Arch Intern Med 127:116-119, 1971.

9. Julius S, Pascual AV, Abbrecht P, London R: Effect of beta-adrenergic blockade on plasma volume in human subjects. Proc Soc Exp Biol Med 140:982-985, 1972.

# A Few Deviations From the Plan

Soon after his arrival, Murray Esler established himself as a mentor to the other fellows; he had a clear research plan and I knew that under his leadership the hemodynamic laboratory would function smoothly. So I took six months sabbatical leave of absence, and in the fall of 1973 the Julius family - Susan, our 8 month old, first born, Nicholas, and I left for Sweden in a roundabout way. We flew to Rome, where we picked up our prepaid new Fiat 1300. The deal included shipping costs from Europe to the United States and if we drove it for six months, the car would be considered used for tax purposes. We found the Italian distributor and, after a half hour roadside stop to figure out how to shut off the stifling heater, we drove to Ancona and took a ferry to Dubrovnik. Nick took it all in stride and having just mastered the art of scooting with his mobile walker he used his new skill to conquer guests in the hotel dining room. He'd drag himself to the table, look up, and stand there expecting to see smiling faces and an avalanche of complimentary petting noises. Boy, was he ever disappointed when we reached Germany! He'd scoot to the table only to find out he had been roundly ignored. Being a true Julius (the gene comes from my mother's side) he could not tolerate such an insult, and he scooted to the serving trolley, pulled out two small bottles and decorated the carpet of the Frankfurt Hilton with red vinegar and olive oil. There it was; a clash of cultures in a nutshell.

I mailed my bibliography to a few German professors interested in hypertension and asked whether I could visit their departments. A positive response came from Professor Peter Wolf in Frankfurt on Main and from Professor A.W. von Eiff in Bonn. Dr. Wolf, a slim, tall and elegant figure, was a superb host. He frequently vacationed on the Dalmatian coast and we exchanged anecdotes about beautiful small places only insiders would know. He proved to be a wine connoisseur and arranged for us a tour of nearby vineyards. I gave a small presentation to his staff but Dr. Wolf had also secured for me a spot at a larger meeting in Wirzburg. I cannot quite remember whether this was a cardiology or internal medicine meeting but it was a large recurring event and about 200 people came to hear my lecture. That single lecture seemed to open doors; the next few years I had frequent invitations to speak in Germany and German physicians started to come for training in the hypertension division in Ann Arbor.

Incidentally, this was the next to last lecture I delivered in German. As my mastery of English improved, my language skills in German deteriorated. About 15 years later, I was invited to give a lecture in Vienna and the host indicated that he'd prefer if I were to speak in German. I translated all the slides into German but the lecture was an absolute disaster. I mixed German and English and just could not express myself the way I wanted. There are three elements in this sad devolution of one's language skills: lack of practice, lack of discipline, and change of background thinking. Lack of practice is self evident; if you don't frequently speak a language you forget some words. The skeleton of basic and often used words never disappears but you cannot discipline yourself to express ideas in a less ornate and simpler way. And that is when you fail. You get into the middle of an elegant sentence but do not know how to proceed; rather than focusing on delivering the thought you are looking for ways to extricate yourself from the linguistic trap. Finally, the biggest disaster occurs when you think in English, form your sentence in English and try to translate them. It just does not work. Here is an example: when our son was eight and our daughter, Natasha, six years old, we took them on a tour of Croatia. Natasha was not quite sure she could manage the large amount of whipped cream on the top of the cake and I told the waiter to bring the cream on the side. Back came the cake smeared on both sides with the cream! I should have asked him to "bring the cream separately" as the term "on the side" in Croatian means literally "put it on the side of." When this can happen in your mother tongue, imagine how difficult it gets when you deal with foreign languages.

Our next stop was in Bonn where my brother Djuka was stationed as the foreign correspondent of the Belgrade newspaper "Politika." The brotherly visit went less than well as I had to tell Djuka I decided to stay in the United States and he got very upset. At that time Djuka was still a rigid leftist; politics was his life line, and he simply could not comprehend that I sided with the "enemy." My argument that by advancing knowledge in medicine I might be working for the entire humankind fell on deaf ears.

"Sure, you are above it all. That is what all defeatists say. My little brother is saving the world while idiots like me work hard to assure that social justice prevails in the world. "

Frankly, that was exactly what I believed but I tried not to argue.

"Okay," I said, in the biting spirit of our immature childhood fights, "if this is really a struggle between two gigantic forces, isn't it better for our family to have one guy on each side?"

Djuka found nothing humorous in my retort, became livid and stomped out of the room to take a headache pill. I think my intonation of "gigantic forces" told him what I thought about the class warfare nonsense and that got to him.

Poor Susan had never seen anything like that in her life. We were both yelling at the top of our lungs. I had to keep my appointment with Professor von Eiff and Susan was left alone in the poisonous atmosphere. My sister-in-law tried her best but she did not speak English and could not be particularly effective. Djuka came into the room to announce he would never again speak to me and asked Susan to convey the message. Susan did the right thing; she told him something to the effect that she could not be a messenger between fighting brothers and that we would have to settle this among ourselves. By evening Djuka had cooled down a bit and we had a "civilized" dinner. I tried to lighten the tension by telling how much Professor Wolf knew about the Dalmatian coast.

"And what do you think he was doing there? Vacationing? My foot!" snarled Djuka, "He was drawing maps for the German Army. That is what they were doing - sending 'tourists' to map out every corner of Yugoslavia. As you know, the Germans had much better maps of the old country than the Yugoslav Army."

I got the message; Djuka's little brother was not only a defeatist but he also socializes with former enemies. I let that one go by. For all I know, Djuka might have been right but the thought never came to my mind. The war was long over and I couldn't care less.

It took a few years to patch things up. The altercation in Bonn was the nadir of our relationship and from then on to the end of Djuka's life our relationship continued to improve. Eventually we forgot politics and remained good brothers.

Professor von Eiff was a man of great energy and remarkable enthusiasm. When he spoke about the effect of stress on the autonomic nervous system his protruding eyes would light up and his white hair seemed to stand on end. He had a great idea which I still think is very valid but his instruments of measurement were not accepted by others. And his passionate conviction interfered with a wider acceptance of his results. There is something funny about research and researchers. If you show your enthusiasm, you are perceived as not being objective. Of course, we are all enthusiastic, otherwise we wouldn't spend endless hours in research, but we learn to hide our feelings behind a stone wall of indifference. If you care, don't show it is the name of the game. Dr. von Eiff focused on the effect of stress on muscle tension and measured the muscular tone with a device called "integrated electro-myogram." The reproducibility and reliability of results with this device were not well known but I have absolutely no doubt that von Eiff was right; muscle tension and muscle mass are important determinants of the resting blood pressure. And nervous tension doubtless causes muscle tension as the body prepares to react to the perceived danger. In fact, some of our later research showed that muscles can be the site of very strong blood pressure-raising reflexes.

About two weeks after the start of our trek through Europe we reached Goteborg in Sweden. I wrote to Professor Folkow in the Department of Physiology and he invited me to work for six months in his laboratory. Len Hansson found us a nice furnished apartment in a downtown area reasonably close to the health complex around the Sahlgrenska Hospital. Len warned me in advance that winter in Goteborg was not much fun. However, the situation in Ann Arbor dictated that I take the sabbatical right then. Had I waited, the opportunity to leave might have evaporated. And a good part of not waiting for the spring was my own bravado; if you Swedes can take it, so can we! Well, the issue wasn't the low temperature but the geographic latitude of Goteborg and the irritating lack of daylight. Don't get me wrong, Goteborg is a really nice town with a great, well-protected harbor, lovely parks, nice buildings, including Opera and Concert Hall houses, but you have to be able to see them to enjoy them. Most of the time it was dark and the daylight at its sharpest point resembled a foggy early morning in Croatia. It eventually gets to you. There is scientific evidence on the influence of the eternal night (or day) on mood, outbreak of psychiatric problems, and suicides. Eventually we adjusted but this was reasonably easy for me as I worked in the laboratory. Susan had to take care of Nick and she kept it all together without complaint. Under some conditions, a good dose of the British stiff upper lip can be very useful! Of course we had friends and that helped a great deal. Swedes are very hospitable people.

After a few days, I reported for work in Bjorn Folkow's laboratory. The purpose of a sabbatical leave is to acquire new skills and I could not have asked for a better mentor than Bjorn Folkow. Tall and slim, in his mid-sixties, Folkow was in incredibly good physical shape; he was a mountaineer, an avid fisherman and a soccer player. This remarkable scientist had a wide range of interests but this did not interfere with his research. He worked tirelessly on providing further evidence for his basic concept of structural reinforcement in hypertension. Bjorn tackled the issue of progression of hypertension. Under progression I mean that if hypertension is not treated with anti hypertensive medication, blood pressure will steadily increase to ever higher levels. In the early 1960's, before effective drugs for blood pressure lowering were available, this vicious cycle of blood pressure self-acceleration was the curse of the medical profession. Physicians could do nothing but passively watch the worsening of the condition and dread the day when a patient might develop "malignant hypertension." They knew how to diagnose the malignant phase of the disease; all it took was a peek at a patient's eye-grounds where one can see naked blood vessels on the surface of the *retina.* When retinal arterioles are seriously damaged, they bleed and leak plasma into the retina (retinal bleeding and *exudates*). Once the patient reached that phase the clinical outcome was sadly predictable; a galloping 25% mortality per year, a median life expectancy of 2 years and not one survivor past the fifth year.

Folkow performed a series of convincing animal and human experiments to show how hypertrophy of the arteriolar wall accelerates the increase of the blood pressure. Both the *smooth muscles* in the arteriolar wall and *striated skeletal muscles* respond to repetitive increased work load with hypertrophy; they become stronger and larger. For small arteries, the steady maintenance of high blood pressure level is an increased work load; an equivalent to heavy lifting during athletic training. The hypertrophy of skeletal muscle improves the performance of an athlete but hypertrophy in the arteriolar wall is bad news. The thick wall protrudes into the arteriole and diminishes the width of its lumen which, in turn, greatly increases the resistance to the flow. To cope with the high resistance and continue providing adequate blood supply, the heart must increase its ejection pressure. Folkow showed that, even when all muscle fibers in the vascular wall are fully relaxed and the arterioles are maximally dilated, the lumen of arterioles in hypertensive animals remains smaller than in animals with normal blood pressure. This is bad enough but the real damage comes when hypertrophic arterioles start to constrict; the wall protrudes even more into the lumen and the vascular resistance increases exponentially.

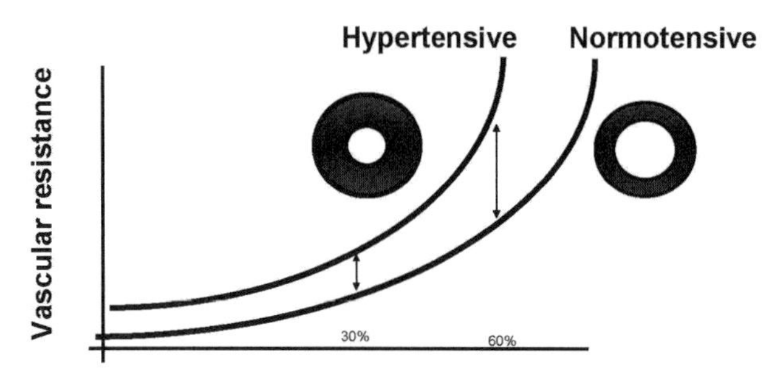


**Figure 1**

Note in Figure 1 that the gap between the resistance in hypertensive and normal blood vessels widens with increasing degrees of vascular constriction. As the pressure which the arterioles must sustain increases, the wall of the arterioles thickens further and the heart keeps increasing the pumping pressure. A particularly bad feature of this vicious cycle is that, due to their geometry, the hypertrophic vessels will over-respond to all vasoconstrictors and under-respond to every vasodilator. In addition to structural changes, the high blood pressure eventually damages the inside layer of the artery (*endothelium*). Endothelium produces nitric oxide, a short lived but extremely potent vasodilator. So the deck is stacked to favor ever increasing constriction and diminish the effect of vasodilatation.

Folkow's concept strongly supported my notion that the pathophysiology of hypertension ought to be studied early before secondary - blood pressure induced - damage alters the picture and beclouds the interpretation of findings. Furthermore, if this structural reinforcement plays an important role in the upward spiraling of blood pressure then early treatment might prevent the vascular hypertrophy and interrupt the vicious circle of hypertension. We all agree that as things stand now, once treatment has been initiated patients must continue taking pills for the rest of their lives. Otherwise hypertension will return. But what if the treatment was started very early and the development of vascular hypertrophy and endothelial damage were prevented? Would the hypertension return immediately or could the reoccurrence of hypertension be postponed? It is a long shot; but could early treatment "cure" some people with hypertension and prevent the later development of established hypertension?

Folkow was also interested in the physiology of the autonomic nervous system. He studied in detail the structure and organization of the "defense reaction" and in a seminal experiment was able to cause hypertension by electrical stimulation of the midbrain in experimental animals. The defense reaction is a state of alertness which prepares the body to cope with perceived threat; there is an increase of blood pressure, heart rate, cardiac output, and blood flow through skeletal muscles. By activating the defense reaction, the body is primed for an efficient "fight or flight" response to danger. Students with prehypertension in Ann Arbor sure looked as if they had a permanently activated defense reaction.

So Folkow's and my area of interest overlapped and there was plenty I could learn from him. His laboratory was very active and superbly organized. We had a weekly research session to discuss the progress of projects, listen to planned public presentations, or review the literature. English or according to Folkow "Swenglish," was the official language in these sessions as Folkow felt that mastering the universal language of science would be useful to trainees. We also had a luncheon break in the small cafeteria of the department and again everybody spoke in English. Of course, almost everybody in Sweden speaks good colloquial English but that was not the point. Folkow believed that his coworkers wouldn't break out from the local scene and be heard in the world unless they learned the disciplined and somewhat formalistic scientific English.

I tried to participate in animal experiments but it soon became clear I was too clumsy to be of any use. Part of the problem was that my vision rapidly deteriorated; I got my first reading glasses and within six months had to change to stronger ones. A serious message from Mother Nature that I wasn't young anymore! Eventually we agreed that I should spend most of my time reading and discussing with others.

Bjorn's thought processes were very different from other physiologists and I greatly enjoyed talking to him. He'd put every hypothesis through a test of physiologic feasibility - asking whether the proposal could be supported by quantitative data. When I first spoke with him about the decrease of plasma volume in hypertension and proposed that this might be due to a slight increase of the capillary pressure, he took a piece of paper and started to calculate. The premise was that the increased capillary pressure would filter more fluid into the extra vascular compartment but this would eventually be offset by the *oncotic* pressure of proteins in plasma. As the pressure-induced fluid filtration increases, the concentration of proteins, which cannot pass through the capillary membrane, also increases. The higher concentration of proteins exerts the oncotic pressure which tends to suck the fluid back into the capillaries. After a while Bjorn calculated that a one millimeter of mercury increase of capillary pressure would filter out about 300 milliliters of plasma before a new equilibrium is established. Our observation of an average loss of about 600 milliliters of plasma was fully explainable by a two millimeters increase of the capillary pressure. In other words, our hypothesis was physiologically feasible. If a loss of 600 milliliters required, let's say a 10 mmHg increase of capillary pressure, this would be hard to believe. Such a measurable large increase in the capillary pressure has never been documented in hypertension.

I was fascinated with Folkow's approach and I wish I could emulate him but the fact is that I never mastered that part of the art. However, I frequently call on advisors to ask whether some of our findings or ideas make physiological sense to them. Another beauty in Folkow's thinking was that he evaluated, to the extent that it was possible, the meaning of things. *Teleological* thinking, the idea that in biological sciences one can presume that some interrelationships serve a specific purpose, is much maligned from many different quarters. Marxists embraced Darwin's theory of evolution as a biological proof of their belief that progress is made through a continuous clash of opposing forces and insisted that assigning a purpose to observations was presumptuous. Behind that was their fear of religion. If you think everything has a purpose, than you might conclude that something or somebody is directing the development. The modern empiricists in science are also concerned about teleological reasoning as they dread the thought that in trying to find the purpose of things you might lose scientific objectivity. And indeed, if you overdo it, and first ask the "what does it mean?" question, you might reject some findings out of hand. But one should not become doctrinary about it. As Sir Roger Bannister, the British neurophysiologist famous for being the first to run a mile under four minutes, once told me: "If the data do not fit the question you might be asking the wrong question." One has to be flexible about the purpose of things. But it certainly is much more fun to think about what some finding might mean than to keep assembling a large list of hard facts. Better to build a house than to just look at a pile of bricks. The trick in science is to first have enough bricks rather than to design a house and then search for the bricks. But it is definitely great fun to think about the meaning of your findings. Bjorn Folkow had a wide knowledge of the evolution across different species, of nature, of the autonomic nervous system and of circulatory physiology, and he did not hesitate to use this knowledge to evaluate the results. It was a sheer pleasure talking to him and to watch him reasoning.

After about three month in Goteborg, we fully adjusted to the new circumstances and started to enjoy our stay. When the six month term expired, we were quite sorry to leave. But all good things come to an end. We packed up and returned to Ann Arbor without a hitch. In due time our Fiat arrived and seemed to have shrunk; what goes for a midsize car in Europe is a mini-car in the USA. Like true Americans, we started to fret about its size and its presumed weakness in case of an accident. Nevertheless, we hung onto it a few years but the pervasive salting of streets in the winter resolved the dilemma; in two years the rust literally ate the car.

Everything went very well in the lab. Murray Esler almost completed his first project and everybody realized how good he was. I had no difficulties convincing Dr. Robinson to appoint him as an Assistant Professor. Murray accepted the position but made it perfectly clear that he intended to return to Australia. And after three years he returned to stellar productivity and an outstanding career in the Baker Institute in Melbourne. His focus remains on the hemodynamics of hypertension and the role of the sympathetic nervous system. He developed his own widely accepted methods to measure catecholamine turnover in the entire body and also in various organs. His excellent relationship with patients and the trust his Institution has in him enables him to do things nobody else can. He measured catecholamine turnover from the brain, across coronary arteries and in the kidneys and he has truly advanced our knowledge in the area. He is a Professor and Director of the cardiovascular neurosciences in the Institute; and has trained a large number of world class investigators from all around the globe.

Before he left, Murray published a very interesting paper in the British journal Lancet[1] in which he found decreased sympathetic activity in patients with low plasma renin levels. He documented the subnormal sympathetic activity with an array of tests. Levels of plasma norepinephrine were low, release of norepinephrine in response to tilt was decreased and injection of tyramine to release norepinephrine from nerve endings elicited a subnormal norepinephrine increase. Murray was very careful in drawing conclusions and suggested that the low renin hypertension was hemodynamically a heterogeneous condition but in some such patients the central (cardiopulmonary) blood volume was increased. In those patients he suggested that "*sympathetic nervous under-activity would be expected to follow from stimulation of cardiopulmonary volume receptors by the higher central blood-volume.*" As in our previous work on borderline hypertension, the total blood volume in these patients was not decreased and the high central volume was due to a redistribution of the blood from the periphery to the cardiopulmonary area.

It was well known that the increased sympathetic stimulation of beta receptors in the kidneys enhances the renal releases of renin. Myron Weinberger and his group in Indianapolis have shown that volume expansion results in decreases of sympathetic and renin activities. However, the low renin state was always considered to be an endocrine condition caused by an excess of *mineralocarticoids* from the adrenal gland. The important message from Murray's work was that decreased sympathetic activity, no matter what might be its origin, can induce inadequate renin release from the kidneys. And in some patients the low renin state might not be induced by the expansion of the total blood volume but be a redistribution of the blood from peripheral veins to the cardiopulmonary space. In such patients a large cardiopulmonary volume causes a larger stretch of cardiac atria, this sends a signal to the midbrain to decrease the sympathetic tone and the low sympathetic tone suppresses the release of renin from kidneys.

I loved that. "Horror of horrors," I though, "one day the endocrinologists might have to learn a bit about hemodynamics!" And of course the whole story made a lot of sense. If atrial receptors are overly stretched, they perceive this as an excess of total blood volume and try to get rid of the fluid by shutting off two important water-retaining mechanisms, the renin-aldosterone system and the sympathetic stimulation of kidneys. In response, the kidneys increase the production of urine (*diuresis*) and some fluid is lost.

Eventually, I started to talk about Murray's results and the hypothesis that in some patients the low renin state might be due to a decreased sympathetic tone. I thought Dr. John Laragh might be interested in this somewhat new concept. He discovered the relationship between renin and aldosterone and later proposed a classification of patients based on the relationship between blood renin levels and the amount of sodium in the urine. In his concept, patients with low renin had expanded total blood volume and those with high renin-angiotensin levels suffered from excessive vasoconstriction. This classification promised to help physicians in deciding which treatment to use; the volume expanded patients should be treated with diuretics and those with vasoconstriction should receive anti renin-angiotensin or vasodilating drugs. Dr. Laragh repeatedly emphasized that to correctly classify the patient one must stimulate the release of renin by drawing the blood when patients are in the upright posture. Of necessity, Dr. Laragh did not study the overall regulation of renin; he investigated patient's adjustment to upright posture. And cardiopulmonary receptors play a major role in that adjustment. I therefore wanted to make the point that low renin in upright posture was not always a sign of the blood volume expansion and that under some circumstances it can reflect an abnormal distribution of an otherwise normal blood volume.

When the opportunity came to talk in front of Dr. Laragh at a scientific symposium I carefully crafted the presentation. To take out any possible edge, I showed the "Swiss cheese" slide. The slide pointed out that what I presented was a hypothesis and that a good new hypothesis should be like Swiss cheese; a) have enough solid material to keep it together, b) taste good to some and be detested by others, and c) it must have great holes which can be patched up or widened in the future. It didn't work and Dr. Laragh was not amused. He was not irritated but repudiated the idea by enumerating all arguments in favor of volume expansion in low renin hypertension. I did not say that all patients with low renin have a normal blood volume. To the contrary, I pointed out that as an exception to the rule, some patients might have a normal blood volume which is abnormally distributed and in them the low renin state may stem from inhibition of the renal sympathetic tone. Laragh based his classification on the outcome of one single type of test (renin-sodium relationship) and I have never seen a hundred percent correct medical test. It just does not happen. But Dr. Laragh did not care for exceptions; he had already built his house and had no use for poorly fitting bricks.

More flack came from an unexpected source. Allyn Mark and Francois Abboud at the University of Iowa used lower body negative pressure to alter the function of cardiopulmonary receptors. They would put the lower part of a subject's body into an airtight chamber, seal it above the pelvic bones and then suck out the air to create various degrees of negative pressures in the chamber. The negative pressure in the chamber would impede the venous return to the heart and decrease the right atrial pressure. A 10 or 20 mm mercury decrease of the pressure in the chamber failed to increase the renin levels. Renin increased only when the negative pressure in the chamber was so high that it also decreased the arterial pressure. At that point, the decreased arterial pressure also activated the aortic baroreceptors. They therefore concluded that in humans the cardiopulmonary receptors do not regulate the renin release, and that the release of renin is under the control of arterial baroreceptors. I never doubted their results; both are superb investigators. But their interpretation bothered me. There was a great deal of literature to show that the cardiopulmonary receptors control the release of renin in experimental animals. If in any other aspect the renin-angiotensin regulation in humans was similar to that of a wide range of mammals, why would human cardiopulmonary stretch receptors function differently? The whole thing did not make sense from a physiological perspective. The renin-angiotensin system is one of the important factors for the control of fluids and electrolytes in the body. Angiotensin I stimulates aldosterone release from adrenal glands and the aldosterone increases sodium and fluid retention by the kidneys. Angiotensin also acts on the brain to increase thirst. Sure, angiotensin increases the blood pressure but that is also a meaningful response from the standpoint of fluid control. If the body is dehydrated the blood volume shrinks, the blood pressure starts to decrease, and angiotensin tries to limit further reduction of the pressure. In the Mark Abboud scenario, the renin-angiotensin would come into play only in a very late phase of fluid loss, when the blood volume was so contracted that the blood pressure dropped markedly. That is too darn late! As I explained earlier, mechanisms to regulate blood volume must be placed in the distensible venous side of the circulation where even the smallest volume changes can easily be sensed.

At that point I made a mistake. Instead of sitting down to sort things out with Allyn, whom I knew well, I challenged him in public. I do not like it when others do this to me and he certainly did not enjoy my question. Just think about it. You have exactly ten minutes to deliver your lecture and five minutes to respond to questions. Within such a time constraint, you can neither fully present your findings nor can you process the questions and give a reasonable reply. To my statement that renin regulation is usually similar in all mammals and to my provocative question whether he wasn't concerned about this presumed difference between humans and animals, he responded with a curt "No." And he was right to do so. Oral presentations at large meetings are not proper forums for debates. Since then, I have tended to restrict my questions to issues of methodology or clarification of results.

Anyhow, the whole thing continued to bother me. In fact it troubled me so much that we took a three-year detour from the borderline hypertension master plan to study the physiology of renin response to upright posture in humans. What I thought would be just one single definitive experiment turned into a lengthy preoccupation. Once you enter a new territory you have no idea where you will end. No political pun intended!

At the time when we took this detour, the hypertension division was very strong. Our staff expanded and we had an extraordinary group of trainees and visitors to the division. Gary Nicholls, a New Zealander, joined us from Glasgow where he had worked in the prestigious British Medical Research Council Blood Pressure Unit. I had an open position and had written to Professor Ian Robertson to ask whether he could recommend someone and back came laudatory letters from Robertson and Professor Tony Lever. I spoke with the new chairman of medicine, William Kelley, and in 1977 he appointed Gary as an Assistant Professor. Kelley was a good judge of talent and the next year appointed Gary to the post of Associate Professor and Assistant Director of the Clinical Research Center in Ann Arbor. Akin to Esler, Gary also made it clear that he wanted to return to his homeland. Gary was a superb physician, a mature investigator, and a great teacher. In Ann Arbor, he pursued an independent line of research on various aspects of the renin-angiotensin system. This wise and eternally optimistic man provided valuable critical input to our research. Unfortunately within a year, in 1979, a position opened in Christchurch, New Zealand and Gary returned home. He became professor of Medicine in Christchurch. Later he was invited to be the Chair of Medicine at the Chinese University of Hong Kong where he reformed the teaching of medicine. After a return stint to Christchurch he was again invited to a foreign Medical School, this time in Abu Dhabi, United Arab Emirates where he was Chairman of Medicine. He has now returned to New Zealand once again.

After Gary's departure, I recruited Alan Weder from the University of Chicago where he had trained in clinical pharmacology under Professor Goldberg. I was very impressed with Alan's sharp mind and his objectivity in research. Alan critically assessed the scope of his own research and that of other investigators. Most striking was his interest in wider issues. Doctors, and that includes me, tend to lose themselves in professional issues and have little time for other things. Alan read non-medical literature, was very well versed in various aspect of evolution and followed developments in other branches of science including philosophy. He brought to the division a new line of research; studies of electrolyte transport in red blood cells of hypertensive patients. He later developed an interest in the genetics of hypertension. Alan advanced through the ranks to full professor of medicine, and in 1999 when I decided to step down as the division chief he inherited the job. He wished to reorient the division towards genetics and vascular biology, an area in which we did not have a particular strength. Presently the hypertension group is a part of the Cardiovascular Division in Ann Arbor. The future will tell whether this friendly merger was useful. I sure hope so - there is much history to preserve.

In 1976, I got a letter from Professor Hans Losse in Munster, Germany. He introduced Dr. Wolfgang Kiowski, described him as a promising young scientist interested in hypertension and asked whether I could offer him a fellowship in Ann Arbor. I responded positively and Wolfgang soon showed up. As soon as he and his wife Monika settled, I started to discuss with Wolfgang the unusual findings from Iowa. I thought their conclusion that the cardiopulmonary stretch-receptors had no role in the regulation of renin release in humans must have somehow reflected their experimental setup. One problem might have been that they collected renin samples 10 minutes after each intervention and this might have been too soon to detect peak increase in renin. The other question was whether the lower body negative suction changed the intra abdominal pressure. A small explanation is needed here. There are well documented arterial stretch receptors within the kidneys. When the intra arterial pressure increases and the receptors are stretched, renin secretion decreases and vice versa; lesser stretched receptors increase renin secretion. If the negative pressure in the abdomen were to distend the kidneys, this would alter the stretch of the intra-renal receptors and renin secretion would decrease. Allyn Mark vigorously defended the lower body negative pressure technique stating that only "a little bit of abdomen" was involved in the chamber. The chamber was sealed at the level of the iliac crest - roughly where men wear belts. When you take a look, a fair bit of abdominal cavity is under the belt. And the abdominal cavity is just that, a cavity. Change of the pressure in one part of the abdomen would quickly distribute throughout the entire cavity. "Yeah," I though, "there is no such thing as a 'little bit' of abdomen in the negative pressure chamber. This is a yes or no thing; it's like telling a woman she is just a 'little bit' pregnant."

Theoretically there was a reasonable objection to our way of thinking. Kidneys reside in a site behind the abdomen called the *retroperitoneal* space. What if changes of intra abdominal pressure are not transmitted to the retroperitoneal space? We kept reading the literature and not surprisingly nobody had measured the retroperitoneal pressure. However, the more we read the more we became convinced that lower body negative pressure is a major intervention which affects many organs. You would not believe what sort of crazy things were done; somebody measured whether lower body negative pressure changed the pressure in subjects' rectums and found a considerable dilatation of the rectum! Well if it does that to the rectum, it might hit the kidneys, too.

Anyhow, we decided to look for a way to reduce the stretch of cardiopulmonary receptors without changing the intra abdominal pressure. Eventually we hit on the idea of inflating blood pressure cuffs around subjects' thighs. We would inflate the cuffs to a pressure about 10 mm Hg below an individual's diastolic blood pressure. This ought to impede the venous return to the heart, but since the external pressure was lower than the diastolic pressure the arterial inflow into the legs would not be reduced.

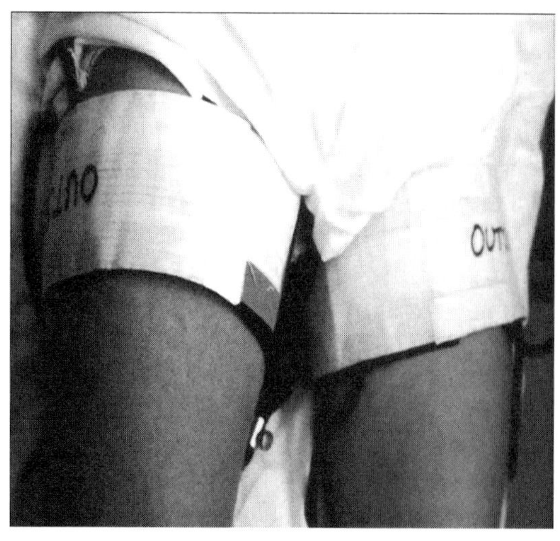

Besides being very smart, pleasant and easy to work with, Wolfgang was also a "can-do man," and he set out to develop the device. Normal blood pressure measurements cuffs were too wide for our purpose, and within a few days Wolfgang brilliantly patched up with adhesive tape a bunch of narrower pediatric blood pressure measurement cuffs and connected them with tubes. His improvisation was so durable that we used the original device once more about 15 years after the first experiment. In the picture you can see the original contraption wrapped (for historical purposes!) around Wolfgang's own thighs.

We designed a two step protocol and submitted it to the Ethics Committee. We would first inflate the cuffs and observe the renin response. If the inflation increased renin levels, we would do two additional things, a) inflate the cuff around subject's ankles to account for the possible effect of discomfort and b) we'd repeat the inflation after an injection of propranolol. It was known that the sympathetic stimulation of renin release was transmitted by beta adrenergic receptors. If beta blockade abolished the renin response, this would prove that the response was indeed mediated by sympathetic nerves. The Committee approved the proposal.

Everything came out fabulously well and knowing we had done a decent piece of work, we sent the manuscript to the toughest periodical in the business. The Journal of Clinical Investigation (JCI) routinely rejected eight out of ten papers. They did not reject the manuscript out of hand but raised a number of objections to our interpretation of the results. We were bitching about the "unreasonable" objections from the reviewer when Oscar Carretero came to visit from Detroit. Oscar, a good friend from the Romero times, is one of the outstanding basic scientists in our field. For more than three decades he has continued to attract steady support from the National Institutes of Health for his large research team at the Henry Ford Hospital in Detroit. We were dejected but Oscar quickly understood the overall situation and proposed a very logical remedy. We ought to find some patients whose kidneys have been surgically removed (*nephrectomy)* and who received a successful kidney transplant. Their new implanted kidneys are *denervated,* that is to say they have no connections with the nervous system and therefore they ought not to respond to the cuff inflation.

Brilliant, but very hard to do! Where could we find such patients? Well, I had forgotten our "can-do man!" Within a few weeks, Wolfgang lined up four such patients, convinced their physicians to let him approach them and convinced the patients to cooperate! Six months after the original submission the paper was accepted[2] and below are the results.

As Figure 2 shows, sham inflation did not cause a change in renin levels. In 24 volunteers, thirty minutes after the inflation the renin increased by an average of 120%.

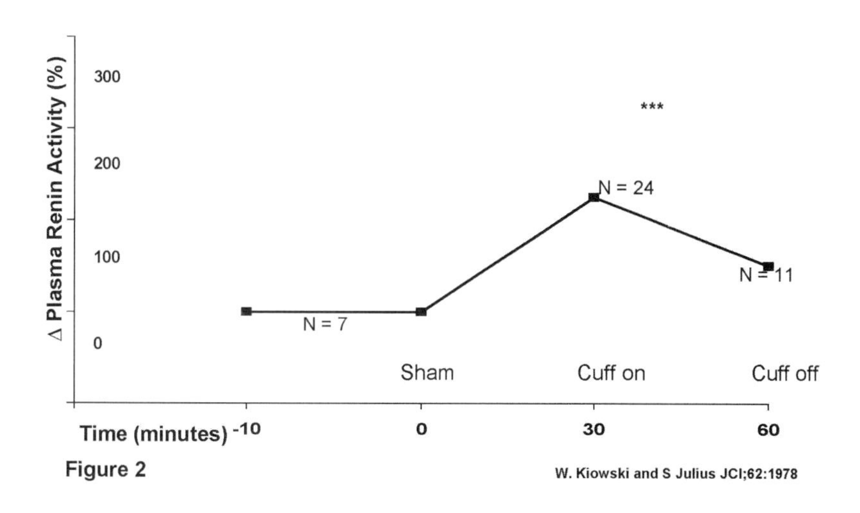

Figure 2 W. Kiowski and S Julius JCI;62:1978

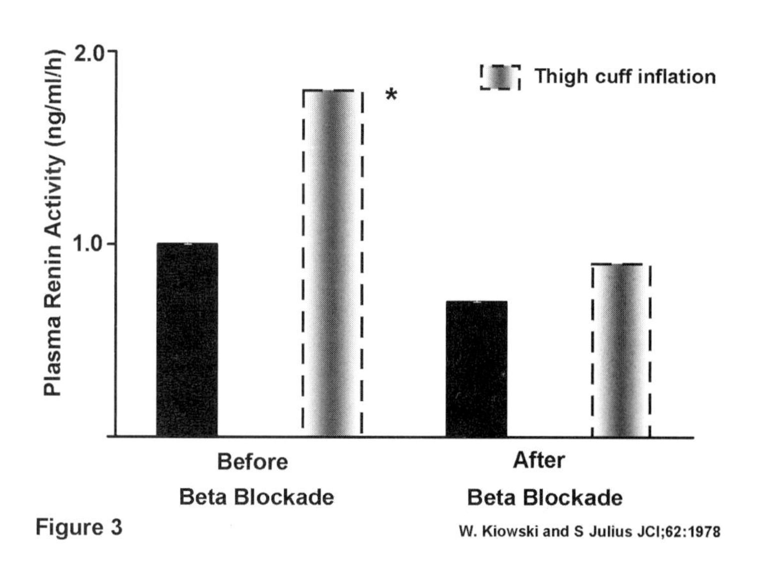

Figure 3 W. Kiowski and S Julius JCI;62:1978

In 11 subjects the cuff was deflated and the renin levels substantially declined. In this study we also measured central hemodynamics and the increase of renin was associated with decreases of the central blood volume and atrial pressure. Upon deflation, the renin went down while the cardiopulmonary blood volume and atrial pressure increased.

In 9 subjects, we repeated the thigh cuff inflation after an injection of 0.2mg/kg of the beta blocker propranolol. (Figure 3) The renin response was greatly suppressed suggesting that the increase was mediated by renal beta adrenergic receptors.

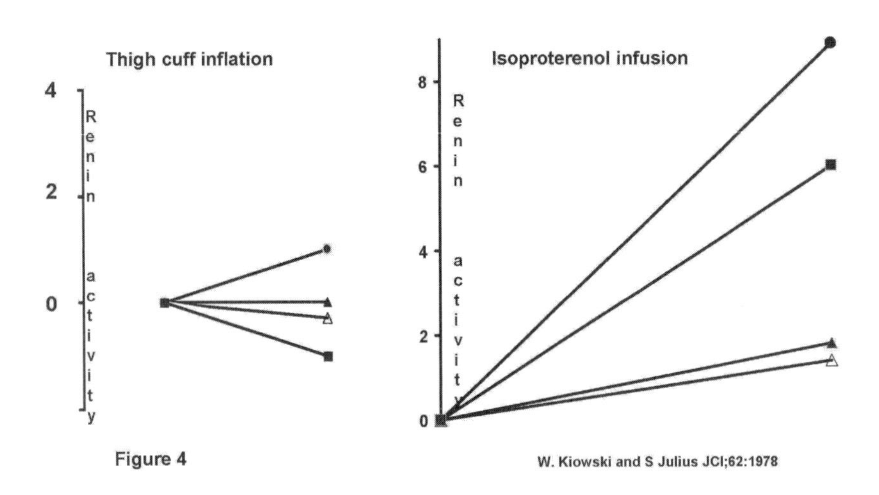


Figure 4

In 4 patients with bilateral nephrectomy and a functioning renal transplant, thigh cuff inflation caused a slight increase of renin in only in one case (Figure 4 left panel).

**Wolfgang Kiowski**

The patients had viable transplants capable of responding to beta adrenergic stimulation; infusion of the beta agonist isoproterenol caused a substantial increase of renin. So the absence of a renin response to thigh cuff inflation was not because these kidneys (right panel) couldn't respond but because their connection with the autonomic nervous system had been severed by the surgery.

At that point we lost Wolfgang. Fritz Buehler came for a visit, had a short chat with Wolfgang and on the spot offered him a staff position at the University Hospital in Basel, Switzerland. Fritz, a trainee of John Laragh, was assembling a strong research team in Basel and he easily recognized Wolfgang's potential. Wolfgang had an outstanding research and teaching career in Basel and later in Zurich. More recently he moved to the Klinik im Park Hospital in Zurich where he founded and developed the Center for Heart and Vascular diseases. Under his leadership, the Center has evolved into a leading institution in Switzerland. We remain good friends and stay in touch. This year, up in the Swiss Alps, we reminisced about the important role Oscar Carretero had in our JCI paper. When I reread the old paper I realized that we did not mention Oscar in the acknowledgments. Sorry for that, Oscar!

**Brent Egan**

Wolfgang's was a slam-dunk paper. We proved our point as convincingly as you ever could. I should have stopped right there. However, I wanted to resolve the last remaining question; do the arterial stretch receptors (baroreceptors) have any role in the control of renin release? This was really a minor issue and in retrospect I wonder whether the intent to study the issue reflected my honest interest or whether I wanted to prove that the Iowa group was wrong in suggesting that the arterial baroreceptors control the renin release. Science is not as unemotional as it appears, and sometimes you get carried away.

**Hans Ibsen**

After Wolfgang's departure, Brent Egan, Hans Ibsen, Chris Cottier and Carl Osterziehl joined the hypertension division. Egan completed his residency in medicine in Michigan and accepted a fellowship in hypertension. Soon he obtained independent research support from the National Institutes of Health. In recognition of his talent and dedication to science, we appointed him as an Assistant Professor. Brent stayed with us eight years and designed a number of his own studies of hemodynamics in hypertension. As I mentioned earlier, he is now professor of medicine and pharmacology at the University of South Carolina in Charleston.

Hans Ibsen from Copenhagen, Denmark was an independent investigator with a long list of important research papers. He decided to take the Danish equivalent of a sabbatical leave and spend some time in Ann Arbor as a visiting professor. Tall, wise and calm, this straightforward Dane fit beautifully into our social and research environment. Though he was an advanced scientist, he never pulled rank and loyally worked on all projects. I often benefited from his insights. Upon his return, Hans established himself as a leader of Danish hypertension research and became the head of internal medicine in the Glostrup University Hospital in Copenhagen. He trained a number of talented hypertension investigators in Denmark and when he retired his colleagues organized a Symposium to highlight his contributions to Danish medicine. Hans and I are lucky to have been involved in many overlapping international research projects and we never miss the opportunity to meet, chat and have good time.

**Stevo Julius, Chris Cottier and Paul Cottier at the 40 year Hypertension Celebration**

Chris Cottier came from Switzerland to continue a family tradition; a few decades earlier his father Paul worked with Sibley Hoobler. Chris is now head of the Department of Medicine in the Ementhal regional hospital in Switzerland where he focuses on the organization of health care in his Canton. Chris recently returned to Ann Arbor for a refresher. Time has not changed him. He remains a lively, interested and pleasant friend. Karl Osterziel came from Germany. He returned to Germany but I have lost contact with him. However, I know he still practices medicine.

About that time Ramiro Sanchez, a hypertension specialist from Buenos Aires came to Ann Arbor. During his stay, he worked on the hind quarter compression model of neurogenic hypertension in animals. He was a knowledgeable person, a hard worker and everybody enjoyed his company. After one year he returned to Argentina. Due to salary structure, Argentine physicians must devote a great deal of time to private practice and there are no adequate mechanisms to protect research time. Though Ramiro works in the Fundacion Favarolo University Hospital, one of the most prestigious institutions in Buenos Aires, the general circumstances there are only a shade better. Nevertheless, Ramiro heads a very active hypertension group in his hospital and they are scientifically very productive. He frequently returns to Ann Arbor to spend a few weeks discussing research and writing papers. Ramiro is a gentle and thoughtful person, a good friend and I always look forward to his visits.

Together with Egan and Ibsen, I continued to work on cardiopulmonary receptors. I wondered whether we could manipulate the distension pressure in the atrial (cardiopulmonary) and arterial stretch receptors (baroreceptors) in different directions. Eventually I decided to use a tilt table. The idea can be best understood from the following cartoon.

Imagine that the heart is somewhere between the arms of the figure in the cartoon. When you tilt the subject (left side of the panel)

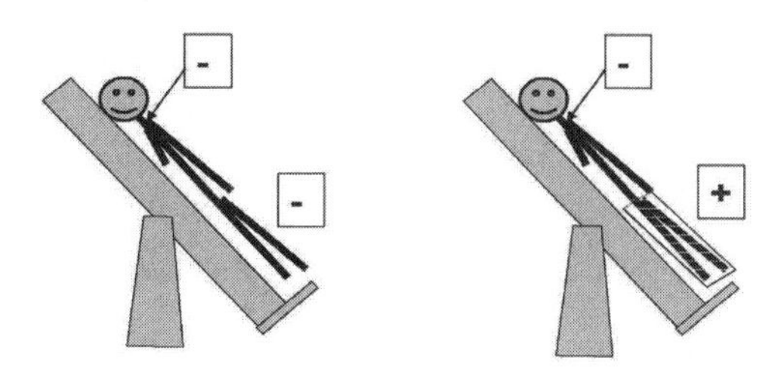

the legs are below the heart, the gravity pools some blood into the legs and the venous return pressure decreases. At the same time, the major blood vessels containing baroreceptors (aorta and the carotid artery) are elevated above the heart and their hydrostatic pressure decreases. So in this condition both the cardiopulmonary and arterial stretch receptors are less distended, and both will send signals to the brain to increase the sympathetic tone. If we were able to find a method to return the blood from the legs to the heart in the tilted position (right panel), this would offset the low atrial pressure. However, the arterial baroreceptors would remain less distended since they would continue to be elevated above the heart. That way one could assess the relative contribution of arterial baroreceptors to the control of renin release. But how could we return the blood from the legs to the heart while the subject still remains tilted?

I looked into pressure suits used by emergency services in people who lost a large amount of blood. The theory was that outside pressure on the lower part of the body decreases the overall capacitance space so that the remaining small amount of blood does not get lost in the large capacitance space. In other words, the external pressure was used to improve the venous return to the heart. But the overall validity of that approach had been challenged in the surgical literature. It was reported that, depending on an individual's anatomy and the design of the suit, in many instances the suit actually decreased the venous return. I then hit on the idea of using fisherman's waders and filling them up with water while the subject remains tilted. The weight of the water would exert a graded external pressure on the lower part of the body; highest at the feet and lowest towards the abdomen. This would offset the gravitational pooling of the blood in the legs. Theoretically the idea was sound but it was nearly impossible to execute it in practice. We'd have to put the waders on naked volunteers, their skin would shrivel from the extended stay in the water, we'd have to heat the water, the water would splash all over the lab and the entire experiment would be a colossal mess. However, if we could use a wader-like suit with double walls our problems would be solved. You'd fill the space between the inner and outer layer with water, and the volunteer would be dry and comfortable.

On a hunch, I called the nearby Jobst Company in Toledo, Ohio. They had considerable expertise in developing gradient compression garments for treatment of venous insufficiency and for management of people with low blood pressure in the standing position (*orthostatic hypotension*). These garments were useful but one garment would not fit all; every patient had to be individually measured and thereafter Jobst would produce an individual suit. As soon as I called I realized I was dealing with an unusual company. For one thing they did not ignore my call and passed it in an orderly way to ever-higher levels within the company. At each step I explained my interest and nobody said no. I think I reached the level of vice president who suggested that I speak to Mrs. Jobst, the president of the company. She inquired in some detail about the science behind the project, understood the problem and connected me with the designers. Within two months we had the double wader ready for use. And the Jobst company did it free of charge!

In 1983, Brent Egan published our report about the effect of posture on renin release including the water suit experiments.[3] In that

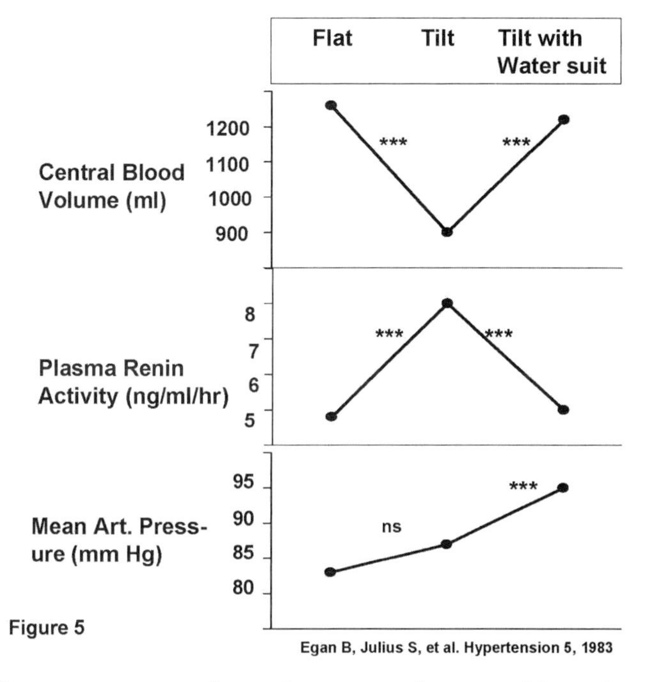

Figure 5

paper, he gave an overview of our experiments with various posture manipulations to document the relative role of cardiopulmonary (atrial) versus arterial stretch receptors. We used a number of imaginative and precise maneuvers and the overall conclusion was that atrial receptors control the renin release in response to upright posture. For the purpose of this book I will focus only on the water suit experiment. (Figure 5) The top panel shows the cardiopulmonary blood volume. As expected during the tilt, the cardiopulmonary blood volume decreased well below values seen in recumbence.

After filling of the suit with water, though the subjects remained tilted, the cardiopulmonary volume was restored to similar levels as in recumbence. So the water suit trick worked as intended! Not shown in the graph are trends in another reliable measure of the stretch of cardiopulmonary receptors, the right atrial pressure. The atrial pressure mirrored the behavior of the cardiopulmonary volume; a decrease with tilt and an increase to recumbent levels after filling of the water suit. The plasma renin (second panel) increased with tilt and after filling the water suit it fell back to levels seen in recumbence. Using a very different technique than the thigh cuff inflation, we again confirmed that cardiopulmonary stretch receptors control renin release. Incidentally, a few years later using the thigh cuff inflation Brent Egan showed[4] that the cardiopulmonary receptors also regulate the release of the pituitary *antidiuretic hormone* further evidence that receptors for regulation of hormones involved in the maintenance of fluid balance are located in the cardiopulmonary space.

But the whole purpose of the tilt-water suit experiment was to separately evaluate the influence of cardiopulmonary and arterial stretch receptors and in that regard we failed miserably! As the bottom panel of Figure 5 shows, after filling of the suit the mean arterial pressure increased. This increase of the arterial pressure was sufficient to increase the stretch of baroreceptors in the aorta and carotid arteries. Nice try but no cigar - we did not succeed to separately evaluate the functions of cardiopulmonary and arterial stretch receptors.

And that is when we took yet another large detour from our planned research. As I said, if you enter new territory you have no idea what will happen. The increase of blood pressure with the water suit was totally unexpected and this became a research topic of its own. It is known that increased muscle tension (isometric exercise) causes a huge increase of the blood pressure. The probable purpose of the blood pressure increase with isometric exercise was to preserve the blood flow through the muscles when increased muscle tension threatens to collapse the blood vessels. I will admit this speculation was a totally unsupported flight of fancy. But that was the way I reasoned at the time and it made sense to me. I wondered whether in order to preserve the muscle blood flow the external pressure on the legs triggers a response similar to isometric exercise. We found a short report in literature from NASA that inflation G suits increased the blood pressure of astronauts.

With this bit of support I discussed the problem with David Bohr in the Department of Physiology. David, who at that time worked with pigs, got interested and suggested that we get a pressure suit for animals, inflate it and see what happens. I called our friends at the Jobst Institute, they came to measure the pigs and soon we had an inflatable suit encompassing the pig's hindquarters. David's pigs had arterial catheters inserted for blood pressure measurements and readings could be taken with animals only slightly sedated with valium. Initial results were encouraging and we decided to continue experiments on dogs. Again, the Jobst folks promptly provided us with two dog suits, we inserted catheters into the aorta and the right atrium to monitor the blood pressure and measure the cardiac output with dye dilution.

The blood pressure increased both in pigs and dogs[5] and this was not just a bit of blood pressure rise; a 50 millimeters mercury (mmHg) increase in the pressure suit elicited a 50/30 mmHg increase in sedated pigs and 44/53 mmHg in anesthetized dogs. The blood pressure increase in animals was rather similar to the pressure in the suit. This nicely fitted the teleological hypothesis that the blood pressure rises in order to offset the external pressure. But it was far too early to discuss this in the paper. Instead, we focused on the mechanism and the duration of the blood pressure response. The pressure remained elevated as long as we maintained high pressure in the suit. In the longest experiment hypertension was unchanged throughout three hours of compression. The high blood pressure was due to an increase of the vascular resistance. Plasma norepinephrine increased 450% above starting values. There were two plausible mechanisms for the increased resistance, a) the mechanical compression of blood vessels under the pressure suit or b) the hindquarter compression elicited a reflex vasoconstriction. Next, we showed that chemical interruption of autonomic nervous signals by the ganglionic blocker trimetaphan almost completely abolished the blood pressure response. Nevertheless, we conservatively concluded that "*...a portion of vasoconstriction occurred in areas outside the suit*" and that "*Lower body compression is a new model to cause prolonged blood pressure elevation by noninvasive and nonpharmacological means.*"

Everything looked great. If we could prove beyond a doubt that the blood increase was neurogenic and if the blood pressure remained elevated for hours on end, we could challenge the prevailing dogma. At that time, the capacity of the nervous system to regulate the blood pressure was investigated by disruption of the negative feedback from arterial baroreceptors. Chronic baroreceptor denervation induced only mild and labile hypertension in animals and Dr. Guyton concluded that the autonomic nervous system has a limited capacity to cause a long term blood pressure elevation.

Karl Osterziel provided the undisputable proof that the blood pressure response to hindquarter compression of dogs was neurogenic.[6] As before, the vascular resistance and blood pressure increased. In parallel, the compression resulted in huge increases of plasma norepinephrine and renin. Local anesthesia of the spine at the lower lumbar level totally abolished the increase of norepinephrine and renin and reduced the blood pressure response by 70%. Karl added an extra touch to these experiments by tying up the lower abdominal aorta and vein which caused a small blood pressure increase. He then inflated the suit and the blood pressure increased in the expected fashion and decreased after deflation. This proved that that the reflex originated in the compressed area and the blood pressure increase was not due to the distant effect of some blood borne factor.

We were lucky to have Mr. David Brant as our laboratory technician. He was extremely capable, meticulous and patient. We discussed whether it would be possible to conduct these experiments in conscious animals. David thought he could familiarize the animals with the procedure and was confident that dogs would cooperate. At that point, Dr. Ying Li, a trained surgeon from China, joined the effort. He and David installed subcutaneous ports connected to the abdominal aorta of dogs. When required, they could monitor the intra-arterial blood pressures and collect blood samples from the port. It took Dr. Li and David on average three weeks to train the animals to rest in the sling and another week for them to adjust to the compression with the suit. After that, it was easy going. We wanted to learn whether repeated and long lasting bouts of high blood pressure would cause permanent hypertension and how such episodic pressure increases would affect the size of the heart. We measured the ventricular mass by echocardiography and our echo technician, Mrs. Lisa Krause, was remarkably skillful; the correlation coefficient between calculated and measured left ventricular mass in six sacrificed dogs was 0.98 – the highest I have ever seen in biological measurements. We designed the experiment[7] to last nine weeks and during that period we applied six hours of compression (three hours AM and three PM) every day. (Figure 6)

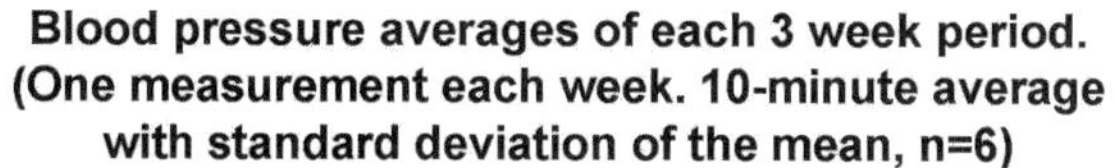


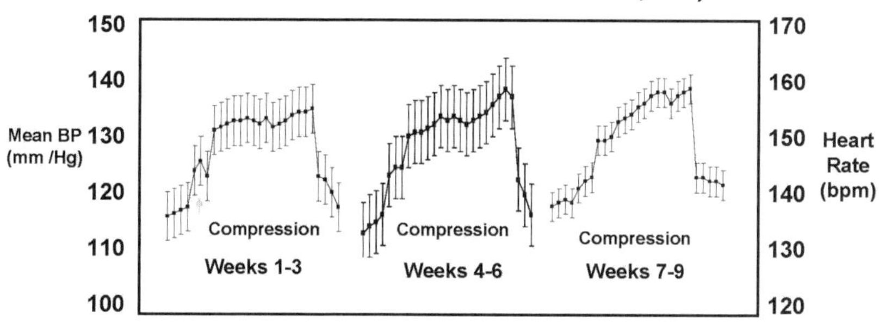


**Figure 6**

Julius et al. Hypertension 13, 1989.

The pressor response was preserved and unchanged throughout the nine weeks but immediately after decompression the blood pressure would readily return to baseline values. However, a concentric left ventricular hypertrophy was already detectable at three weeks, and at the ninth week, the left ventricular mass was 28% above the baseline value.

Bronwyn Kingwell closed the chapter of pressure suit experimentation in Ann Arbor. In a letter this trained physiologist who worked in the renowned Baker Institute in Melbourne, proposed to come to our lab and work with "pressurized" dogs. On occasions I can be a bit formal and I hesitated to respond. I figured out that Bronwyn must be the first name but I had no idea what gender it represented. What do I say; "Dear Sir?," "Dear Madam?," "Dear Mr?," "Dear Mrs, Miss, Ms?" Soon letters of recommendation came and Bronwyn was a she. So in due time this gorgeous, pleasant, steel-eyed, no-nonsense, and goal-oriented young woman came to Ann Arbor. We thought we had learned what we needed to know from the dog model and were about to abandon it. Bronwyn proved to be an excellent person capable of networking with others and soon she revitalized our moribund dog program. The question she wanted to answer never occurred to me, but that is exactly why you want to have guests with different frames of reference. Bronwyn mastered the sophisticated technique of evaluating the baroreceptor function by injecting pressor and depressor substance and assessing the heart rate response. In a new series of experiments, she again proved that intermittent pressure elevation causes left ventricular hypertrophy and after six weeks of seven hours per day (except Sundays) of suit inflation she studied arterial baroreflex responses in conscious dogs. The question was whether changes in cardiac structure affect arterial baroreceptors. This certainly was an appropriate physiological basic science subject. We knew from previous experiments that the hypertrophic hearts of these dogs were stiffer and that such a lack of elasticity would diminish the sensitivity of atrial stretch receptors. Nothing in the body works in isolation and it was logical to assume that the cardiac and arterial baroreceptors “talk” one to another. Bronwyn proved her point[8] and the arterial baroreceptor function was diminished. The degree of baroreceptor abnormality correlated with the degree of cardiac hypertrophy. In clinical medicine we have come to increasingly appreciate that left ventricular hypertrophy is one of the strongest risk factor for future negative cardiovascular events. The mechanism by which a larger ventricular mass affects such distant events as strokes is not clear. Bronwyn’s findings strongly suggest that a primary cardiac lesion can also negatively affect the overall autonomic nervous control of the circulation. Bronwyn returned to the Baker Institute in Melbourne where she is the head of the clinical physiology laboratory and an associate professor. She is a respected scientist and in 2006 she was elected as the president of the Australian Society of Medical Research.

Hindquarter compression was the only animal experiment I ever designed and the results were very important. First, we dispelled the notion that the autonomic nervous system is not capable of maintaining a long term blood pressure elevation. Second, we provided another piece of evidence against the blood pressure hyperreactivity concept of hypertension, i.e., that established hypertension evolves from repeated excessive blood pressure increases. At that time various blood pressure reactivity tests were used to predict future hypertension and our results challenged the basic validity of the concept. If repeated episodes did not translate into hypertension, then the magnitude of the pressor response to a test could not possibly predict future hypertension. Third, the results dispelled yet another myth: that “labile” hypertension is an innocuous condition. The division into “labile” and “established” hypertension was used to decide whether to initiate pharmacological treatment of hypertension. It was accepted that only patients with established hypertension ought to receive pills and that probably was a justifiable yardstick. Unfortunately, by deciding not to treat, most physicians automatically assumed that patients with “labile hypertension” (today we would call it "prehypertension") had an excellent prognosis. Our results proved that temporary blood pressure elevation can cause left ventricular hypertrophy and it is well known from longitudinal epidemiological studies that cardiac hypertrophy is one of the strongest predictors of future negative cardiovascular outcomes.

I realize that some of my description of former fellows might sound repetitious. But the truth is that most of our trainees and visitors were very nice people. What am I supposed to do? Ignore them because they were nice?

So, for fairness sake, let me admit that we occasionally ran into a bad egg. I once hired a fellow who fiercely competed for the position and seemed to be the best of all applicants. When he arrived, the man had an opinion on everything and everybody. In our research conferences, he criticized other people's work and proposed what ought to be done to rectify the situation. However, we simply could not get him to join any of our research projects. Working was not his thing. He'd complete the minimal required patient-related assignments and disappear out of sight. And he remained a virtuoso of passive resistance. You could criticize him as much as you wanted but he was not perturbed and continued to implement the policy of not dirtying his hands with work. Accustomed to interact with ambitious people, I became royally irritated by this obstinate non-performer. Finally I told him not to ask for letters of reference as they would be rather negative. He was not moved and nothing changed! What do you do with such a character? Firing somebody within the University system is almost impossible. I would have to write repeated letters of reprimand, he'd go through counseling sessions, I'd then write that he hadn't changed and just before his one-year contract expired we could initiate the actual firing process. So I was bracing myself for a year of misery. Then I learned that my father-in-law had a nice relationship with his neighbors who knew Dr. X, and over dinner they had mentioned that Dr. X planned to marry and would be soon leaving to return to his homeland. While I weighed whether it was worth embarrassing my in-law's neighbors by confronting the guy, I received a handwritten note that a family emergency forced the good doctor to depart for home immediately. This was a gross breach of contract and we could not find a replacement for him in the middle of the academic year. On the other hand, I got rid of an unpleasant person. I was as ambivalent as the man who watched his meddlesome mother-in-law driving off the cliff in his brand new Cadillac.

We also had a bit of a problem with another very nice fellow. He was pleasant, dynamic, and exuded charm. As they say, everything should be used in moderation and his excessive charm got this slim and elegant tennis player in trouble. This happened at the end of his year in Ann Arbor and we never got directly involved nor could I fully ascertain the facts of the case. Soon after our fellow's departure, I started to receive calls from various lawyers. Apparently he spoke to a patient about his plans to return to the country of origin and improve the health service in his impoverished homeland. His patient was so overwhelmed by this noble enterprise that she practically forced a brand new Porsche on him. Somewhat impractical, an ambulance would do better, but for sure the Porsche would greatly increase the speed of house calls in his homeland! And for all I know she indeed might have forced him to have the fast car. As I said the guy was really charming and we all liked him.

The last of the bad eggs was not a bad egg at all; he was an egg that did not fit into our nest. And I think we could have avoided trouble if we had been just a tad bit more sensitive. He came from Peru and completed medical residency in the USA. He was knowledgeable and a good doctor. I should have known troubles were on the horizon when he vehemently reacted to a perfectly innocuous joke. Both our children read books about the Paddington bear that came to London from "darkest Peru." They loved the book and when he came for dinner I told them our guest came from Paddington's land.

"You mean," said Natasha who has a great memory, "he is from darkest Peru?"

Our guest was visibly offended. Unfortunately, if someone doesn't have a sense of humor it is no use to explain your intent; the entire evening became a tense unpleasant affair. The man was a good fellow but rather withdrawn and not given to socializing. When his term came to end we had a farewell picnic in a riverside park in Ann Arbor. As was our custom, at the end of the meal we handed him a farewell present. A few participants goodheartedly asked him to open the parcel and show them what he got. He refused. I actually knew what this was about; in some countries, including where I came from, it is impolite to open a gift lest others think you are greedy. A few people had more beer than they needed to combat the heat and started to egg him on shouting: "Open, open!" I could not shush them down and eventually our departing fellow grabbed the present and ran away in visible panic.

What they had asked of him was a huge taboo- something never done in his country. We have never seen him again. I hear from him only when he needs a reference letter as he moves from one job to another. I believe he is a respected and successful physician. We offended him and he never forgot the insult.

I guess this ought to be enough to meet the criteria for "fair and balanced" reporting in this book!

**Bibliography**

1. Esler M, Zweifler A, Randall O, Julius S, Bennett J, Rydelek P: Suppression of sympathetic nervous function in low-renin essential hypertension. Lancet 2:115-118, 1976.

2. Kiowski W, Julius S: Renin response to stimulation of cardiopulmonary mechanoreceptors in man. J Clin Invest 62:656-663, 1978.

3. Egan BM, Julius S, Cottier C, Osterziel KJ, Ibsen H: The role of cardiovascular receptors on the neural regulation of renin release in normal man. Hypertension 5:779-786, 1983.

4 Egan B, Grekin R, Ibsen H, Osterziel K, Julius S: The role of cardiopulmonary mechanoreceptors in the ADH release in normal man. Hypertension 6:832-836, 1984.

5. Julius S, Sanchez R, Malayan S, Hamlin M, Elkins M, Brant D, Bohr DF: Sustained blood pressure elevation to lower body compression in pigs and dogs. Hypertension 4:782-788, 1982.

6. Osterziel KJ, Julius S, Brant D: Blood pressure elevation during hindquarter compression in dogs is neurogenic. J Hypertension 4:411-417, 1984.

7. Julius S, Li Y, Brant D, Krause L, Buda A: Neurogenic pressor episodes fail to cause hypertension, but do induce cardiac hypertrophy. Hypertension 13:422-429, 1989.

8. Kingwell BA, Krause L, Julius S: The effect of hypertensive episodes and cardiac hypertrophy on the canine cardiac baroreflex. Clin and Exp Pharm and Physiology 21:31-39, 1994.

# Tecumseh Here We Come
# The 1980's

Throughout our entire work in prehypertension we faced a difficult question. In fact it was more than a question. Frustrated with his WW II ally, Sir Winston Churchill is quoted as saying: "Russia is a riddle wrapped in a mystery inside an enigma." Well, we had our own riddle; we think we resolved the mystery but nobody has yet penetrated the enigma.

The riddle was pretty simple: since in prehypertension the cardiac output is high, and in established hypertension the vascular resistance is elevated, there are two possibilities: a) prehypertensives do not develop established hypertension, or b) if they do cross the border to established hypertension then their circulation must change from a high cardiac output state to a condition of high increased vascular resistance. It is fair to say that when I started to work with prehypertension most of my professional colleagues believed that people with hyperkinetic prehypertension are only "nervous" individuals who will never develop "true" hypertension. That opinion prevailed despite substantial evidence to the contrary. In 1939, Robinson and Brucer, working for the life insurance companies, coined the term "prehypertension" for people with marginal elevation of blood pressure. In 1945, P.D. White and his colleagues showed that a fast heart rate is a very strong predictor of future hypertension and cardiovascular events; and in 1971, Tony Schork and I [1] reviewed a large body of evidence that borderline hypertension is a precursor of hypertension. We might have been somewhat alone at that time but the evidence kept growing and after reviewing the field the most recent report of the Joint National Committee on Hypertension recognized marginal blood pressure elevation as a state of "prehypertension" which requires physicians' and patients' attention.

In short, the evidence against the answer "a" to the riddle is overwhelming and one has to conclude that "b" is correct. As "prehypertension" evolves into established hypertension, the cardiac output decreases to normal values and the vascular resistance increases. Let's call this process the "hemodynamic transition." In physiological terms the outstanding questions were how does such a transition evolve, and what is the mechanism of the change in the hemodynamic profile of hypertension?

From our previous work with the subgroup of borderline hypertensives that had normal cardiac output,[2] we had good insight into the cardiac output side of the "mystery" of hemodynamic transition. The group with normal cardiac output resembled the high cardiac output (hyperkinetic) subgroup. They were overweight, they had tachycardia, and their blood pressure was similar. However, the mechanism of tachycardia in the two groups was different. In the hyperkinetic group, the fast heart rate was due to increased beta adrenergic stimulation and decreased parasympathetic inhibition of the heart. In the normal cardiac output (normokinetic) group, the tachycardia was caused only by a withdrawal of the parasympathetic inhibition of the cardiac pacemaker.

So if they had a faster heart rate, why didn't normokinetic subjects have high cardiac output values? There are two answers to that question. Withdrawal of the parasympathetic inhibition of the cardiac pacemaker speeds up the heart rate (positive chronotropic effect), but does not increase the strength of the cardiac ejection (no inotropic effect). A better explanation is that subjects with normal cardiac output had a fixed decrease of the stroke volume. As shown in Figure 1 below (left side), the resting stroke volume in the normokinetic group was significantly lower than in the control group.

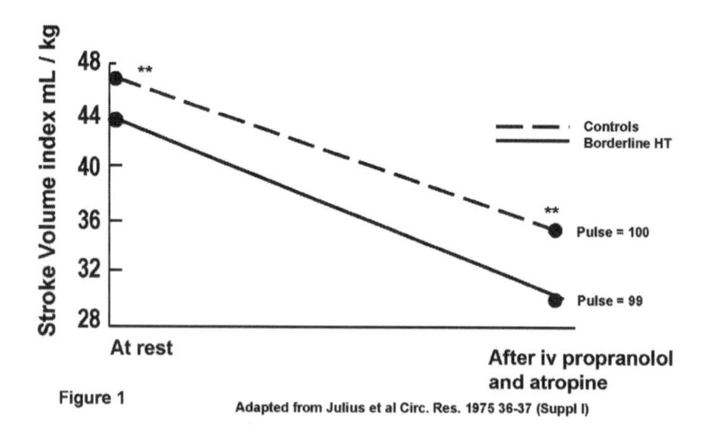


Figure 1
Adapted from Julius et al Circ. Res. 1975 36-37 (Suppl I)

The low resting stroke volume might have been a direct result of the tachycardia; a faster heart rate shortens the time between two heart beats and this may impede the diastolic cardiac filling. However, after complete cardiac blockade with atropine and propranolol (right side) the heart rate of the two groups was similar, but the stroke volume in borderline hypertensives fell even more.

Since the venous filling of the heart (assessed by cardiopulmonary blood volume and right atrial pressure) was normal, the only explanation for the low stroke volume was that borderline hypertensives had stiffer hearts and could not sufficiently activate the stretch-dependent Starling mechanism to improve cardiac inotropy and expel more blood in one heart beat. However, that was not all.

Figure 2 shows the increase of cardiac output during the infusion of a high dose of the beta adrenergic agonist isoproterenol. The

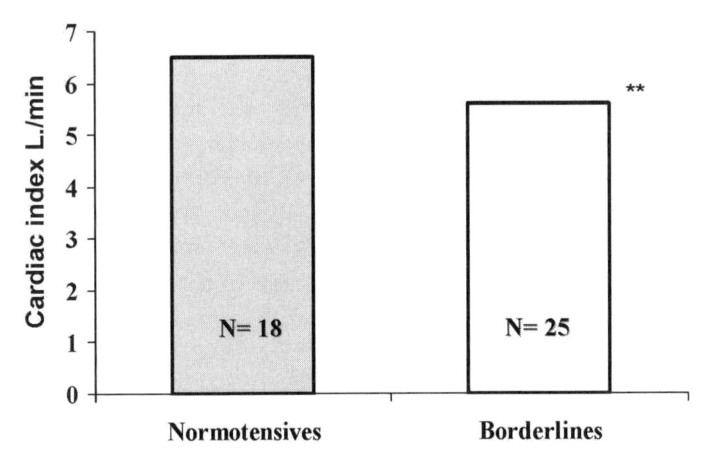


Figure 2

Adapted from Julius et al, Cir Research 1975; v. 36-37

increase of the cardiac index in normal cardiac output type prehypertensives was substantially reduced. Underlying this defective increase of cardiac output was a decrease of both the heart rate and stroke volume responses to isoproterenolol. Such a decreased response tends to occur after a prolonged stimulation with agonists and is technically called receptor *downregulation*. To me this suggested that the normokinetic group might have started as hyperkinetic but after a period of excessive sympathetic stimulation their receptors have downregulated, and they ceased to respond. Since we had no longitudinal data, this remained an unsubstantiated speculation. Nevertheless, we described two different mechanisms by which an initially elevated cardiac output could have returned into the normal range in the course of hypertension: a) structural changes of the heart (increased stiffness) limit the stroke volume, and b) functional changes (receptor downregulation) decrease the beta adrenergic responsiveness. In doing so, we started to remove the wrap of mystery from the cardiac output part of the hemodynamic transition riddle. But why would the vascular resistance increase? After all, Murray Esler showed that in neurogenic hypertension the sympathetic tone to vascular alpha adrenergic receptors is increased. If the beta adrenergic receptors downregulated, why wouldn't the vascular alpha receptors follow suit? That response was provided by Brent Egan's excellent study.

Until the mid-1980's, I personally participated in all hemodynamic experiments. However, as my work increased I kept spending less time in the hemodynamic laboratory. Only rarely, when things did not go well, would I be called to try inserting a needle into the artery or pass a catheter. Luckily, Brent Egan had the interest and energy to complete a study that fitted very well into the old master plan. He designed an elegant and difficult-to-execute protocol and in 1987 the results were published in the toughest of tough periodicals, the Journal of Clinical Investigation.[3] Brent undertook to test Folkow's theory of amplification of vascular resistance responses in the course of hypertension. He decided to investigate young, weight-matched volunteers and patients with "mild" hypertension. (By current standards these patients would be classified as having Stage 1 hypertension). In all our previous papers, the prehypertensives or hypertensives groups were on average about 25 pounds heavier than the control subjects. By undertaking to weight-match the two groups, Egan gave himself a really difficult task; it took him an entire year to find overweight volunteers with normal blood pressure.

In this study, he infused increasing doses of norepinephrine and angiotensin into the brachial artery and measured the forearm blood flow by *plethysmograhy*. If you infuse the same amount to all participants, then for those who have a larger extremity this amounts to a relatively smaller dose (and vice versa). Consequently, Brent adjusted the dose to each subject's forearm volume. To measure the volume he used a neat trick, he'd immerse the subject's forearm and record the volume of displaced water. This attention to details was typical of Brent's patient and meticulous approach to research. Since we measured the forearm blood flow and the intra-arterial blood pressure, it was easy to calculate the vascular resistance at each point of the protocol. The first point of measurement was after ten minutes of *anaerobic* forearm exercise during which the blood flow to the forearm was occluded. As soon as the tourniquet was released, the blood flow increased dramatically in order to bring more blood to oxygen depleted tissues. At that point, all small arteries are maximally dilated and the residual resistance (minimal forearm vascular resistance) reflects the anatomy of the vessels; narrower vasculature offers more resistance. In Brent's experiment, after anaerobic exercise the blood flow in both groups was equally high but the resistance to flow in hypertensive patients was significantly higher than in control subjects.

Figure 3 shows responses to increasing doses of norepinephrine injected into the forearm artery. Clearly patients with hypertension responded with a steeper increase of vascular resistance than control subjects.

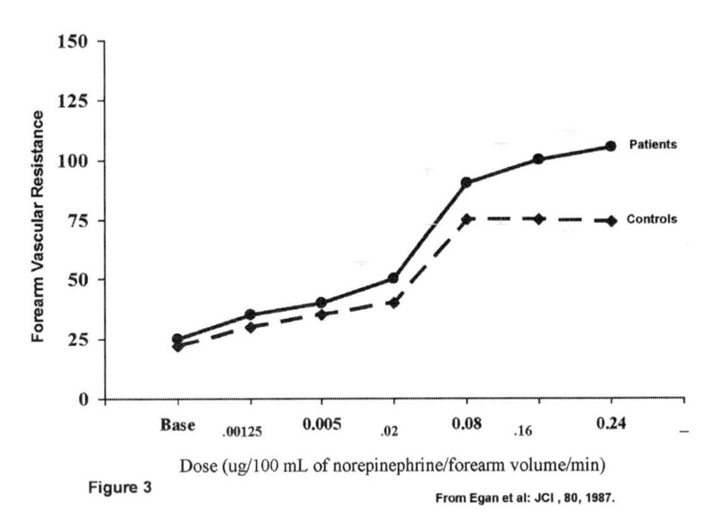


Figure 3

A very similar response was observed also with infusion of increasing doses of angiotensin into the forearm despite the fact that norepinephrine stimulates alpha adrenergic receptors and angiotensin acts on angiotensin receptors.

Taken together, these findings strongly supported Folkow's theory as to why and how hypertension accelerates. Due to their structural properties, hypertensive vessels offer more resistance when they are fully dilated and they over constrict in response to different agonists. Increased vasoconstriction calls for a higher pressure to maintain normal blood flow, the higher pressure enhances incremental vascular structural changes, and the vicious circle of a steady increase in blood pressure continues.

In the last two decades, Folkow's theory has attracted some opposition. It has been reported that the muscle wall in resistance arteries is not always hypertrophic and a process called *eutrophic inward remodeling* has been detected in some resistance arteries. When eutrophic remodeling takes place the lumen and the external diameter of the blood vessel are smaller but the muscular component of the vessel wall is not thicker. This is probably true but the physiological meaning of this difference escapes me. Brent Egan recognized the problem and inserted the following disclaimer in the paper: *"While our data strongly implicate a structural reduction in luminal cross-sectional area as an important contributor to resistance responses, the nature of the vascular abnormality is not proven. The possibilities include increased wall thickness,. . . a reduction in vessel size, or a reduction in vessel number. Since we have not anatomically examined vessel structure we cannot definitively exclude any of the three."* Right, but does it matter? We had documented a decreased minimal resistance and hyperresponsiveness to different agonists. This pretty much "nailed" our conclusion that the underlying mechanism was a structural alteration of resistance vessels; no other explanation fitted the data. What counts, regardless of the anatomic substrate, is the enhanced vasoconstriction which accelerates hypertension.

I think what happened to Bjorn Folkow is quite similar to the troubles Freud had with his pupils Adler and Jung. They diverged from the original theory but both did not alter the division of the psyche into conscious and unconscious and both accepted the idea that suppressed yearnings somehow are not good for you. As far as I am concerned, the concept of structural reinforcement is alive and well.

With these studies we removed the wrapping of mystery from the processes that cause the hemodynamic transition in the course of hypertension. (Figure 4) The cardiac output decreases back to a normal level due to a decreased stroke volume. The fall of stroke volume reflects two joint processes: increased cardiac stiffness and decreased beta adrenergic receptor responsiveness. In parallel, changes in the structure of small arteries (arterioles) favor the increase of vascular resistance, a process that enhances vasoconstriction and limits the capacity for vasodilatation. The unfortunate thing about all of this is that the body does not change its mind and all of a sudden decide to normalize the situation. Changes in cardiac output and vascular resistance in the course of hypertension are pathological processes that

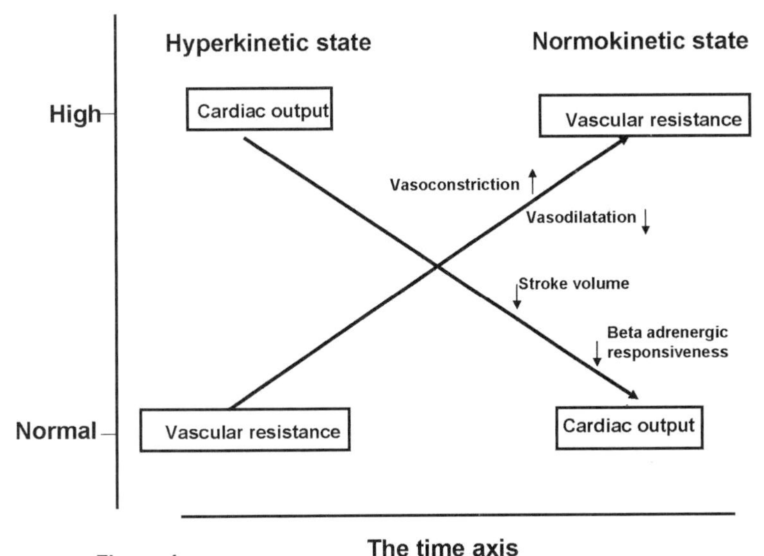


Figure 4

reflect underlying damage to cardiac and vascular structure and function. And all of that in a group of people who barely had a blood pressure elevation. What we used to call "mild" hypertension is not mild at all!

Our conclusions were supported by similarities between the high and normal cardiac output sub-types of prehypertension. Akin to the hyperkinetic group, those with normal cardiac output had tachycardia and a decreased parasympathetic inhibition of the heart and we simply assumed that they in the past might have had a hyperkinetic circulation. However my good friend, Per Lund Johanssen, followed a group of people with borderline hypertension in Bergen, Norway over 30 years and demonstrated that they develop established hypertension and that as time passes their cardiac output and stroke volume decreased while the vascular resistance increases. This unique longitudinal study dispels any doubts about the mechanism of hemodynamic transition.

So much about the riddle and the mystery. But the enigma remained untouched and I might as well level with you about it. There is a very unpleasant little fact in the entire story of prehypertension. It is reasonably easy to prove the autonomic imbalance and sympathetic overactivity in prehypertension, but in established hypertension the sympathetic overactivity seems to disappear. So, if sympathetic overactivity initiates hypertension, why would it later evaporate? I cannot say that we resolved the enigma but a decade later I felt brave enough to grapple with the problem and provide a conceptual framework for why this might happen. Since I am trying hard to write this memoir in a reasonably chronological fashion I must restrain myself for the time being. Eventually we'll get there, but for now let's move on.

Already in 1969 I anticipated that at some point we would have to move out of the laboratory into the real world. The volunteers studied in Ann Arbor might have not been representative of the general population. Most were students, and everybody will agree that students are birds of a different feather. The motivation to volunteer for invasive studies might have been very different in the normotensive control and in prehypertensive groups. What if "nervous" self-concerned individuals with borderline blood pressure volunteered for our experiments? Our finding would be skewed towards a group that in real life might be very rare. What if the degree of education, the type of employment, degree of physical activity, or the income status affected the degree of nervousness? To resolve this issue, we had to mount an epidemiological study in a real life setting. But this investigation had to be different from garden variety epidemiological studies. Almost everything that could be done with blood pressure and body size measurements and biochemical tests has already been reported. In the laboratory we studied physiology and we needed to design a study of physiological measurements in an epidemiological environment. Figure 5 is a reproduction of the bottom half of the original "master plan" graph. (E.H., A.Z. and S.J. stand for Ernest Harburg, Andrew Zweifler and Stevo Julius respectively.) From the legend you can see that already in 1969 I started writing a proposal for a study of the natural history of hypertension. I completed the document in 1970 and not surprisingly, that grandiose proposal was rejected. Rightly so on three accounts; the "who are you" issue, the "what is new" principle and the "we've got no money for large studies" operational reality. We had just started working in borderline hypertension and nobody knew us, there was little new in the proposal and completing the study would require an enormous financial commitment. I tried to minimize the cost but that did not fool anybody, and I learned a basic rule for grant applications; do not try to sneak in by the back door. People recognize what you are doing and the review will contain a flat unemotional killing statement that the study can't be executed with the proposed budget. But the real Achilles' heel of the

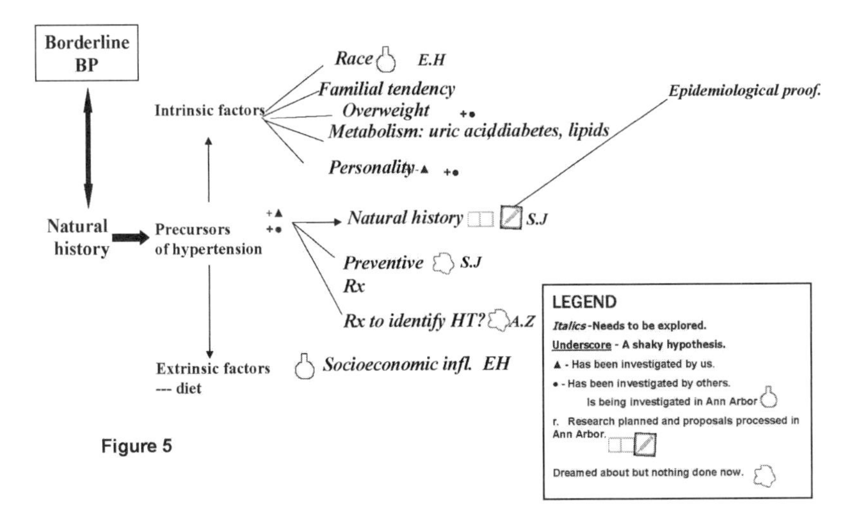


Figure 5

proposal was that it did not include hemodynamic studies. This is where our strength was and this is how our proposal could differ from others. Therefore, the number one task was finding a reliable method for noninvasive measurement of cardiac output.

Unfortunately, an earlier study of a noninvasive method to measure cardiac output was disappointing. Back in 1968, I was part of James Conway's team evaluating the validity of the carbon dioxide (CO2) re-breathing method.[4] James connected with an exercise physiologist and we tested the method in a group of top athletes. Compared to the "gold standard" of dye dilution the method worked very well during the exercise but at rest there was practically no correlation between the two measurements.

Incidentally, I learned something about athletes during these experiments. As physicians we are trained to help people in discomfort and I could barely tolerate seeing what was in front of my eyes. These guys worked up to highest levels of exercise and when they got there, medically speaking, they were in acute distress. Covered in sweat, they breathed quickly and with effort resembling patients in asthma attacks, as in congestive heart failure their neck veins were distended and from their faces you could tell they were in great pain. And just when I wanted to step in and stop the nonsense they'd come off the bike and with the first breath they'd invariably say "that was great," "it felt good" or some other words to that effect.

"Felt good? You must be out of your mind. Just a minute ago you were dying on that bike and I was ready to pick you up from the floor," I thought. "All right, so you completed the task but don't tell me you enjoyed the torture." These guys were trained to cope with pain. Their bodies were sending them all the right signals but somehow they got the wrong message; discomfort became enjoyment. What a masochistic denial!

Anyhow the message was that in the resting state, where we detected difference between prehypertensives and normal controls, the CO2 re-breathing method would not work. In the meantime technological advances made echocardiography suitable for measurement of cardiac output, but it was not clear whether the method was sufficiently sensitive to detect minor differences. Wolfgang Kiowski accepted the challenge to compare echocardiographic with the standard dye dilution measurements of cardiac output and reported the results in 1981.[5] The design of the study was quite neat. Wolfgang measured the output with both techniques two times at rest, once after speeding up the heart by infusion of the beta adrenergic agonist isoproterenol and once after he slowed the heart with intravenous beta receptor blocker propranolol. That way he got a wide range of cardiac outputs to compare results. Both methods were reproducible, that is to say that two times in a row they showed similar average values. There was also a strong correlation between the two resting measurement performed with the same method and both methods produced similar average resting values. However, the correlation between the dye dilution and echocardiographic readings was poor. In other words, the echocardiographic method frequently overestimated or underestimated the resting cardiac output values. With the echo method, one first measures the stroke volume and then multiplies it by the heart rate to calculate the cardiac output. Our heart rate measurements were precise as we derived them from recorded electrocardiograms. There had to be something wrong with the echo measurement of the stroke volume. Wolfgang pinpointed the problem, the echo method resulted in significantly lower values if an individual's stroke volume increased more than 40% over the baseline. Nice science but the gist of it was that the echo method was not good enough for our purposes.

Seven years after Kiowksi's report, we returned to the topic of noninvasive measurement of cardiac output. The leaders of that effort were Alan Hinderliter and Drew Fitzpatick and they had statistical support from Nik Schork, the son of my friend Tony Schork. The apple does not fall far from the tree and young Schork was a positively brilliant statistician. Al Hinderliter completed his residency in medicine in Ann Arbor and wanted to take a research year to round up his training in cardiology. Drew Fitzpatrick, a cardiologist, came from Australia to hone his research skills. Both were energetic, hard working and friendly persons. Both proceeded to careers in academic medicine. Hinderliter is an associate professor of medicine in the cardiology division of the University of North Carolina Medical School where he focuses on hypertension and echocardiography. Drew Fitzpatrick is an associate professor and director of cardiology in the Nepean Hospital, Penrit, Australia. He excels as a medical educator and is interested in improving cardiology health care delivery in the region.

They compared the results of cardiac output obtained with the gold standard of dye dilution to values measured with an improved version of the CO2 re-breathing and with a newly developed echo Doppler technique.[6] This protocol created an even wider range of cardiac outputs than the Kiowski study. There were two resting measurements, a measurement after inflation of thigh cuffs, during 35-degree head-up tilt, during isoproterenol infusion, and after blockade of beta receptors with intravenous propranolol. The estimates of resting cardiac output were reasonably accurate by either CO2 re-breathing or echo plus ultrasound (Doppler) techniques. However, the Doppler technique was more accurate in detecting acute changes in cardiac output.

Previously we used correlation calculations to compare the results with various techniques. These calculations were technically correct but this time around Nik Schork found a way to illustrate the vast superiority of the Echo-Doppler technique in simpler and more understandable way. Using changes in cardiac output that we detected by dye dilution at various points of the protocol, he calculated how many people would have to be investigated to detect such changes by noninvasive means (at a statistically significant level). The table below shows these calculations.

The echo Doppler method was as sensitive as the dye dilution and hands down won over the CO2 re-breathing. As in our first study, the CO2 method worked well when the cardiac output was high (in the first study by exercise and now by isoproterenol infusion), however in conditions with low cardiac outputs CO2 re-breathing failed miserably.

To detect a significant decrease of cardiac output with tilting, one needed only five people with dye dilution or echo-Doppler techniques and over one hundred with CO2 re-breathing.

**Mean cardiac output (Q) changes ( L/min.) with calculated sample size needed to detect observed changes at a 5% significance level (beta set at .2)**

| | Dye dilution | Doppler | $CO_2$ re-breathing |
|---|---|---|---|
| **Cuff** | | | |
| Mean Q change | -0.36* | -0.43* | -0.18 |
| Sample size | < 5 | < 5 | 33 - 40 |
| **Tilt** | | | |
| Mean Q change | - 0.59* | - 0.98* | - 0.06 |
| Sample size | < 5 | < 5 | > 101 |
| **Isoproterenolol** | | | |
| Mean Q change | 1.59* | 1.67* | 1.21* |
| Sample size | < 5 | < 5 | < 5 |
| **Propranolol** | | | |
| Mean Q change | -0.45* | -0.39* | -0.22 |
| Sample size | 5 | < 5 | 22 - 27 |

Figure 6 * = $p < 0.05$ AL Hinderliter, A Fitzpatrick, N Schork, and S Julius Clin Pharmacol Ther 41, 1987

In the echo Doppler technique we finally found what we were searching for, a reliable and sensitive noninvasive method of measurement which could be used in field studies. Echo-Doppler was both noninvasive and non-threatening. When we eventually implemented the field study many participants told us that they liked seeing their beating hearts and hearing the swooshing sound of the blood flow. To each his own, I would have never guessed that any of our procedures would be entertaining but that is what they told us.

When I entered the "Natural History" arm into the master plan (Figure 5), I had the Tecumseh study in mind. In the mid 1950's, the University of Michigan School of Public Health obtained funds to initiate epidemiological studies in the village of Tecumseh, 30 miles to the south of Ann Arbor. Contrary to large cities where people tend to come and go, in Tecumseh the population was stable. The villagers either worked the land or were employed by the Tecumseh small engines company which since 1930 provided a steady source of jobs. As people tended to stay put, it was not unusual to find in the village and its vicinity three to four generations of the same family. That was one of the reasons that the National Institutes of Health (NIH) funded the Tecumseh study along with the famous Framingham study. Both were supposed to investigate long time health trends in the community. From the very beginning, the Tecumseh study focused on two major tracks: infectious disease and pulmonary respiration, and cardiovascular diseases. People in Tecumseh proved remarkably cooperative and did not mind coming for repeated health exams although they received little direct feedback from these examinations. The organizers of the first study understood the local situation and the number one rule was to leave the individual's health care in the hands of local physicians. Tecumseh had a very good community hospital with a capable staff. If you found something during a periodic health exam, you'd send a note to the doctor to follow it up. Local doctors initially had their doubts but after a few examination rounds realized that research teams from Ann Arbor did not in any way compete with them. Knowing how the University Hospital in Ann Arbor worked, I was not surprised that the local doctors had their doubts. To put it as gently as I can, let's just say that communications with local doctors were not our strong suit and that occasionally we'd retain patients for repeat consultations, teaching or research purposes. No wonder then that despite frequent on-site visits from the University of Michigan, our colleagues in Tecumseh had and have a strong preference to refer their patients to specialists at St. Joseph's Hospital in Ann Arbor. Eventually when our team worked in Tecumseh I found that this arrangement of "cooperative detachment" worked rather well. No business relationship - no conflict.

The original Tecumseh study completed repeated rounds of health surveys and published valuable papers in the epidemiology of viral diseases as well as cardiopulmonary and other types of chronic diseases. Most importantly, the early investigators arranged a great data base for each round of exams including a code for family relationships across generations. However, at some point about two decades after the first round of the Tecumseh study, the NIH decided to guarantee automatic continuation of the Framingham Study whereas the Tecumseh study had to compete for individual grants for each of its projects. By-and-by the funding dried up and in 1987 when I proposed to initiate a study of prehypertension there were no on site research activities in the village.

During my first two years in Ann Arbor, Dr. Hoobler asked me to go once a week to Tecumseh to perform physical exams. That helped Hoobler to cover part of my salary but the arrangement was also good for me. I learned the ins and outs of the field work, and was very impressed with the way the original investigators designed procedures and documentation for the Tecumseh study. Knowing that there were no ongoing studies in the village, and hoping that my field experience credentials would help, I thought that my initiative to start a new prehypertension study would be well received. I was dead wrong!

After my presentation to the permanent standing Tecumseh Study Committee in the School of Public Health a chill fell over the room. Nobody was impolite but they sure responded with a stream of stories about obstacles. Everything related to the Tecumseh project seemed to be extremely difficult. It was difficult to set up a field office, it was hard to access the files stored in a former factory in Ann Arbor, the citizens in Tecumseh might be fed up with new studies and so the litany went. My ears perked up when somebody said that the original computer-ready data were kept on paper rolls which had started to deteriorate. That was serious! And the School apparently had no resources to fix that particular problem. Then I got a stern lecture about rules for submitting proposal to the Committee research and the code of behavior in Tecumseh. Finally, it was made clear to me that the primary responsibility of the Committee was to protect the integrity of the study.

Of course, they had to review the proposals, and they had to protect the village population from repetitive exams but what was bothering these folks? I had never before met such a dejected and pessimistic committee. Part of what bothered them soon became evident; they were too polite to outright reject my proposal but approving it would cause them a great deal of internal troubles. In a few days I received a cease and desist letter from a School of Public Health professor. He was preparing a large study of respiratory diseases and should his grant get funded there would be no space for other projects. The letter ended with something that at first glance appeared as a compromise but deep down was plain blackmail. If I were to use his subjects, I would have to financially compensate his project. The letter was hilarious but I lost my sense of humor and fired off a nasty letter in reply. Something about a free society and that he neither owned a group of people nor could he restrict activities of other investigators.

After a while I cooled down and started to think. The confrontational style would get me nowhere. I was the intruder and I had better learn their rules. And I was quite stunned by their ticking time bomb. If it was really true that the original computer input data was about to fall to pieces, this would be an absolute disaster. Yet the Committee was not moved to action.

I decided not to be affected by the morose atmosphere and continued to work on the prehypertension project as if everything was fine and dandy. But I knew the rules. The Tecumseh study was supported by NIH. Data collected by public funds must be made available to the public. If I were to get some funding, nobody in the School of Public Health could legally deny me access to old data. Nobody but a physical problem, if time ate the data everybody would be in big trouble. It was also clear that I could not get a grant from the NIH without showing them some preliminary data about the feasibility of the project. So, steady as she goes and damn the torpedoes! I had to find money to generate feasibility data.

Let me reiterate that to succeed in research one has to have some luck and my lucky star shone again. At that time I was well known in the behavioral medicine community. Our demonstration of sympathetic overactivity in hypertension provided a physiological support for their theoretical concepts, and we also studied the personality and behavior of subjects with borderline hypertension. As Figure 5 shows, we intended to take another look at social and behavioral characteristics in prehypertension in the future epidemiological study. I discussed this with Ray Rosenman of "type A" fame and Charlie Spielberg, a master developer of personality questionnaires. I met them during a Symposium and sought advice about psychosomatic measurements appropriate for use in a field study. I got much more than I expected. After a few months I was invited to give a presentation to the Scientific Research Committee of the RJ Reynolds Company. Yes, dear reader, the tobacco company! And they gave me two years of very solid support for my Tecumseh project! Go figure!

I know purists will be shocked but the truth is that this was as untainted a grant as I ever got. They did not set up any precondition besides stating that studying tobacco ought not to be the primary focus of the research project. Of course, we did not intend to only study smoking but we could not possibly omit collecting data on such an important cardiovascular risk factor as smoking. I also indicated that we'd report the results no matter what came out. That was fine with them.

As they say: "Don't look a gift horse in the mouth." But let's examine the motivation and the ethics of the entire thing. Why would they support me? The most benign explanation is that RJ Reynolds sought to improve their public image and they supported health research to atone for possible negative effects of smoking; I am saying "possible" effects since at that time the tobacco industry insisted that negative cardiovascular effect of smoking had not yet been scientifically proven.

A more devious and Machiavellian explanation is that they wanted to highlight all other cardiovascular risk factors. If sued for damages, they could raise doubts whether in a specific case smoking was the only cardiovascular risk factor. And indeed smokers are usually physically inactive, might have hypertension and so on.

Okay, why did I accept? Let me first say that they gave me no reasons not to accept. They did not ask for progress reports, did not insist on seeing our manuscripts before publication and there was not an inkling of any kind of quid pro quo. During World War II, I had seen many people saying what they would or would not do in a hypothetical situation and when the situation arose they turned 180 degrees and did the opposite. Therefore I cannot say for sure what I would have done had they attached some strings to the project but I prefer to see my self as a highly honorable person on a white horse. So there! I would have rejected it with disgust!

For moralists I have no excuse for the appearance of conflict or for the possibility that I aided a source of evil. May they be purer than I!

But I got a great extra kick out of the whole thing. At the end of the first report from the Tecumseh study,[7] I inserted the following acknowledgement: *"This investigation was supported by grant HL37464 from the National Heart. Lung, and Blood Institute, National Institutes of Health, Bethesda, MD, and in part by the RJR/Nabisco Company, Winston-Salem, NC"* Strange bedfellows! I will admit to having a quirky sense of humor and I still find it funny. Nobody ever complained.

As soon as the RJ Reynolds Company sent the first installment of the grant funds to the University of Michigan, I informed the Tecumseh Committee in the School of Public Health that to help preserve data from previous Tecumseh exams I would send an annual financial contribution to the Committee. I could not tell them outright what to do with the money but the contribution was larger than what was needed to hire a transcriber of old data into a format compatible with modern computers. The Committee accepted the contribution without too much of a comment.

Next, I gave a Grand Round lecture to the staff of the Herrick Hospital in Tecumseh explaining our previous results and informing them of what we intended to do in the village. That looked straight-forward but it wasn't. I knew that anybody coming to Tecumseh from the Ann Arbor University Hospital automatically acquired an onus of suspicion and ambivalence. As a first step, I spoke to Dr. Brandt in the Ann Arbor St. Joseph hospital. He was the cardiology consultant in the Tecumseh Herrick Hospital. Generally there was very little communication between the St. Joseph and the University hospital staff in Ann Arbor. A few of the St. Joseph physicians practiced also in the University Hospital, but this was a one-way street, nobody with a primary appointment in the University Hospital worked in St. Joe's. However, I knew Dr. Brandt socially and after I explained my plans, he graciously agreed to suggest that the Herrick Hospital invite me for Grand Rounds. I cannot tell whether this made a difference but generally speaking one should not elbow his way into a community. It is much better and more appropriate to work through the existing community leadership structure.

The grand round presentation went well and the Herrick physicians seemed interested in the project. Following the lecture, a local physician came to chat and after usual perfunctory compliments told me in confidence that he planned to retire. He and his wife practiced in a free-standing building at the edge of the village. Should we decide to open a field office in Tecumseh, he'd be delighted to rent us his building. I took note and when we eventually got the NIH grant we rented his office. The building was on a two-acre lot. Up front it had sufficient parking space for multiple cars and behind the ranch style house was a spacious remnant of an old orchard. The building was built for a physician's office; it had a waiting room in front of the receptionist space and a corridor leading to multiple examination rooms. A few rooms were larger and could be used for laboratory and medical examination devices space. The smaller rooms had patient examination tables.

The house served us very well and practically no alterations had to be made. The biggest room accommodated the echo Doppler machine, whereas three additional large rooms were used for special measurement equipment and for blood sample processing. We even had a "staff room."

A large study requires a great deal of preparation and a capable staff. The beauty of a large University is that you can easily find capable people and I soon assembled an extraordinarily talented group of coworkers. I already mentioned Nik Schork and he became a crucial member of the team. Nik, the son of my friend Tony Schork, completed his college degree and started taking graduate courses at the University of Michigan.

The University had a very good policy for an employed student. When hiring a student, all employers in the university system had to permit the student to attend classes for a specified number of credits per semester. In addition, the employer had to pay tuition for those classes. While he worked with us, Nik completed master degrees in philosophy and statistics, and obtained a PhD in epidemiology Though he did this over a short period of time, Nik flawlessly handled both his studies and his work. This extremely talented, and occasionally moody young man became an indispensable member of the team. Here is an illustration: At some point during the development of the Tecumseh prehypertension protocol, we got access to source files held in storage but were unable to obtain the code for family relationships among various participants in previous exams. Of course, this was very important information for the conduct of the study. After spending some time in the storage room and looking at numbers on various files, Nik realized that there was a system in those numbers. It did not take him long to penetrate the entire code as some number stood for the family name, another for the relationships within the family, and yet others for when the exam was completed. It was a brilliant code and it took a very smart man to decipher it. I didn't have to go back to the Tecumseh Study Committee to request the code! Nik Schork is now an internationally known expert in statistical analysis of genetic data. He is currently the director of research, Scripps Genomic Medicine, and professor of molecular and experimental medicine in San Diego.

**Nik & Tony Schork in 2006**

One of the early members of the Tecumseh team was Cosmas van de Ven, an optimistic and lively physician who had just graduated from Medical School in Leiden, Holland. He had earlier taken a break in his studies to work with David Bohr in Physiology. After graduation he decided to immigrate to the United States and joined us for a year before he got a residency at the Henry Ford Hospital in Detroit. Cosmas was an excellent organizer and greatly contributed to the development of the documentation (forms, filing system, data transfer and the like) for the study. After his residency, Cosmas returned to the University of Michigan. Presently he is Clinical Professor of Obstetrics and Chief of the Neonatal Unit at the University Hospital in Ann Arbor.

At that time I also hired Eric Grant as data manager. Eric proved to be a capable computer programmer and he particularly excelled in digitalizing biological signals. He also was a hardware man. He'd read the literature and then build our own equipment to perform specific tests. As we worked on the study protocol, I remembered an oddly fascinating study from Italy. Some Italian dentists reported that patients with hypertension better tolerated pain than people with normal blood pressure. To prove the point, they connected an electrode to the tooth, gradually increased the charge until patients reported having pain and sure enough, hypertensives had a higher threshold for pain. Disregarding for a moment the brutality of the test, the finding was very interesting. A few papers in the psychological literature suggested that besides the circulatory response the defense reaction increases the threshold for pain. It sure made a lot of sense to me; if the body prepares itself for a fight and possibly for an injury, suppressing pain helps to continue fighting. Speak to soldiers wounded in a face-to-face combat and they will tell you they didn't notice they were wounded. Anyhow, I got this idea that threshold to pain may be a good way to assess the degree of activation of defense reaction in an individual. Remember that the circulatory pattern in hyperkinetic borderline hypertension resembles the circulatory response during the defense reaction! Anyhow, I thought it would be interesting to test the threshold for pain in a less cruel fashion. Eric found that some investigators in the US developed a tip-of-the-pen like device in which they could gradually increase the heat. They then recorded the heat at which a person reported discomfort. He got in touch with them and soon assembled our own device together with a computerized recorder of the entire experiment. I wish I could tell you that we had great results with this method but we did not. The point here is that Eric was capable of designing and producing medical devices.

Eric was particularly interested in handling large data sets and he streamlined the flow of data from various study parts (field office, laboratory data, and in-house analyzed records from the field) into the master file for computer filing. This expertise served him well when his wife's parents in Japan got sick and he had to move the family to Hiroshima. We gave him a strong letter of recommendation; Jim Neal our colleague in the Department of Genetics also gave him written support. Jim used to work with the United States investigators on the effects of the atomic bomb. Eric applied for and got a job with the USA research center in Hiroshima. Recently he returned to the USA to complete his PhD.

In addition to these three excellent young people at the beginning of their careers, I also hired Dr. Agnes Mejia, a fully trained nephrologist from the Philippines. Akin to Arturo Pascual, Agnes was calm and well-organized. She excelled in coordinating the entire team. Developing a field study is a huge enterprise and Agnes became my right hand. Unfortunately, I was unable to convince the authorities to grant a permanent visa to Dr. Mejia. I wrote a sincere letter explaining how good she was and how important it was for us to retain her. But the authorities were not convinced. Agnes was a proud person and instead of hanging on and coping with protracted insecurity, she returned to Manila. Presently she is professor of Medicine at the University of Philippines Medical College and chairwoman of the Department of Medicine in the renowned Philippine General Hospital in Manila. Agnes and I had to go through every aspect of the study. First we needed to decide exactly which tests to select for the final protocol. Next we determined how much time was needed for completion of each test. This sometimes could be predicted but on occasion time and motion studies were in order. After that we designed a sheet for data entry of each procedure. Once these issues were resolved, we started to assemble and train the research team. Most crucial was to find local people for the field office in Tecumseh. They were our outreach arm and through them we communicated with our research subjects. Luckily, we found some very capable workers with experience in previous Tecumseh exams. Training of the entire team was crucial. Field people visited villagers in their homes to instruct them in home blood pressure technique. Since more than one person went to the homes, it was very important to precisely define what has been taught. We also insisted on explaining standardized blood pressure measurement procedures to all physicians and rehearsing their implementation.

**Hypertension Division Physicians in 1982**
**First row:** *T. Shahab, A. Mejia, R. Gupta,* **Second row:** *D. Bassett, A. Zweifler, M.A. Sekkarie*

We implemented continuous quality control. Laboratory results were monitored by standard procedures but for other components we had to invent our own monitoring system. For example, for plethysmography measurements, which can vary with the season, we compared the average monthly results. For blood pressure measurement, we compared the averages among various physicians. And we were glad to have done so. In the first field office in Tecumseh the temperature inside the building was not well regulated. Despite our rule that everybody must be in the room half an hour before the measurements, the average forearm blood flow in winter was lower than in other months. When we moved to the new well heated and air conditioned building, the plethysmography result became stable. Blood pressure measurement monitoring also paid off. Two weeks after a new physician from Europe joined us, we found his average diastolic blood pressure was about 10 millimeters above average values of other physicians. This was a senior physician and I had the painful task to tell him that his blood pressure measurement technique was not adequate.

Against every literary rule, I have already indicated that we got the NIH grant for the Tecumseh study. As a writer you are supposed to keep the reader curious and not spill the beans in advance. So it might be anticlimactic to now tell you that we got the grant but I did not tell you what sort of a grant we got. The behavioral medicine branch of the NIH chose my grant for the coveted MERIT award! As everything in NIH, "MERIT" is an acronym, here standing for "Method to Extend Research in Time." Normally the maximal duration of a grant was five years but ours was approved for ten years of funding!

Having a decade of financial support was an incredible luxury and we were now ready to implement the study.

**Bibliography**

1. Julius S, Schork MA: Borderline hypertension - a critical review. J Chronic Dis 23:723-754, 1971.

2. Julius S, Randall OS, Esler MD, Kashima T, Ellis CN, Bennett J: Altered cardiac responsiveness and regulation in the normal cardiac output type of borderline hypertension. Circ Res 36-37 (Suppl. I):I-199-I-207, 1975.

3. Egan B, Panis R, Hinderliter A, Schork N, Julius S: Mechanism of increased alpha-adrenergic vasoconstriction in human essential hypertension. J Clin Invest 80:812-817, 1987.

4. Ferguson RJ, Faulkner JA, Julius S, Conway J: Comparison of cardiac output determined by CO2 rebreathing and dye-dilution methods. J Appl Physiol 25:450-454, 1968.

5. Kiowski W, Randall OS, Steffens TG, Julius S: Reliability of echocardiography in assessing cardiac output. Klin Wochenschr 59:1115-1120, 1981.

6. Hinderliter AL, Fitzpatrick MA, Schork N, Julius S: Research utility of noninvasive methods for measurement of cardiac output. Clin Pharmacol Ther 41:419-425, 1987.

7. Julius S, Jamerson K, Mejia A, Krause L, Schork N, Jones K: The association of borderline hypertension with target organ changes and higher coronary risk. Tecumseh Blood Pressure Study. AMA 264:354-358, 1990.

# Thinking Time

The late 1970's and '80's were the golden years of the Ann Arbor Hypertension Division. After I obtained the MERIT award from the National Institutes of Health, the field work in Tecumseh started in earnest. This called for a substantial increase in technical and medical personnel. Eventually the growth of the Tecumseh staff tripled the personnel of the division.

**The Hypertension Division and Tecumseh Staff in 1983.**
**First row (l to r):** *D. Meadows, L. Krause, A. Zemva, S. Julius, A. Meija, L. Earle*. **Second row:** *C. Mead, P. Wlazlo, K. Jones, L. Vadnay, M. Rapai, S. Baker, J. Kneisley, P. Minick, M. Wonderly, M. Borondy*. **Third row:** *A. Sekkarie, R. Gupta, J. Mertens, D. Brant, R. Schmouder, J. Ferraro, N. Schork, S. Kjeldsen, A. Weder*

By some standards this was a fabulous progress, but sudden growth brings its own problems. I very much enjoyed the volunteer spirit in the old, smaller, division. We shared the same interest, we loved our work, and it was easy for me to lead a group of enthusiasts. The Tecumseh expansion threatened to involve me in administrative responsibilities. I knew my science but whether I could handle budgets, accounting, personnel files, and also manage responses to a myriad of inquiries was an open question. Nobody seems to care when you have a small budget but after you cross a certain threshold everybody pays attention. Soon you are inundated with requests for progress reports from the granting agency, the Department Chair, the Personnel Department and all the way down to the lowliest university accountant. Seemingly they work independently, but I swear all of them are part of a well choreographed conspiracy to rob you of free time. As soon as you respond to one request the next pops up with an equally short deadline and with a patently false message of catastrophic urgency. This dynamic of growth eventually got to me.

At that time Dr. Robinson had retired and William N. ("call me Bill") Kelley from Duke University was recruited to the post of Chairman of Medicine in Ann Arbor. He was young, had excellent scientific credentials, and proved to be a strong administrator. On his first day as Department Chairman, Kelley removed three division chiefs and initiated a search for their replacements. The previous leaders could stay on as acting division chiefs if they so desired. If not, one of the division's youngest staff members was appointed as the acting chief. That shock treatment got everybody's attention!

Kelley was the high priest of constant growth. He had an insatiable appetite for expansion and in each annual report he'd proudly point out how much the department's research funds had increased. As did all agents of change, Kelley polarized people around him; some physicians hated every change he introduced, some appreciated the productivity-based incentive system he implemented, but nobody could ignore the new chairman of medicine. You'd think that adult academicians might be sure of themselves and capable of laughing off various attempts to change their ways, but the truth is that the new environment made everybody tense and insecure.

A few years after Kelley came to Ann Arbor our division's research grants exceeded one million dollars. Bill was delighted and invited me to his office to congratulate me.

"This is a great achievement," Bill intoned in his Southern accent, "and I wish to also congratulate you on recruiting three junior faculty members. You've certainly done your part but why don't you push the new faculty a bit. Realistically speaking in three to four years your division could easily double its research funding. Two million dollars would be a great benchmark."

Typical Kelley, nothing was good enough and there was no end to expansion. Driven by some internal demons, Kelley simply could not stop the mad race to the top. If I took his bait he would hold me to the new benchmark. I had to register some sort of disagreement.

"I am not good at pushing people," I said, "I push myself and hope others will follow suit. Our new guys are a bit too young to bring huge grants but let me see what we can do."

"So you don't want to commit to two million?" Kelley got the message but did not seem to be particularly irritated.

I tried to finesse my way around the issue: "Oh, I just read a booklet entitled 'The Peter Principle.' In this witty, tongue-in-cheek critique of the industrial society, the author describes how contemporary competition motivates people to aim at ever higher goals. An ambitious person is often promoted to an ever more demanding job until he reaches his own level of incompetence. At that point promotions come to a screeching halt and the individual remains forever frozen in his last post. Consequently the higher the position in a company is, the more likely it will be occupied by an incompetent individual. The same might apply to the academic environment."

Hearing myself talk, I became concerned about whether I may have gone too far. But Bill was perfectly at ease. He knew he had not yet reached his own level of incompetence!

"I might be able to handle a million dollars but should our research grants grow to two millions I might fail," I said after a short pause.

"Oh come on, you can do better. Try it. You might like it," said Kelley in a benevolent tone.

The issue of a two million dollars goal never surfaced again. Of course Kelley did not change his point of view but he was content to let me do it my way.

Kelley occasionally needled people just for the fun of it. One day out of context and for no particular reason he asked me to explain why he should not make the Hypertension Division a section in the Cardiology Division.

"Because I would not like it and you have nothing to gain from such a merger," I said laconically.

"I guess you are right," Kelley laughed, "If it ain't broken, don't fix it."

One of Kelley's effective innovations was the external review board, an impressive group of eminent national leaders who once a year visited Ann Arbor to evaluate the performance of the department. The review board served a dual purpose. It recorded the department's progress and thereby gave Kelley ammunition against in-house criticism, and it provided him with a rationale for future personnel changes. If a division lagged, the chairman could use the findings of the board for further decapitations. Bill Kelley kept inviting three to four division chiefs on a rotating basis to give a progress report to the review board. Given the history of Kelley's famous first day division chief massacre, nobody was looking forward to these presentations. Luckily, when my turn came the hypertension division was still "on the up." We had our million dollars, we were in the midst of the long-term NIH grant for the Tecumseh study, and we had good staff. Our only serious problem was the organizational oddity I inherited from Hoobler; other departments of medicine did not have a hypertension division. But from previous conversations I knew I had Bill's support and that was all that counted.

After Kelley introduced me to the visiting group of remarkably inexpressive and deadly serious individuals, I first informed them about the increase in our clinical activities, the staff development, and major lines of research in the division. My report seemed to have no affect at all, the visitors just sat there in silence as if they were mourning some deceased dignitary. I found the funereal atmosphere quite irritating and tried to lighten the situation a bit. The last slide of my presentation was a graph of our research funds which, due to the recent Tecumseh grant, showed a steep increase in the last two years. No reaction!

"Gentlemen," I said, "you are all eminent scientists and you can recognize an exponential curve when you see one. This is a serious situation and I hope you will recommend that we slow down. If we were to continue at the present pace, within four years we might have more funds than the entire Medical School. Five years after that we might be in a position to buy the entire University."

No reaction again! Okay, even if the joke was inept or sounded disrespectful I'd expect to see a smirk or a frown. But no, they just sat there in the mummified decorum of self-perceived elegance.

"Any questions?" Kelley asked.

No questions.

"I guess that is it. Thank you," Kelley said.

As I stepped out of the room I heard Bill saying to the group "Well, Julius is a bit different." I did not hear the rest of the conversation but am sure that Bill defended me with his favored "if it ain't broken" argument.

Kelley was a good leader and I was sorry that he left Ann Arbor. When he arrived, Ann Arbor ranked forty-second in research grants among departments of medicine. Fourteen years later when Kelley left, the department had become the fourth largest recipient of research grants in the nation. With time I came to appreciate Bill Kelley's modus operandi. He managed to set clear goals for everybody and as long as one met the expectations everything was fine. Bill was remarkably objective. It did not matter whether you played cards with him or stayed out of his sight; you were measured with the same yardstick. He was self assured and openly judgmental. You did not have to worry what he thought about you, Bill was more than willing to give you a "to your face" straight positive or negative assessment. Devoid of any visible emotions, he could congratulate you or fire you in the same measured even voice. Since it did not bother him a bit to give you bad news, Kelley had no use for the usual elliptical academic backstabbing. And I found that very refreshing.

The real problems arose when Kelley underestimated a person's potential. He wanted to recruit junior faculty capable of landing huge research grants. Bill's formula for evaluation of someone's prospects relied more in an individual's pedigree than in his present-time personal characteristics. No matter whether a person was bright and hard working, if he was not trained in a leading academic institution or did not acquire some "cutting edge" methodological skills, Kelley would not take chances. Unfortunately Kelley was also dead set against in-house promotion. Graduates from Michigan programs were advised to leave the institution, acquire new skills and then compete for a staff position at their Alma Mater. The intent of this policy was laudable. Inbreeding and development of a rigid "school of thoughts" are threats to academic institutions and a steady infusion of "new blood" is the best antidote against narrow minded provincialism. However, Kelley was too rigid in applying his rule. In the case of our division only a few other centers in the USA offered training in hypertension. Consequently, we had to bring up our own trainees and promote them to faculty positions. Needless to say, this occasionally put me on a collision course with Kelley, and in only one instance did I succeed to have him bend the rule.

The exception to the rule was Kenneth Jamerson whom I first met in the early 1980's on internal medicine wards. I was immediately attracted to this tall, slim and talented native of Detroit who was a graduate of the University of Michigan Medical School. You had to be good to be a resident in Michigan and Ken was no exception; he was knowledgeable and an avid reader of current medical literature. Whereas most residents relied heavily on laboratory findings and performed only a perfunctory physical examination, Ken was interested in the old fashioned art of bedside diagnosis. At that time I still had considerable expertise in patient examination, and Ken and I had a great deal of fun auscultating, percussing and palpating our ways to a diagnosis.

Ken's family emphasized the importance of education but had very limited financial means. He took his schooling very seriously and graduated from an exclusive Detroit high school. It must have been very hard for him to cope with inequities and prejudice but he never complained. Ken trained himself to respond with laughter to not-so-infrequent intended or unintended insults. Having watched other people emerging from hard times as bitter and suspicious individuals, I understood the emotional cost of Ken's optimism. You might finesse your way around an incident but deep down it hurts. It is profoundly unnatural not to redesign someone's nose when he well deserves a punch. I know how hard it is to swallow abuse, and Ken deserves compliments for being who he is. It is fun to be in the company of this witty and sharp man. In due course, we became good friends and Susan and I consider him a close member of our family. Whereas Ken acknowledges me as his mentor, I have learned a great deal from him. He has a keen understanding of people and circumstances. In Yugoslavia, I knew whence the wind blew from but in America everything appeared fair and square and I lost my bearings. However, instinctively I know that everything is not as nice as it seems and Ken's understanding of American ways and his recognition of hidden obstacles is very helpful to me.

**Ken Jamerson**

Ken was instrumental in the implementation of the Tecumseh study and in designing subsequent experiments to investigate the mechanisms of insulin resistance in hypertension. Ken is a "people person" and his exceptional ability to work with others proved to be a great asset; he has assumed leadership positions in many large epidemiological studies. He is a superb speaker and educator and is widely recognized as a leading expert in hypertension. Ken is a productive scientist and in a span of ten years he has deservedly advanced from fellow to full professor of medicine at the University of Michigan.

At about the same time Ken joined the division, another tall and slim man came to work in Ann Arbor. Sverre Eric Kjeldsen, an instructor in medicine in the Ulleval University Hospital in Oslo, Norway, had already written a considerable number of research papers on hypertension, including some interesting work about the autonomic nervous system in hypertension. Len Hansson considered him one of the most promising young medical scientists in Scandinavia and thought a stay in Ann Arbor might be very useful for his further development.

**Sverre Kjeldsen**

I met Sverre during some scientific meetings, knew what he looked like, was awed by his height and his ramrod straight posture, but knew very little of him as a person. He turned out to be as resourceful and precise as Len Hansson once was. Akin to Len, Sverre settled all administrative issues (medical license, driver's license, social security number and the like) already in the first week, found an apartment for his family and was ready to go. He was incredibly well organized and disciplined; he never forgot a task and always lived up to every self-imposed deadline. Sverre has a legendary memory, he never forgets a paper he read. I tend to remember contents of articles but have little recollection when and in which Journal a paper was published. It was a pleasure and a luxury having Sverre around. Instead of searching the literature I'd simply ask him and presto, he'd come up with the reference. Sverre is one of the kindest and most patient individuals I ever met but even he had his limit. One day when I asked for a reference he gently reminded he had given it to me a week earlier. I got the message! Sverre was also a quick writer and he could finish a high quality paper in half the time it would take me to prepare the first draft. Not surprising then that Sverre had a stellar academic career. He is now a professor of medicine in Oslo. A professorship in Norway is very different from a similar appointment in the USA. Whereas the Department of Medicine in Ann Arbor has over 40 full professors, the entire country of Norway has just handful of professors of medicine. Sverre's talent and hard work was recognized by his peers and he was elected as the president of the European Society of Hypertension. Though he returned to Norway, Sverre did not completely leave Ann Arbor. I continued to cooperate with him in a number of research projects and he remains a visiting professor in Ann Arbor. In fact since he left Ann Arbor, Sverre and I have been coauthors on 58 peer reviewed papers!

The two tall men, Jamerson and Kjeldsen, were a great help to Agnes Mejia, the diminutive dynamo responsible for the Tecumseh field work and the future chairperson of Medicine at the University Hospital in Manila, Philippines. This pleasant triumvirate of hard working motivators interacted very well with the rest of the Tecumseh team. Eventually the study ran on autopilot and my only role was to regularly review the progress of the study. This, in turn, left me with plenty of thinking time. But before I tell you how I used that time let me give the credit where it is due.

Other physicians and scientists working in the first five years of the Tecumseh study were R.Schmouder, J. Ferraro, J. Weisfeld, R. Gupta, E.M. Sekarrie, and E. Johnson, PhD from the U.S., J. Petrin, A. Zemva and R. Accetto from Slovenia, and P. Nazzaro from Italy. Ales Zemva and Rok Accetto are professors at the University of Ljubljana, Slovenija, and Jurij Petrin is president of Pharmaceutical Regulatory Services, Inc, Princeton, NJ. Pietro Nazzaro is associate professor, Stress Research Center University of Bari, Italy. Robert Schmouder works in the Exploratory Development, Novartis Company, New Jersey, and Abed Sekkarie is a clinical associate professor of medicine, University of West Virginia, Bluefield, WV. Unfortunately, I have no information on the present status of other American Tecumseh study investigators.

Freed of daily oversight of the Tecumseh study, I wrote a number of reviews of our previous work and each time I realized we were left with a huge unresolved problem. You might remember from the previous chapter the "riddle wrapped in a mystery inside an enigma" story. The early phase of hypertension looks very different from the later established hypertension. Many patients in the prehypertension phase exhibit a "hyperkinetic state" of a high cardiac output and a fast heart rate. In later established hypertension the cardiac output is normal and the vascular resistance is high. The "riddle" was whether the hyperkinetic state is a different entity than the high resistance hypertension. The literature provided ample evidence that the hyperkinetic state eventually evolves to high resistance hypertension and the "mystery" of how this hemodynamic transition occurs had to be explained. Our studies removed the wrapping of mystery from that process; the transition is due to the secondary effects of high blood pressure and increased sympathetic activity on the heart and blood vessels. Due to the high pressure the heart become stiffer whereas sympathetic overactivity renders the heart rate less responsive to sympathetic stimulation. Combined together, these processes are responsible for the decrease of cardiac output. Simultaneously the high blood pressure thickens the walls of resistance vessels and they over-respond to constrictive stimuli. Furthermore, the pressure damages the endothelium of arterioles and this renders them less responsive to stimuli that widen the arteriolar lumen (vasodilatation). The end result is an increase of vascular resistance. As simple as that! But the literature strongly indicates that the sympathetic system is overactive only in early phases of hypertension whereas patients with advanced hypertension seem to have a perfectly normal sympathetic tone. Why and how would that happen? That is a tough one!

Whereas I think I finally understood why the sympathetic tone might "evaporate" in the course of hypertension, explaining this to others proved very difficult. I hope you can stay with me through the next few graphs but if you want to skip this part I will not complain. And should you decide to pass, rest assured you will be in good company. I published my thoughts about the enigma of sympathetic tone in hypertension as a front editorial in a very reputable Journal,[1] but the paper failed to excite most of hypertension specialists. Probably the topic was too complicated or, maybe, I should have used the J.S.Bach technique and written a myriad of variations on the same theme until it finally "sinks in." Certainly other people proposing new concepts in hypertension followed up the initial salvo with a continuous barrage. What works in artillery seems to work also in science; hypertension experts might agree or disagree with the authors but repeated papers on the topic made the concepts of J. Conn, J. Laragh and A. Guyton known to just about everybody.

Anyhow, this book gives me another chance to explain the enigma. Let's start by repeating that the autonomic nervous control of the circulation is arranged as a thermostat-like negative feedback. A negative feedback tends to stabilize the variation around a set point. If you set the thermostat at 72 degrees, a lower temperature in the room will fire up the furnace to increase the heat, and vice versa, if the heat exceeds 72 degrees the furnace shuts down.

The negative feedback of a thermostat is straight forward, it senses the temperature and regulates the temperature. However, regulation of the circulation of fluids is more complex. The flow in a river or in any system of pipes depends on the pressure, the amount of water and the resistance to the flow. (In human circulation this translates into blood pressure, cardiac output and vascular resistance.)

Let's now assume that an increase of the central nervous autonomic discharge (tone) alters the circulation and that this then activates a negative (restraining) feedback signal to the brain (Figure 1). Which signal does the brain sense; the pressure, the flow, the resistance or a combination of these three components of the circulation?

To answer this question, one has to observe responses to induced changes of the circulation. In the course of our hemodynamic studies we repeatedly altered the circulation on purpose and there is no doubt in my mind that the central nervous system tightly regulates the blood pressure while at the same time it permits wide fluctuation of the

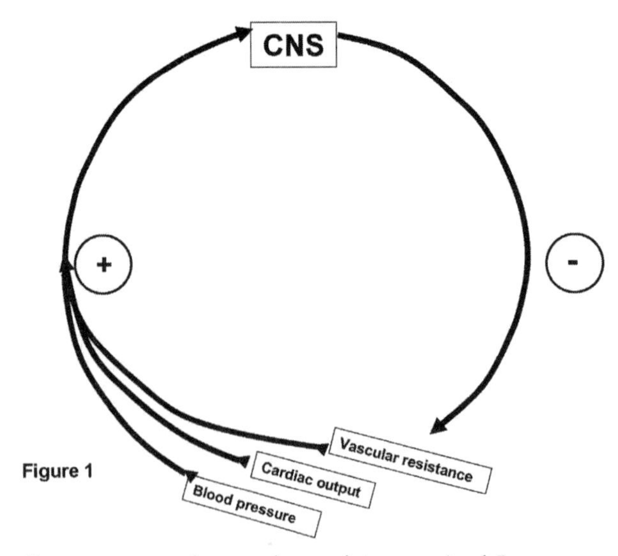

Figure 1

of the cardiac output and vascular resistance. And I can say even more; the autonomic nervous system on purpose changes the cardiac output or vascular resistance in order to preserve the "set" blood pressure high in hypertensives and normal in normotensive individuals.

To make myself clear, I must introduce to you the basic hemodynamic grid. (Figure 2)

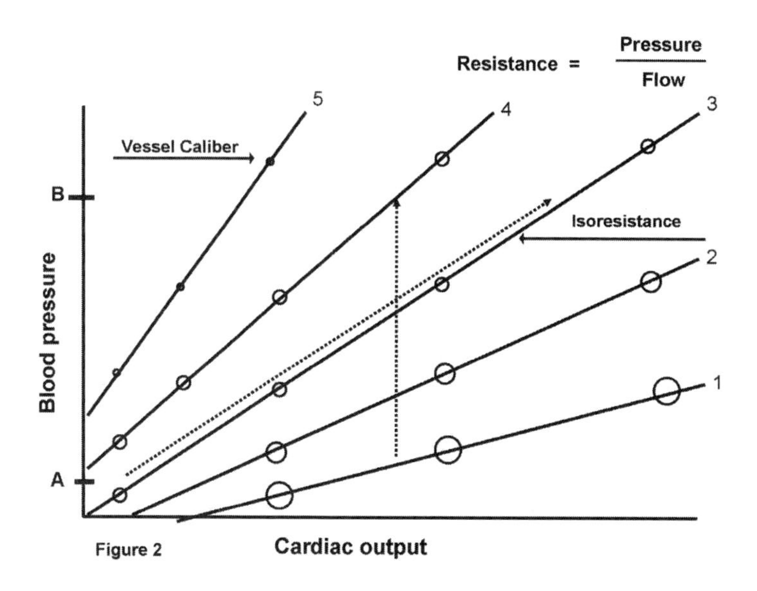

Figure 2

On the vertical axis is the blood pressure and on the horizontal axis is the cardiac output whereas the diagonal lines represent vascular resistance. Let's start with these diagonal "isoresistance" lines. The lowest line represents the lowest calculated vascular resistance. Circles along each resistance line depict (not very successfully) the arteriolar diameter (lumen). The diameter is widest at the lowest level of vascular resistance and as the resistance increases (upwards from line 1 to line 5) the lumen of the arteriole becomes narrower. There are a few points to remember about these lines. First, the same level of vascular resistance ("isoresistance") can be seen at lower or higher levels of cardiac output and blood pressure. Second, at lower levels of resistance, when the diameter of the resistance vessel is wider, the slope of the curve is flatter so that a small blood pressure increase (vertical axis) results in a large increase of cardiac output (horizontal axis). Conversely, at higher resistance level (line 5) it takes a big increase of blood pressure to elicit a small increase of the cardiac output. The graph permits you to follow the hemodynamic underpinnings of any change in flow or pressure. Broken arrows in the graph depict two different scenarios for the same increase of blood pressure (from point A to point B; vertical axis).

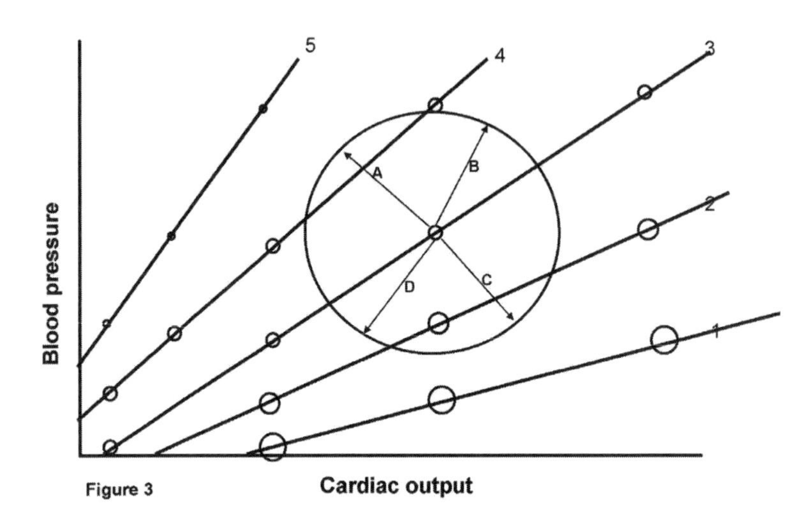


Figure 3

In one instance (vertical arrow) the cardiac output did not change but the vascular resistance increased from line 1 to line 4. The diagonal arrow shows the same rise of blood pressure but in this instance it is entirely due to an increase of cardiac output without any change of vascular resistance.

Before I present some specific experiments using similar graphs, some additional points about how to read them may be helpful. An upwardly-directed arrow indicates an increase of the blood pressure, downward-oriented arrow a decrease and a horizontal arrow no change of the blood pressure. In regards to the cardiac output (or cardiac index), an arrow pointing to the right shows an increase, to the left a decrease and vertically no change. Of course the diagonal lines permit you to evaluate whether the vascular resistance changed, that is to say whether the arrow moved towards another isoresistance line. Unfortunately in real life changes are rarely at right angles. Figure 3 might be helpful as a training graph. The arrows within the circle point in different directions and each of them describes a different hemodynamic trend. I would not be a good university professor if I could not ask a few multiple choice questions. So here we go:

1. Which arrow shows an increase of the blood pressure, a decrease of cardiac output and an increase of vascular resistance?

2. Name the arrow showing an increase of blood pressure, increase of cardiac output and increase in vascular resistance.
3. Which arrow shows a decrease of blood pressure, a rise in cardiac output and a decrease of vascular resistance; C or D?
4. What does arrow D show?
(answers at end of chapter)

Knowing that Figure 3 made you an instant hemodynamic expert, I can now proceed telling the story of an experiment with unexpected results. Long term prognosis of a patient who had a recent heart attack depends on the degree to which the pumping performance of the heart was damaged. Assessing the cardiac performance immediately after a heart attack is technically difficult and requires fairly sophisticated equipment. Some physicians hoped that the blood pressure response to handgrip might provide a simple bedside test for assessment of cardiac pumping capacity. When a person squeezes a hand-held device (dynamometer) as hard as he possibly could, his blood pressure increases a great deal. The reflex increase of the blood pressure under this circumstance is associated with a substantial surge of the cardiac output. It was therefore expected that patients who sustained serious damage after myocardial infarction would not be able to raise

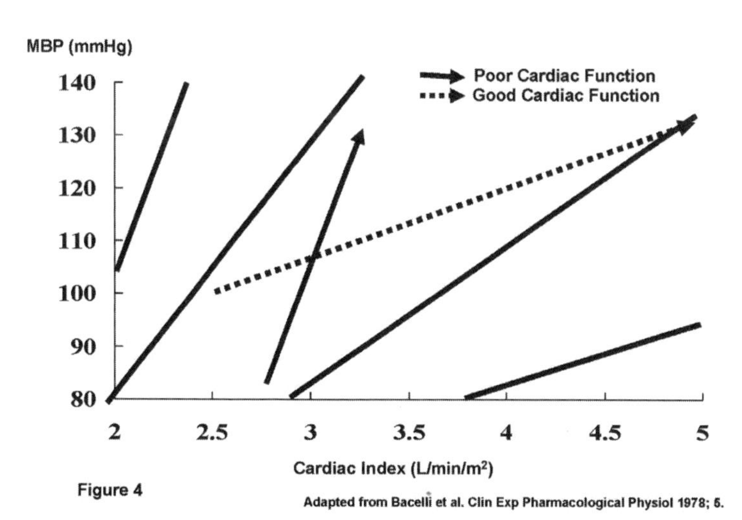


Figure 4

Adapted from Bacelli et al. Clin Exp Pharmacological Physiol 1978; 5.

their cardiac output and their blood pressure would increase less. A group of Italian physicians decided to test the idea and their results are presented in Figure 4. As expected, patients with a poor cardiac function responded to handgrip with a lesser increase of cardiac output (7%) than patients with a good cardiac function (50%) but their blood pressure increased to exactly the same levels as in the good cardiac function group.

To understand what happened let's for a minute speculate why a handgrip causes a high cardiac output type of blood pressure elevation. Handgrip is a typical example of isometric exercise; the muscle tenses up but produces practically no external motion. The increased tension threatens to compress blood vessels and thereby cut off blood supply to hard-working muscles. Whereas muscles can work without oxygen (*anaerobic exercise*) for a while, a steady blood supply is necessary to sustain a prolonged exercise performance. So the brain responds with increased sympathetic drive, which increases the intravascular pressure and thereby prevents a collapse of muscle blood vessels. It makes sense! But it would not make sense if the blood pressure response were an increase of vascular resistance (vasoconstriction) without an increase of cardiac output. After all, the working muscles need more blood and vasoconstriction diminishes the blood supply. All of this works beautifully as long as all organs in the circulation respond in a normal fashion. But when the heart is not able to increase its output, as was the case in the Italian patients with poor cardiac function, the brain increased the blood pressure by vasoconstriction. We call this "hemodynamic plasticity." In certain situations, the brain is programmed to increase the blood pressure to a given level and will achieve that level with whatever is available. If it can not be a high cardiac output, it will be an increase of vascular resistance but the blood pressure response will be the same.

The response to handgrip is rather short-lasting and the question arises whether this has any relationship to long term regulation of the blood pressure. However, subjects with borderline hypertension who have a continuous mild elevation of blood pressure also show considerable hemodynamic plasticity in maintaining their higher than normal blood pressure level. In an early hemodynamic study, we compared the response to upright posture, mild exercise, and blood volume expansion in borderline hypertension and normotensive control subjects. The effects of beta adrenergic blockade with propranolol and of complete autonomic blockade of the heart with propranolol plus atropine were also evaluated. Across these various conditions, the cardiac output varied from high to low and the vascular resistance from normal to high, but the blood pressure in borderline hypertension remained consistently higher than in the control group. Two examples from this study illustrate the point.

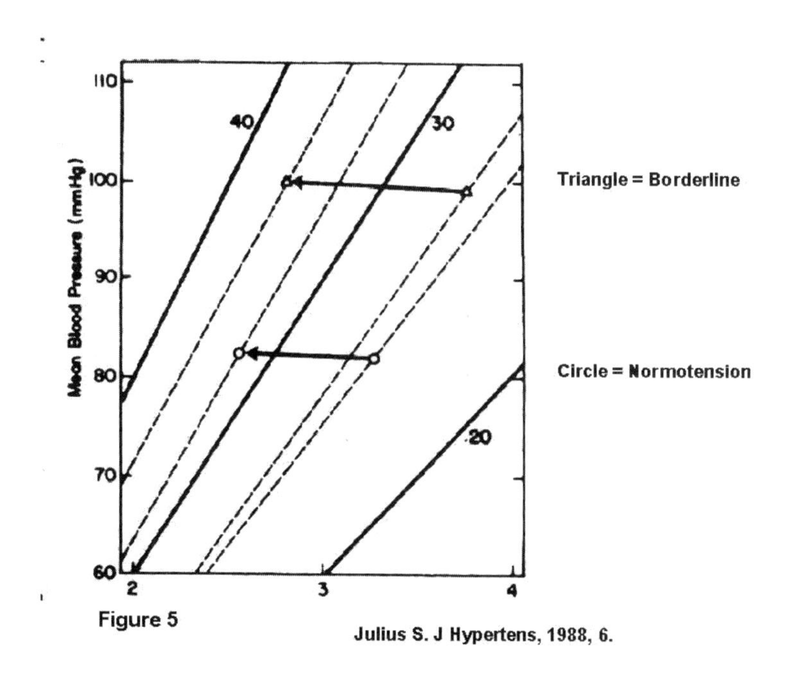


Figure 5

Julius S. J Hypertens, 1988, 6.

Figure 5 shows the responses to blockade of cardiac betareceptors with intravenous propranolol in 77 patients with borderline hypertension (triangles) and 82 control subjects (circles). In the resting position, the borderline hypertension group had a significantly elevated cardiac index and a nominally normal vascular resistance. If an excessive sympathetic drive was responsible for the increase of the cardiac output, then a blockade of cardiac beta receptors should abolish that increase and that is exactly what happened. However, the blood pressure remained elevated and what started as a high cardiac output type of hypertension had been converted into a high resistance kind of high blood pressure. I commend to your attention the precision with which the brain maintained the blood pressure elevation.

Our radical intervention rendered the heart unresponsive to sympathetic stimulation but the blood pressure did not change whatsoever. The baseline mean pressure was set at a higher level at baseline and remained at the exactly same level after the intervention.

Even more impressive is what happened to the blood pressure when we blocked both the sympathetic and parasympathetic receptors and totally isolated the heart from all autonomic nervous stimuli (Figure 6).[2]

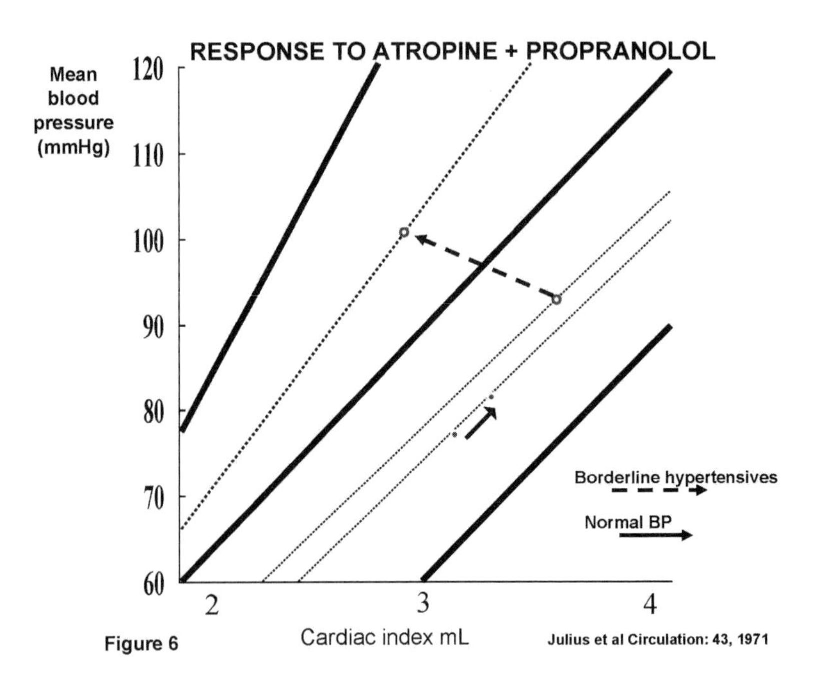

Figure 6

In this subgroup of experimental subjects (18 borderline hypertension, 17 controls), the borderline hypertensives had a higher baseline cardiac index. After the blockade both groups experienced the same minor, but significant, increase of blood pressure. However, the hemodynamic underpinning of this equivalent increase of pressure was very different in the two groups. In control subjects, the cardiac index increased and the vascular resistance was unchanged. In borderline hypertension, the cardiac output decreased and the vascular resistance increased. (Why the cardiac output in borderline hypertension decreased is not relevant to this topic. Nevertheless, let me remind you of our previous work showing that a chemical "denervation" uncovers a decreased stroke volume in borderline hypertension.) In other words, the relative between groups difference in blood pressure was preserved and to achieve this, the brain relied on two entirely different hemodynamic mechanisms.

The common feature of the experiments in Figures 4-7 is that the brain tightly regulates the blood pressure to reach a certain goal or maintain a certain existing level of blood pressure. I call this "the blood pressure seeking property of the central nervous system." In people who perform handgrip after myocardial infarction, the brain seeks a certain increase in blood pressure and uses whatever is available. If the heart is normal, the cardiac output will increase and if myocardial damage impaired the cardiac performance, the vascular resistance will increase. Similarly, in borderline hypertension the brain seeks to maintain a higher blood pressure under all circumstances and uses whatever it can; when we prevented the increase of the cardiac output with chemical means, the brain turned to vascular resistance to keep the blood pressure high.

Okay, we are getting pretty darn close to discussing the enigma of why the high sympathetic tone can be seen so readily in early stages of hypertension only to disappear in later phases of the disease. But before we get there, I feel I must show you yet another graph. Call this a tactical decision; a preemptive strike. There is a weakness in the argument that I presented so far and I'd rather talk about it than let you discover it by yourself. All examples of hemodynamic plasticity I showed so far dealt with the conversion of a high cardiac output into a high resistance state. But are there any circumstances in which the nervous system increases the blood pressure through a primary increase of vascular resistance and can such a reaction be converted into a different hemodynamic pattern? The unequivocal answer to both questions is yes. A. Andreen in Len Hansson's laboratory in Goteborg demonstrated that background noise causes a high vascular resistance type of blood pressure elevation in humans. He then prevented vasoconstriction by treating his subjects with the alpha adrenergic blocker prazosin. This did not diminish the blood pressure response to noise but the hemodynamic background changed from a high resistance to a high cardiac output state. In his experiments the increase in blood pressure was modest and the periods of blood pressure elevation were short. I will therefore show you our experiments with the hindquarter compression of dogs which elicits long lasting and potent neurogenic blood pressure elevation (Figure 7).

In this experiment we investigated the hemodynamic response in 10 anesthetized dogs before and after compression of hindquarters with a pressure suit. The suit was first inflated without pharmacological intervention (Baseline A). After one hour of compression the suit was

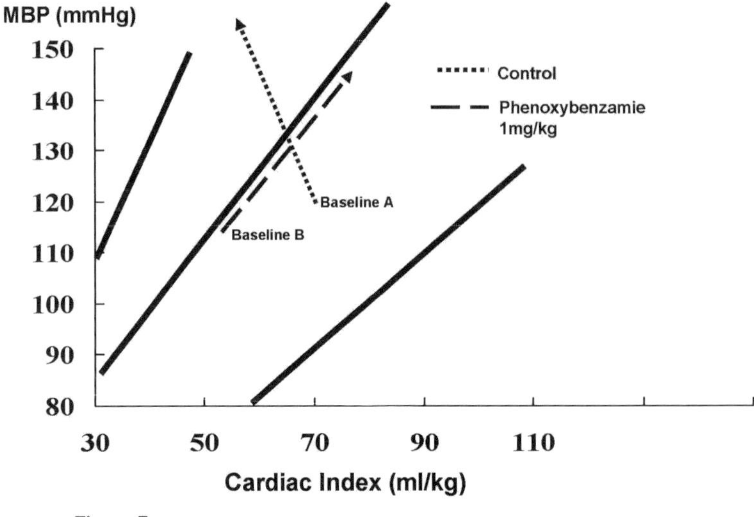


Figure 7 Adapted from Julius,Sanchez,Brant J.Hypertension. 4 suppl. II. 1986

deflated and the alpha adrenergic blocking agent phenoxybenzamine was injected. One hour later (Baseline B) the suit was inflated again for another hour of compression. Arrows show the response to compression. The first compression elicited a decrease of the cardiac index and an increase of vascular resistance resulting in large blood pressure increase (+34 mmHg systolic and +30 diastolic pressures). After we prevented vasoconstriction with phenoxybenzamine the baseline blood pressure was about 10 mmHg lower but the compression nevertheless elicited a 27/21 mmHg increase of the blood pressure. The increase of blood pressure after the pharmacological blockade was entirely mediated by an elevation of the cardiac index.

Hopefully by now you are convinced that under all these circumstances the brain strives to preserve the blood pressure level or the pressor response and when an organ is impaired it shifts the autonomic nervous drive to the system which is still capable of responding. Since the blood pressure is so tightly regulated the negative feedback loop to the brain must be related to the blood pressure. Otherwise how would the brain know whether and when to stop? Remember, I've never given you an example of a blood pressure overshoot in patients with hypertension. Not because I've hidden something but because we could not find any evidence of pressor hyperactivity in borderline or mild hypertension! In hypertension the feedback systems functions beautifully but it operates at a higher set point. To put it differently, the brain efficiently manages to keep the blood pressure lower in people with normal blood pressure and higher in patients with hypertension.

So far I showed how the brain reacts when one of the organs in the circulation is dysfunctional and unable to respond. From there it takes only the smallest leap of imagination to visualize what would happen if an organ were hyper-responsive. In Chapter 7, I already mentioned Brent Egan's finding that intra arterial infusion of different constricting agonists, be they norepinephrine or angiotensin, elicit excessive increases of vascular resistance in patients with stage one hypertension. We view this hyper-responsiveness as a consequence of a higher blood pressure which elicits structural alterations of the vascular wall and also causes endothelial damage. These processes then favor increased vasoconstriction and limit the capacity for vasodilatation. When such an amplification of vasoconstriction sets in, the brain can raise the blood pressure more efficiently, that is to say it needs less sympathetic drive to maintain the increased blood pressure level. Within that framework, the "disappearance" of sympathetic tone in advanced hypertension is not an enigma anymore, but rather, an expected outcome.

Despite the finding that plasma norepinephrine or some other index of sympathetic tone is not elevated anymore, the brain and the autonomic nervous system have not abandoned their role in advanced hypertension. The feedback loop which brings the blood pressure back to a higher set point works perfectly well also in advanced hypertension. The only difference is that the autonomic nervous system quietly withdrew to its new role of a back-stage conductor of circulation.

I have never been as sure about anything as I am about the mechanism of withdrawal of the sympathetic tone in the course of hypertension but for now this remains a hypothesis. The hypothesis can be tested only with longitudinal observation and I hope that a younger, energetic investigator will set up a study to investigate hemodynamics in the course of evolution to test the hypothesis that over time changes of vascular reactivity would negatively correlate with changes of the sympathetic tone.

Oops, as soon as a research project idea pops up I fall back on scientific lingo. What I said above means that simultaneously with the ensuing increase of vascular responsiveness there will be a decrease in the sympathetic tone. It should not be difficult to complete such a study, all it takes is perseverance.

The period of the Ann Arbor Division expansion coincided with a substantial recognition of our work. In 1979 I had a very pleasant surprise. Out of the blue the University of Goteborg informed me that I had been selected to receive an Honorary Doctor of Medicine degree. I must hand it to Len Hansson; together with Bjorn Folkow he quietly processed the nomination but didn’t give me a hint what might be coming. When it appeared that the nomination might be successful, he asked me to reserve a date in November for a “Symposium” in Goteborg. A clever move; when Goteborg informed me, I already had a reserved time for the event.

Good news tends to generate more good news and Susan and I were delighted that my dear old friend Tony Schork decided to join us and attend the ceremony in Goteborg. We arrived a few days earlier to measure my head and finger for the symbols of a Goteborg doctorate; a traditional top hat and an elaborate gold ring. As is the tradition in Europe the formal proceedings in Goteborg were conducted in Latin and when the Promotor Professor, Harje Bucht, read the announcement I could barely restrain myself from crying. The announcement formula was almost identical to Latin words my father used twenty five years earlier when he conveyed on me and my class the degree of Doctor of Medicine in Zagreb. That year the University of Goteborg chose six individuals for various honorary doctorates and I was the only recipient of the Doctor of Medicine degree.

I also had the honor of responding in Goteborg with the same Latin words I used to express our thanks in Zagreb.

“Domine Promotor meo colegarumque meorum nomine gratias vobis ago.”

After the ceremony there were numerous receptions, and above is a picture of Len (on the right) and me taken at the dinner in his house. Twenty three years later during his funeral Len's hat was placed atop of his casket. I love the European tradition, a doctorate is a meaningful achievement and the symbol of that achievement shares the fate of its owner.

**Answers from page 239:**

1. A
2. B
3. C
4. Decrease of cardiac output, blood pressure and resistance

**Bibliography**

1. Julius S: Editorial Review: The blood pressure seeking properties of the central nervous system. J Hypertension 6:177-185, 1988

# Tecumseh Studies: Two Rounds in the Forest 1990 - 1998

In internal conversations I frequently referred to the Tecumseh blood pressure study as our "forest expedition." This moniker did not make too much sense to others but to me it had a special meaning as it reminded me of the novelette "The Birch" by Slavko Kolar, the Croat writer and a friend of the Julius family. In this short piece a well educated forester, for reasons known only to him, tries to convince a villager to marry a girl whom he finds to be as stunning as a birch in the forest of regular trees. The villager does not get it; to him a girl is a girl and that is all there is to it. This eventually leads to all sorts of complications but that is beside the point. To me the question was whether the students we earlier selected for invasive hemodynamic studies of borderline hypertension were a bunch of rare birches. Everybody who has spent even a short time on a university campus would agree that the lively, weirdly dressed, and still growing youngsters, who at traffic lights routinely challenge all motorists, are probably not typical average USA citizens. We had to go to a regular forest to validate our previous findings.

In 1989 when I took the first look at the results of the Tecumseh study, I knew we had a winner. Everything seemed to line up beautifully and I quickly prepared a manuscript which was published in the Journal of American Medical Association (JAMA) on July 18th 1990.[1] I knew the article was good but I could not possibly anticipate the frenzy it created. It all started with a New York Times front page paper about the study which was published on the same day as the JAMA presentation. JAMA had already then developed what is now a routine for all major medical Journals; the journalists are given advance copies of the yet to be published papers with an "embargo" date after which they are permitted to publish comments.

The New York Times gave a balanced overview of our findings with a focus on the evidence of slight cardiac damage in people with borderline hypertension. In the words of the New York Times we found that "people with borderline high blood pressure pumped less blood with each beat of the heart than average and (their hearts) …did not relax properly in between contractions. Both conditions are considered very early indicators of potential heart problems." The reporter, Elizabeth Rosenthal, also mentioned in passing some other findings in the study but the main message was that borderline hypertension is not an innocuous condition. The New York Times is in essence a local paper with a huge national readership and as is their traditional practice, Ms. Rosenthal asked two local hypertension experts to comment. The experts' opinions were divided. John Laragh at Cornell found our evidence of organ damage "fragile" and considered the risk to borderline hypertensives as "generally trivial." Marvin Moser at Yale took our findings more seriously. "People with borderline hypertension should not be patted on the back and told, 'Don't worry about it - you'll be fine." He then suggested that treatment could be tried in some of these patients and warned that "Treatment of this group should not mean tons of tests and taking three pills a day." His assessment was much more in line with our own suggestions in the JAMA paper: "*We generally do not advocate pharmacological treatment of borderline hypertension,* we said and added that if the nonpharmacological treatment (life style modification) has failed and repeated blood pressure readings suggest that treatment may be needed a "*One year trial of antihypertensive treatment with small doses is suggested. After a year a full assessment of blood pressure levels and of organ status is in order. Regression of abnormalities or a return of the high blood pressure justifies continuation of the treatment.*"

Though we were careful and our comments were rather conservative, the notion that treatment with medication might be appropriate for some borderline hypertensives was a stark break with tradition. I hoped the New York Times article would stimulate further scientific discussion but nothing of the sort happened. What ensued was sheer madness and I found out what the proverbial 10 minutes of fame really means. The next day there was an avalanche of telephone requests for interviews and among them most striking was an invitation by NBC to come to New York for a "live" interview with Deborah Norville, the co-host of the Today television show. And live it was; without preparations or a pre-interview conversation, I was ushered onto the stage where I sat on the designated chair and off we went. Ms. Norville's questions were easy and the interview went quite well but it would have been even better had I not decided not to be impressed by my interviewer's quite obvious beauty. I sat there in front of her with my sight fixed at the level of her thyroid gland. That way, I thought, I could converse with her without getting blinded by her gorgeously youthful aura. Of course that made me look stiff, but even worse, it made her quite uncomfortable. Somewhere towards the end of the interview, thinking something might have been wrong with her appearance, she started to fidget and rearrange her blouse. I sensed what was going on and was sorry to have caused her discomfort. The lady later had some career difficulties and I wonder whether a pre-interview chat might have minimized the impact of her natural assets, relaxed the interviewee and alleviated one aspect of the problem. But then again I am a bit weird and my reaction might have been idiosyncratic. Anyhow the interview was seen by a wide audience and contributed to the prolongation of my ten minutes of fame.

Until the publication of the Tecumseh results I did not properly understand the media hierarchy. Something interesting that appears in New York Times gets quickly recycled in numerous regional and local papers. That cycle passes quite rapidly but soon the secondary recycling starts. In my case, the Tecumseh story was retold with or without interviews for quite a while. It started with national illustrated magazines and trickled down to niche-oriented smaller journals. A short story about Tecumseh appeared in the July issue of Time magazine and in September in People magazine. At least the one in Time appeared in the "Medicine" section but the placement in the "Body" section of People magazine was a bit less dignified. I mean that magazine is all about owners of beautiful bodies and whereas the section "Body" refers to physiological aspects of the organism this tends to get lost in the sea of gossip and personality interviews.

The article in People was quite an experience. Overall they did a good job but it had to be on their terms. In addition to informing the readers, the piece had to have some entertainment value. To make its articles more interesting, People magazine illustrates everything with pictures and I was told that, separately from the interviewer, their best photographer would come to Ann Arbor to shoot some pictures. That was fine but I wondered whether it would not have been better for him to first hear the interview and then choose the topic for photography. Poor me; always an educator, I tend to believe in the power of information. No, that was not necessary said the interviewer and explained that the two teams work separately quite on purpose. A third editor then takes the text of the interview and matches it with pictures. I wondered how that works but when the famous photographer arrived I understood their modus operandi. The photographer's technique was brilliant, while incessantly talking with me in the office or in our house he'd constantly click in such a quick succession that it felt as an irrelevant background noise and I forgot he was taking photos. Some of literally hundreds of pictures he took were bound to be good. And he did his own interviewing not about hypertension but about my family and my habits. He seemed fascinated with my hobby of collecting very old geographic maps, asked to see the entire collection and then arranged for my entire family to sit on the floor and pretend we were examining a huge old map. Of course this totally artificial picture became the centerpiece of the printed interview. We never before or after sat on the floor to enjoy a map simply because I am the only one that really likes them. But I understood the intent to show the readers "the human side of a scientist."

According to a description in Wikipedia, Ms. Norville, the NBC interviewer, was later herself interviewed by People magazine. Her session with the photographer was not as benign as mine. Apparently keen on presenting the human side, the photographer took some shots while she was breastfeeding her newborn child. As was the case with me that uncharacteristic picture found its way into the magazine and the purists in the NBC headquarters were scandalized by what they viewed as a breach of etiquette. Television must be a very difficult métier; Wikipedia maintains that this innocuous picture further contributed to Ms. Norville's separation from the big company!

Various interviews and articles about the Tecumseh study continued to trickle in. I knew we were close to the end when a report appeared in the Dec 1990/January 1991 issue of Bride magazine. There were not many more niches to fill! I think the final, almost official, announcement of the end of the season was when Ms. Rosenthal published an abbreviated version of her New York Times article in the December 1990 issue of the Readers Digest.

So what did we find to create such a stir? To me the most important discovery was that hyperkinetic neurogenic borderline hypertension was also found in the Village of Tecumseh[2] despite the new and more objective definition of the hyperkinetic state. Earlier in the hemodynamic laboratory, we defined individuals whose cardiac index was two standard deviations above the mean of normotensive control subjects as hyperkinetic. This arbitrary division changed with time as the size of the control group increased and, importantly, we were not blinded. Blinding, that is to say classifying without knowing to which group an individual belongs, is a very important precaution against investigator bias. It is amazing how our brains work. I consider myself as honest and honorable but am not sure that I fairly classified individuals whose cardiac index values were close to the cutting line. Knowing that we expected to see more hyperkinetic individuals in the borderline hypertension group, I could have told myself "here is another one" if in a hypertensive individual the cardiac index was marginally elevated. Conversely, in the normotensive group I might have moved a "fence-sitter" downwards. That tendency is so inherent, so much of part of human nature, that blinding is the only way to overcome it. However, in Tecumseh we had accumulated cardiac index measurements in 691 persons, a group large enough to take a look at the distribution of the cardiac index in the population and to develop an objective computerized definition of the hyperkinetic state. It is expected that most variables in a large group of people will be "normally" distributed in a bell shaped fashion; most observations will aggregate in the middle (the top of the bell) and from there the less frequent high and low values will accumulate in an orderly, symmetrical, descending fashion (the two sides of the bell). In Tecumseh, the distribution of both the blood pressure and cardiac output was skewed towards high values on the right side of the graph. (Figure 1, ovals on both panels).

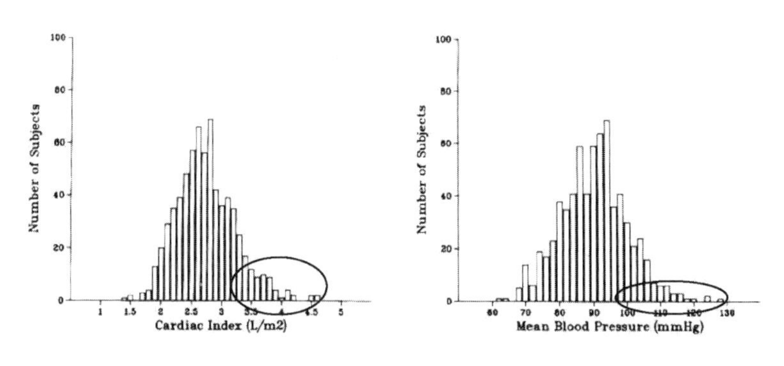


**Figure 1** Julius S, et al. J Hypertension 9, 1991

The left panel shows the distribution of cardiac index and on the right panel is the distribution of the mean arterial pressure. (Mean arterial pressure is an average of the "area under the curve" of diastolic and systolic blood pressure). Prior to the analysis of the Tecumseh data, our statistician, Nik Schork, developed a method to investigate whether the skewing might reflect an overlap between two normally distributed subpopulations; the larger one on the left and the smaller one on the right. It is difficult to simplify intricate issues and what Nik developed was a rather complex and lengthy computerized statistical program. To avoid possible misrepresentation, I prefer to use the description which we used in the paper and to which Nik Schork agreed: "*Normal mixture analysis is a statistical technique for isolating homogenous subpopulations within the distribution of a variable collected from an apparently heterogenous population.*" So, there!

The beauty of the program was that it had fixed objective criteria to first search whether there might be two subpopulations and then to determine whether an individual belonged to one or the other subpopulation.

In Figure 2, we used an objective bivariate analysis (of both the blood pressure and cardiac index) and found a larger group (594 subjects) who had both lower cardiac index and blood pressure values (normokinetic group) as well as a smaller group (97 individuals) with both higher cardiac index and blood pressures (the hyperkinetic group). In the next step the two groups (normokinetics and hyperkinetics) were divided into normotensive and hypertensive using the traditional cutting point of 140 mmHg systolic and or 90 mmHg diastolic.

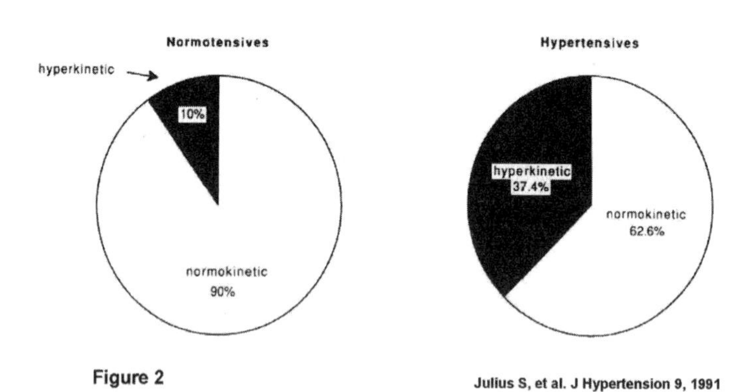


Figure 2

Julius S, et al. J Hypertension 9, 1991

Only 10% of normotensive subjects had hyperkinetic circulation compared to 37% hyperkinetics in the hypertension group. So, we had no reason to worry, the hyperkinetic state which we found among selected students with borderline hypertension existed also in the real life situation in the village of Tecumseh and was even more frequent among villagers than in university students.

In invasive studies in Ann Arbor, we used receptor blockers to prove that the hyperkinetic state was neurogenic. We could not possibly repeat such investigations in a field study and eventually settled on collecting blood samples for determination of plasma norepinephrine levels. Whereas this measurement generally reflects the level of sympathetic activity, it tends to fluctuate a great deal and we were not sure whether norepinephrine measurement would give us reliable results.

To our surprise, hyperkinetic borderline hypertensives in Tecumseh had significantly elevated plasma norepinephrine levels. (Figure 3, "Hypertensives").

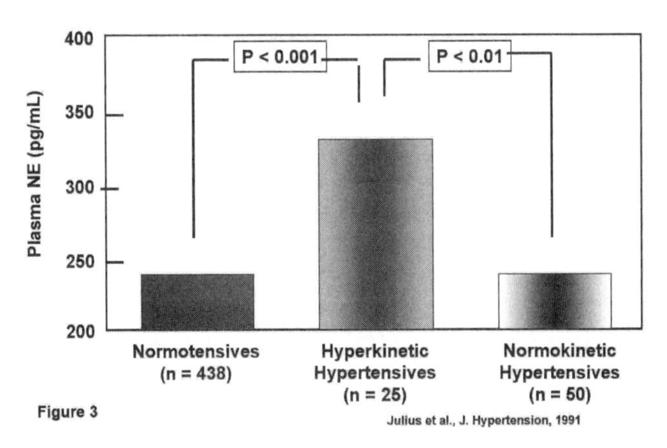

Figure 3

By itself this finding would not persuade people who believed that the hyperkinetic borderline hypertension is a benign condition. Just as James Conway did many years earlier, they would view this as a proof that we had found a group of "anxious" people who are unlikely to ever have hypertension. After all, John Laragh thought that the cardiovascular risk found in the Tecumseh study was "generally trivial." He and others with a similar mind-set were likely to view our new finding as another proof of a trivial pursuit. But we succeeded to uncover some really powerful data to persuade the doubters.

When we studied them, the hyperkinetic subjects in Tecumseh were on average 32.6 years old. We were able to locate records from previous exams when they were 5, 8, 21 and 23 years old. In all these exams the blood pressure of the hyperkinetic group was significantly higher than in the normotensive group. The blood pressure of children "tracks" and those with higher childhood pressure are very likely to end up as hypertensive adults. Furthermore, the hyperkinetic group had significantly faster heart rate when they were 21 and 23 years old, suggesting that the sympathetic overactivity was a life long characteristic of these individuals. Since hypertension is a heritable disease we took also a look at records of mothers and fathers of our Tecumseh volunteers. When they were 33.2 years old (a similar age to our hyperkinetic and normokinetic subjects), fathers of hyperkinetic subjects had significantly higher blood pressure than fathers of the normotensive group. In fact, the average systolic blood pressure of fathers of hyperkinetic individuals was in the hypertensive range. A similar statistically significant blood pressure elevation was also seen in mothers of the hyperkinetic group but their average blood pressure did not yet reach the level of hypertension. Fathers of hyperkinetic subjects also had an elevated heart rate in the fourth decade of their lives. Finally, it is important to point out that non-hyperkinetic borderline hypertensives in Tecumseh had elevated childhood blood pressure similar to the hyperkinetics, and their parents' blood pressures were also elevated. And, just as in our previous studies, the normokinetic borderline hypertensives had significantly faster heart rates than the normotensive control subjects.

Though we had no space to discuss this in the published paper[2] these similarities between the two hypertensive groups were very important to us. As I discussed in Chapter 7, the medical literature and our early hemodynamic work supported the concept that the hyperkinetic state eventually evolves into a typical high resistance established hypertension. The important elements of this hemodynamic transition (from high cardiac output to high vascular resistance) were the ensuing decrease of both the stroke volume and heart rate and an acceleration of hypertension due to structural changes in arterioles. All these elements were again found in Tecumseh; the normokinetic group had significantly decreased stroke volume and their heart rates, though lower than in hyperkinetics were significantly higher than in normotensive subjects. Furthermore, Tecumseh subjects with borderline hypertension had increased minimal forearm vascular resistance[1] which, according to Folkow, is a reliable sign of vascular hypertrophy. Apparently the small but persistent blood pressure elevation in borderline hypertension was sufficient to induce structural arteriolar changes which render the arterioles over-responsive to vasoconstriction and under responsive to vasodilation. Or, to put it differently; borderline hypertensives in Tecumseh had already stepped on the escalator to higher blood pressure levels.

The Tecumseh findings in both borderline hypertension groups were strong and should have moved the doubters to recognize that the hyperkinetic state is not an innocuous condition. But it is difficult to change old habits and there is no evidence that the Tecumseh paper had an impact on clinical practice.

Luckily soon after we published the first Tecumseh report I met Paolo Palatini, the bicycle enthusiast from Padova, Italy. Paolo, a professor of medicine at the venerable University of Padova, had designed two longitudinal studies of the natural history of hypertension. Paolo has a keen intellect, a lovely hidden sense of humor and an insatiable appetite for hard scientific work. His studies have been inadequately funded and were it not for his ability to project enthusiasm on others they'd probably fail. But Paolo and his crew of volunteers have kept working and publishing important papers along similar lines, but in a different age range than the Tecumseh study. Paolo and I became very good friends partially because I understood his whimsical, self-deprecating style. The first time he gave a lecture in Ann Arbor, Paolo projected a slide with a big ship named Tecumseh next to a small boat carrying the name of his study. Only a self assured person fully aware of the importance of his data could pull off such a trick. It did not take us long to see the beauty of his findings but I did not fall into the trap. Had I jumped to assure him his data were important, I'd be acting as a senior mentor commenting on a junior's performance. And very likely Paolo would in private laugh his stomach off.

Anyhow, as we walked though a beautiful park of a decaying but still luxurious old palace in the Riviera de Brenta near Padova, Paolo told me he'd like to take a year sabbatical leave in Ann Arbor. I was delighted and suggested that he ought to write a review paper on tachycardia and cardiovascular disease. Soon after he came to Ann Arbor, Paolo searched the literature and produced a well written and convincing paper about heart rate and hypertension.[3] We were amazed by the sheer volume of evidence and by the strong consistency of findings across different papers. In numerous large populations, heart rate correlated positively with blood pressure and in prospective studies tachycardia was a strong and statistically independent predictor of future hypertension. Furthermore, a fast heart rate forecast future sudden death and coronary heart disease. There was also sufficient evidence in the literature about the possible mechanism of the tachycardia related cardiovascular damage.[4] Part of the effect could be explained by the ensuing hypertension, and a part reflected the association with sympathetic overactivity possibly leading to later arrhythmia and sudden death. But most fascinating was the evidence that the frequent back-and-forth flow in *coronary arteries* might mechanically damage their walls and thereby cause future heart attacks. Scientists working in the primate center of the Wake Forest University campus in Winston Salem proved that if monkeys are fed a high cholesterol diet they will develop coronary atherosclerosis. If they decreased the heart rate in these monkeys either surgically (by creating an *atrio ventricular block*) or by giving them beta blocking agents, the animals with slower heart rate had significantly less coronary atherosclerosis than their counterparts with faster heart rates.

Paolo's first paper caused a minor boom as many investigators went back to their epidemiological files and uniformly confirmed the important predictive strength of tachycardia. New epidemiological studies also paid attention to heart rate and reported that tachycardia is indeed a strong predictor of poor cardiovascular outcomes. Three years after Paolo's paper, the publication boom resulted in twelve new heart rate-related articles. Paolo and I continued to publish reviews and original papers on tachycardia and over a decade co-authored 19 papers on the topic. Paolo networked with investigators in France and Belgium to publish a few more papers about the ubiquitous association of tachycardia with a poor cardiovascular prognosis. Recently Professor Palatini hosted a conference under the aegis of the European Society of Hypertension. Upon reviewing the literature a consensus emerged[5] that tachycardia is a strong risk factor which ought to be considered in clinical practice. Various health authorities are producing a steady stream of cardiovascular treatment guidelines for practicing physicians but none of them had yet listed tachycardia as a cardiovascular risk factor.

Remember the Pettenkofer story in Chapter 3? It is hard to overcome inertia! The prejudice that tachycardia is a sign of benign "nervousness" still prevails and we will have to reiterate our message until it finally takes root.

With the Tecumseh study we met with a goal outlined already in the 1969 "master plan;" individuals with borderline hypertension indeed had a strong familial background of hypertension (Figure 4). Similarly the issue of overweight was also listed as one of the goals in the "natural history" section of the graph.

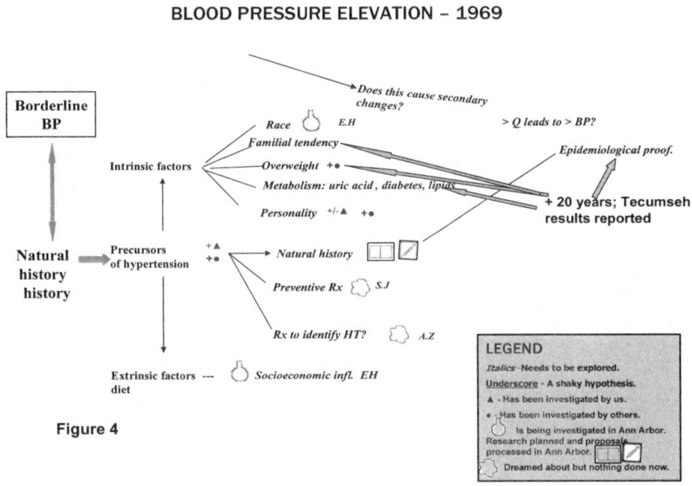

Figure 4

**Figure 4 is a reproduction of the lower part of the "master plan" which was posted in the old hemodynamic laboratory in Ann Arbor.**

In all our previous studies the borderline hypertension group was consistently heavier than the control group and we were not surprised to find this also in Tecumseh. But the sheer degree of overweight in Tecumseh was stunning. As Figure 5 shows it did not matter whether we looked at total weight, the percentage in which a person was overweight or how much fat was under their skin (skinfold thickness), hypertensives in Tecumseh were grossly overweight.

All that was important, but I do not want to leave you with the impression that in Tecumseh we just kept confirming previous findings; the study produced a number of new very important observations. One of the most interesting was the close association of borderline hypertension with other cardiovascular risk factors as shown in the table below.

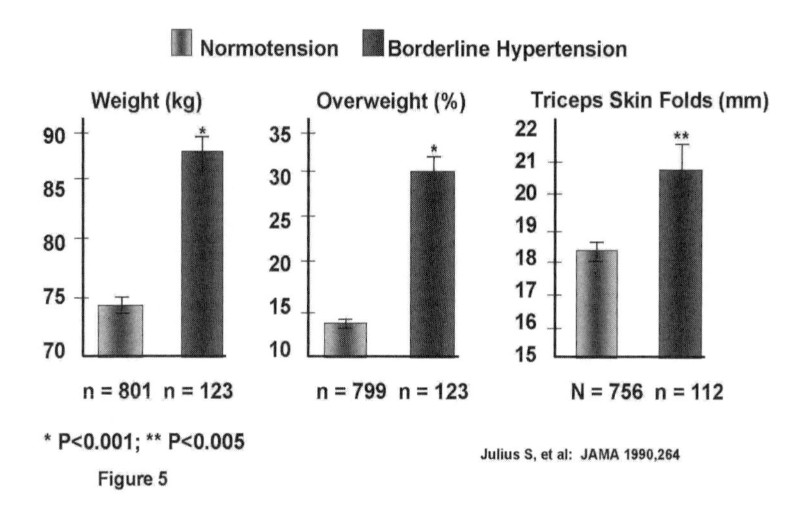

Figure 5

| | Normotensive (n=822) | Borderline (n=124) | p |
|---|---|---|---|
| Cholesterol (mg/dL) | 176 | 190 | <.0001 |
| HDL (mg/dL) | 43 | 40 | <.001 |
| Triglycerides (mg/dL) | 95 | 135 | <.0001 |
| Insulin (μU/dL) | 12 | 18 | <.0001 |
| Glucose (mg/dL) | 92 | 95 | <.001 |
| Insulin-glucose ratio | 0.155 | 0.200 | <.0001 |
| Waist-hip ratio | 0.83 | 0.86 | <.005 |

Julius S, et al: JAMA 1990; 264.

In every category the group with borderline hypertension was significantly different from the normotensive group. The values in borderline hypertension were higher in categories in which a higher value meant an increased risk (cholesterol, triglycerides, blood sugar and a measure of obesity called waist to hip ratio) but lower in the *HDL* (high density lipoprotein) category where a lower value signifies a higher risk. And of course being overweight is also a cardiovascular risk factor. So, as I like to say on the lecturing circuit, borderline hypertension comes as a "package" of cardiovascular risks competing one with another to hurt the individual. There is nothing particularly benign or fleeting about prehypertension and this condition requires much more attention than it receives in clinical practice and in research.

In the table above there was also a brand new finding; subjects with borderline hypertension have significantly higher levels of *insulin* in the blood. In 1969 I included in the master plan investigations of "metabolic factors' including diabetes but, frankly, I knew next to nothing about the topic. Ten years later when I wrote the first version of the Tecumseh grant proposal Brent Egan suggested I ought to include measurements of plasma insulin in the protocol. I complied but the truth is that at that time I did not quite understand the potential importance of this measurement. As it turned out, the finding of elevated insulin levels in borderline hypertension was one of the most interesting features of the Tecumseh study. Thanks to Brent this new finding forced me to learn more about the insulin/blood pressure relationship. As I read the literature, I became so fascinated that I eventually redirected my research toward a new unanticipated line of work.

To understand the importance of insulin one has to appreciate that the human body has certain priorities, and one of them is not to permit a high concentration of glucose in the blood. High glucose concentration, as is the case in diabetes, can cause an "internal dehydration" of the body. The osmotic pressure in the blood increases and this sucks fluids from tissues into the blood. To compensate for this diabetics get thirsty and drink more water, but if the glucose is not quickly cleared from the blood bad things start to happen. The glucose passes into the urine and the kidney draws a huge amount of fluid from the body into the urine. If not stopped in time, a condition called "hyperosmotic syndrome" ensues; the patient may become unconscious and the situation can become life threatening.

After each meal a large amount of glucose passes from guts into the blood and we are temporarily in a *hyperosmolar* state. True to its priority of preventing *hyperglycemia* the body responds with increased secretion of the *pancreatic* hormone insulin which quickly clears the blood by stimulating the chemical transformation of the sugar in the liver to glycogen and by directly storing the glucose in the skeletal muscles. In some people the blood clearing action of insulin is less efficient and it takes a larger amount of insulin to rid the blood of glucose. We call this condition *insulin resistance*. Thus the finding of high insulin and elevated but not abnormally high blood glucose meant that the borderline hypertension group was insulin resistant. To make it simpler, because they are insulin resistant, subjects with borderline hypertension must secrete more insulin to maintain a near normal glucose level. But there is a long- term price to pay for this prevention of hyperglycemia; the capacity of the pancreas to produce insulin can become exhausted and diabetes, a state of high glucose and low insulin, develops. So, borderline hypertensives are both pre-hypertensives and pre-diabetics and are heading towards the double trouble of hypertension and diabetes. Diabetes and hypertension accelerate atherosclerosis and, when both conditions occur in the same individuals the cardiovascular risk increases dramatically.

Bear with me; I am not yet done with the story of convergence of multiple cardiovascular risk factors in Tecumseh. The graph below (Figure 6) shows in some detail an observation we hinted at in the original JAMA paper.[1] In an infallibly dense scientific fashion we reported that the correlation between "*systolic blood pressure and various risk factors was found across the whole range of blood pressure distribution in Tecumseh.*" Translated in English this meant that a normotensive individual with a lower blood pressure was likely to have lower cholesterol, lower triglyceride, lower blood sugar, and lower insulin levels than a person with a higher but still normal blood pressure. To a physiologist such a constellation means that factors which normally regulate the blood pressure must be also involved in the regulation of these other risk factors. Extended further, this suggests that one underlying mechanism might be responsible for the association of all these risk factors both in hypertensive and normotensive individuals.

For Figure 6, I used statistical output based on diastolic blood pressure and decided to place the diastolic value as the central ring of the Venn diagram but it could have easily been systolic blood pressure or any of the variables. I also added, for convenience, a ring with hematocrit which is yet another well-documented but less-appreciated cardiovascular risk factor.

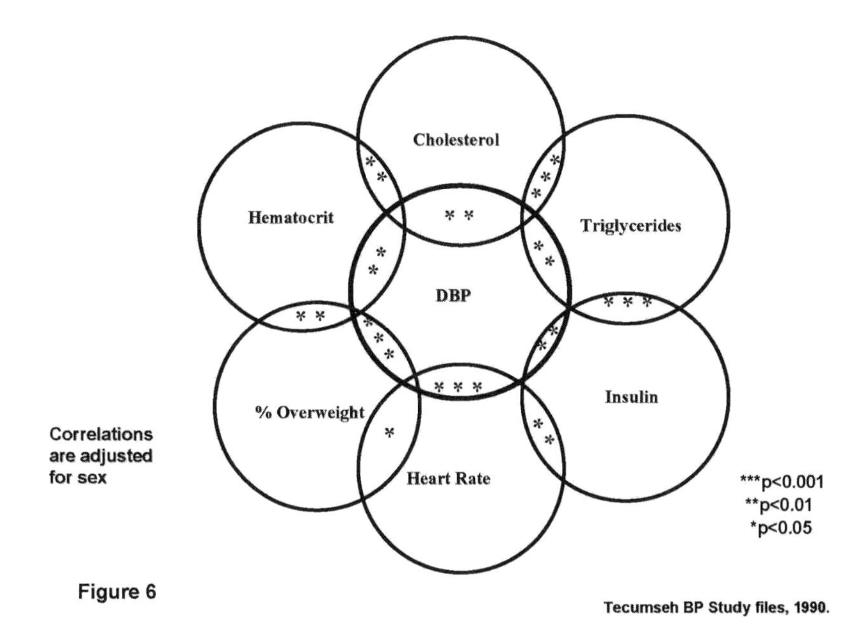


Figure 6

**Pairwise correlations among diastolic blood pressure and metabolic and hemodynamic measures**

When you take a long look at the interrelationships in the graph, a natural question will arise: why are such diverse factors so closely interrelated? Why would a host of metabolic variables relate to hemodynamic variables (blood pressure and heart rate) and to a hematological variable (hematocrit)? And you can guess what I will say: all of this reflects the degree of sympathetic activity - if the autonomic activity decreases all these apparently diverse factors will settle at a more favorable level. However, I am not suggesting I've resolved the chicken and egg question. I think that sympathetic overactivity might connect all these factors but I have no idea when and why the sympathetic overactivity activity kicks in. The sympathetic overactivity may be triggered for different reasons but the overall picture will end up looking the same. Let me give you some examples. Dr. Lewis Landsberg, a distinguished scientist and presently the Dean of the Northwestern University Medical School in Chicago, offered a calorie rich "cafeteria food" to his experimental rats and they started to overeat. Next, their plasma insulin increased and the high insulin stimulated the brain to increase the sympathetic tone. Under these circumstances the eating behavior elicited a high sympathetic tone secondary to high insulin levels. Some patients with hypertension develop *sleep apnea,* a problem with breathing while asleep. Because they fail to properly ventilate the lungs, their arterial blood is insufficiently oxygenated and this causes the brain to increase the sympathetic tone. This in turn increases the blood pressure. But in hyperkinetic individuals the excessive sympathetic tone appears to be the primary driver of the process.

With Tecumseh data in hand, we tried to resolve the question whether overweight or primary sympathetic overactivity was responsible for the association of high blood pressure with other cardiovascular risk factors in prehypertension. This turned out to be a lesson in how difficult it is to make such judgments. In the JAMA paper,[1] we first used the statistical method of analysis of variance with "adjustment" for other variables. This technique takes into account preexisting conditions which might affect the outcome. In our case, because the blood pressure in women is lower than in men, this might affect the relationship between blood pressure and other variables.

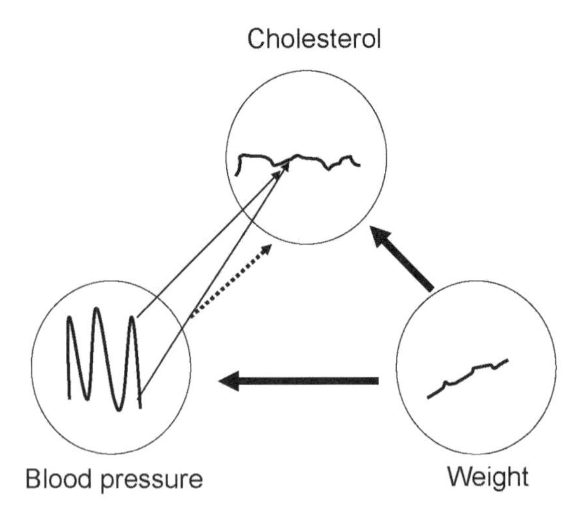


**Figure 7**

When we mathematically adjusted for gender (sex), the systolic blood pressure still correlated strongly with cholesterol, HDL, triglycerides, and insulin. Next we adjusted both for the gender and the percentage overweight and the remaining correlation of the blood pressure with other variables became substantially weaker. From this we could have concluded that being overweight was the essential element in the inter-relationship between various risk factors and we would have been wrong. The issue is the relative stability of weight and the well-known variability of the blood pressure. (Figure 7)

As anybody who tried to adhere to a diet knows, body weight is stubbornly stable; the scale shows the same value day in and day out. And if you don't do something about it, the weight will ever so slowly but surely creep up. That is not the case with the blood pressure which oscillates widely.

As the lower left circle in Figure 7 shows by chance, depending whether the blood pressure was taken at a high or at a lower point, the relationship between the blood pressure and cholesterol is bound to be weaker than with the stable weight. Moreover, a stronger correlation does not necessarily imply that the stronger variable is biologically more important; in real life blood pressure is a much stronger predictor of cardiovascular mortality than weight.

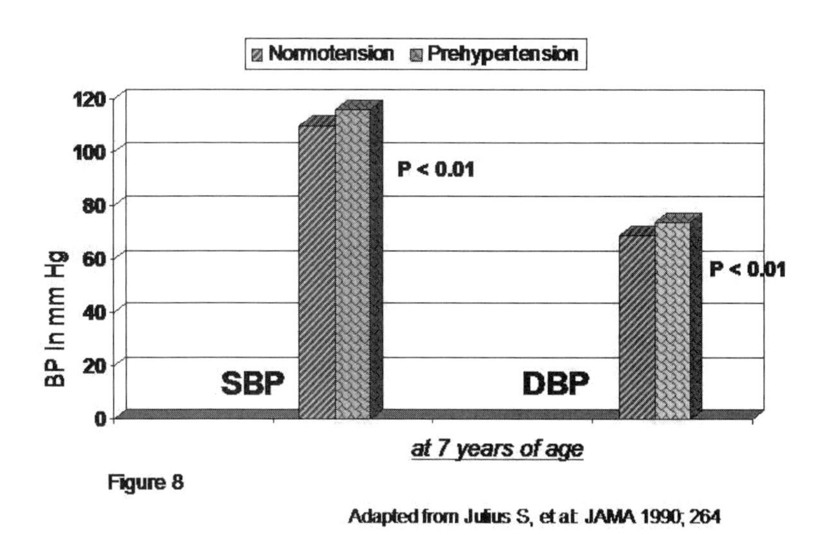


**Childhood blood pressures in subjects classified as prehypertensive or normotensive at 32 years of age**

The covariate analysis we used in Tecumseh was a snap shot at one time point and a very different picture emerged when we analyzed the relationship between weight and blood pressure over a longer period of time. Figures 8 and 9 show measurements recorded during previous health assessments in subjects we examined at an average 32 years of age. Figure 8 indicates that people classified as prehypertensive adults had significantly higher blood pressures already at seven years of age.

Young children grow at different rates and giving their gross body weight would not suffice. Probably the most reliable measure of overweight in growing children is the amount of fat under the skin (skin fold thickness.) The grossly overweight adult (32 years of age) prehypertensives in Tecumseh were not overweight as children. However 14 years later as young adults (21 years of age), the prehypertensives became significantly overweight. (Figure 9)

Taken together, Figures 8 and 9 suggest that high blood pressure is a steady life-time characteristic of borderline hypertensives whereas the overweight develops later. Thus, a primary problem with overeating is not likely the cause of borderline hypertension.

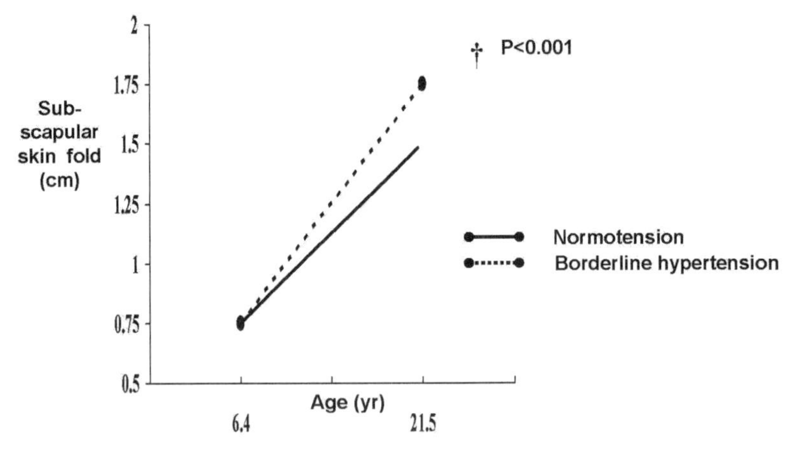


**Figure 9** Adapted from; Julius S, et al, JAMA 1990;264.

Before we move forward, I want to point out yet another important Tecumseh paper[6] in which we compared borderline hypertensives who had elevated blood pressure in the office and at home ("sustained" prehypertension) with those that had high office but normal home blood pressure readings ("white coat" prehypertension). Both groups were very similar; they had elevated blood pressure since childhood, came from families with higher blood pressures, their lipids, insulin and insulin glucose ratios were abnormal, had fast heart rates and they were overweight. By reporting this we hoped to convince the medical community that white coat blood pressure elevation is not necessarily an innocuous condition. Apparently we failed; to this day most physicians consider white coat hypertension to be a benign condition.

In parallel with publication of early papers, we developed and pre-tested the protocol for a new round of exams. Satisfied that everything was in place, we started the new Tecumseh field work in 1993. Simultaneously an almost entirely new team of physicians joined the study. The leader of that round of exams was Shawna Smith who soon married and is now Shawna Nesbitt. Shawna completed her internal medicine training in the Allegheny General Hospital in Pittsburgh, Pennsylvania and applied for a hypertension fellowship in Ann Arbor. During the interview, I felt that behind her very pretty face there was an exceptional person. She was highly intelligent, calm, self-assured but not full of herself, and clearly dedicated to learning more about hypertension. My intuition proved to be correct and after her fellowship Shawna was promoted to the rank of assistant professor in our division. Though she kept a full research and clinical work load, Shawna also managed to take courses in the University of Michigan School of Public Health and eventually obtained a MS degree in clinical research design and statistical analysis. Presently Shawna is an Associate Professor of Medicine at the University Texas Southwestern Medical Center in Dallas.

Shawna led the Tecumseh group with equal precision and skills as her predecessor Agnes Mejia. But the two of them had somewhat different jobs; Agnes reigned over a bunch of men and Shawna worked with a predominantly female contingent of physicians. Silja Majahalme came to us from the University Hospital in Tampere, Finland where she was trained in cardiology. In the introductory letter, her peers described her as a hard working scientist. Despite a heavy clinical load, Silja completed a very well designed study of intra arterial blood pressure monitoring in hypertension in Finland. She presently practices cardiology in Appleton, Wisconsin. Olga Vriz came from the University Hospital in Padova at the recommendation of Paolo Palatini, and presently practices cardiology in the regional hospital in San Daniele, Friuli, Italy. Maria Consuelo Valentini came from Torino, Italy and is now a member of the research team of Professor Gianfranco Parati at the University of Milano. An Ann Arbor fellow in cardiology, Nancy Lewandowski, also worked in Tecumseh and is now a physician in Detroit. Lillian Gleiberman PhD also joined the group and to this day continues to work in the hypertension division. In addition most of our technical personnel were women; Lisa Krause was the echo technician and Michele Wonderley managed field operations in Tecumseh while Mary Rapai and Barbara Toretti did the laboratory work in Ann Arbor. I do not know whether working with a gender-solid group made Shawna's job easier or harder.

**Photo of the Tecumseh second round group. First row (left to right):** *Shawna Nesbitt, Stevo Julius, Silja Majahalme*. **Second row:** *Lillian Gleiberman, Nancy Lewandowski, Olga Vriz, Michele Wonderley, ConsueloValentini*. **Back row:** *Mary Rapai, Eric Grant and John Amerena*

But the fact is that everything went well and under her leadership the second round of Tecumseh study was successfully completed. The group proved to be very compatible, easy to get along with and dedicated to hard work.

Well, of course there were some men too. John Amerena who is now a senior lecturer in the University of Melbourne Geelong Teaching Hospital took a two year fellowship in Ann Arbor. John reminded me of his Australian predecessor Murray Esler. He had the same pleasant

Australian accent and like Murray, John was tranquil, laid back, and had a calming effect on everybody around him. He is living proof that someone can be full of energy without exhibiting the fast-talking, quick-moving and impatient characteristics of a type A personality.

Next to join us was Thorkel Gudbrandsson from Iceland. I first met him in Goteborg where he trained with Len Hansson. In Goteborg, before he returned to Iceland, Thorkel completed an interesting study on vascular damage and hematological changes in malignant hypertension. Back home he continued to focus on high blood pressure and was recognized as one of the foremost Icelandic opinion leaders in hypertension. Presently Thorkel practices medicine in Reykjavik. Historians report that in the early 11th century there was a Danish prince called Thorkel the Tall. Well, Thorkel Gudbrandsson deserved the same nickname; he was slim and very tall. Or so I thought.

**Hypertension division staff in 1992. Front row (from left to right)** *A. Zweifler, S .Julius, S. Nesbit, D Bassett*. **Back row** *K. Jamerson, A. Weder, T. Gudbrandsson*

Delighted that Torkel decided to come to Ann Arbor, I announced to the group that we would get yet another very good coworker. As an aside I teased Ken Jamerson and said something to the effect that he will lose the distinction of being the tallest person in the division. Ken had seen Torkel at an international meeting and confidently said that Thorkel was actually shorter than he. Ken was right but for me Thorkel was tall enough, and as the picture shows for a while we had two guys to look up to.

In the second round of the Tecumseh study we reexamined people we had seen three years earlier. Reading the paper about the second round of exams,[7] I am again impressed with the cooperative, willing spirit in Tecumseh volunteers. Of 946 subjects examined the first time, 79% (735 persons) showed up for the second exam!

Eighty one percent of the people who had normal blood pressure the first time around were normotensive also on the second exam. Of those who had hypertension at the first exam, only 40% remained hypertensive three years later ("sustained hypertension"). The remaining 60% of the hypertensive group became normotensive ("transient hypertension"). However, 10% of normotensives at the first exam were classified as hypertensive the second time around ("de novo hypertension"). We hoped to find some characteristics that might predict who will remain hypertensive and who not. With such information in hand the practicing physician could adjust his/her approach to an individual with prehypertension. Consequently, in the first pass analysis we compared the three hypertension groups and to our chagrin they were not different one from another. This wholesale change of classification was not a surprise; in fact this is why we did the second study. However, compared to the normotensive group each hypertension category had high childhood blood pressure levels. Furthermore, as adults they had high self-measured home blood pressure, were overweight, and had significantly higher levels of cholesterol, triglyceride and insulin.

These findings were a mixture of good and bad news. Bad because the groups were so similar that it was unlikely we could foretell which individual would move in which direction. And good because all three groups had a steady background of higher blood pressures as well as higher levels of cardiovascular risk factors which once again suggested that borderline hypertension is not an innocuous condition.

Luckily not everything was lost. In the first round we undertook a major logistic effort to teach all participants - nearly a thousand people - how to measure their own blood pressure. Our field team visited participants in their homes to instruct them about the measurement technique. At the end of the training session, the volunteers operated a blood pressure measuring device connected to two stethoscopes; one for the visiting trainer and the other for the volunteer. The readings were repeated until the number obtained by the volunteer repeatedly matched values the trainer wrote down. In the past home blood pressure measurement yielded reliable results[8,9] in students but it was not clear whether the measurements would be equally reliable in people with different educational and social backgrounds. It turned out that 20 minutes of training sufficed to obtain reproducible results in the diverse Tecumseh population.[10] It did not matter whether someone was a professional or manual worker, had university or elementary school education; everybody in Tecumseh seemed equally capable of measuring the home blood pressure.

Our effort to obtain baseline home blood pressure readings paid off nicely in the second round of the Tecumseh study. We used a computer program to search for a home blood pressure cut point that would provide maximum sensitivity with at least 90% specificity to predict sustained hypertension. A home blood pressure equal or above 128 and 82 mmHg predicted with 93% specificity that a person will exhibit "sustained" office hypertension three years later. But nothing is perfect and the sensitivity of detection was only 48%. Translated in plain language this method found a group that had a 93% chance to remain hypertensive but missed half of all people who remained hypertensive. A home blood pressure cut off at 120 and 80 mmHg diastolic predicted with 91% specificity that a patient who appeared to have hypertension in a doctor's office will three years later be normotensive. However the sensitivity was only 45%.

According to numerous surveys, about 25 million people in the United States have slightly elevated blood pressure similar to the one we found in the Tecumseh borderline hypertension group. In the fourth decade of life, the incidence of future hypertension in such subjects is about 10% per year. This group has a high risk of future cardiovascular morbidity and mortality but it is presently generally accepted that antihypertensive treatment should be started only if patients have repeated blood pressure readings equal or higher than 140 and or 90 mmHg. Unfortunately hypertension is the "silent killer," patients have no subjective symptoms and detection of hypertension entirely depends on how frequently the blood pressure is measured. If all 25 million people with marginal blood pressure elevation had to be seen frequently just to detect ensuing hypertension this would overwhelm the health care delivery system. Our paper proposed a very reasonable triage scheme for management of these individuals. Based on home blood pressure cutting points, we defined "high risk," "intermediate," and "low" hypertension risk group and proposed how to manage them in clinical practice. The high risk group would receive intensive nonpharmacological (life style modification) treatment and be evaluated each 3 months. If at the six-month point the blood pressure was elevated both at home and in the office, we suggested that pharmacological treatment be started. On the other hand, we felt that in the low risk group the blood pressure could be measured every three years.

This might have been somewhat of a pipe dream and nobody hastened to implement our recommendation. But when you raise a large health issue you must provide a rational approach to it. In the meantime things have changed a great deal. The noise level about the importance of the marginal blood pressure elevation has increased to the point that it might have created an over-reaction. In its seventh and most recent report, the USA Joint National Committee on Prevention, Detection, Evaluation, and Treatment of High Blood Pressure (JNC) introduced a new category called "prehypertension." I am delighted that they highlighted the importance of early phases of hypertension - an area to which I devoted my entire scientific life. But they became over exuberant when they widened the definition of prehypertension to include also people whose systolic is in the 120-129 mm Hg and the diastolic in the 80-84 mmHg range. With this new definition the prehypertension group grew to 65 million people. Technically the JNC 7 was correct; this group indeed has a higher degree of cardiovascular risk. But besides vague statements about life style modification the Committee gave no guidance about what to do with these individuals.

As we kept agitating about the importance of prehypertension, we felt it necessary to also address some practical issues. And the second Tecumseh round gave us sufficient standing to propose a management algorithm for borderline hypertension. Were we right or wrong? I do not know, but at least we tried!

The Tecumseh study generated more publications than I could possibly describe. All together we published 23 papers and I can proudly say that all were original analyses of data. Despite various invitations we never wrote an overview of already published work. And I won't do it this time either; after all I am writing memoirs and not a textbook of hypertension. As a compromise, a full list of Tecumseh publications is attached at the end of this chapter.

The time has come to move on. Let me just add a short historical note. After ten years of field work I decided not to apply for an extension of the Tecumseh grant. The study raised interesting questions which I wanted to investigate with sophisticated invasive techniques. I also got involved in a few large scale trials of long term outcomes of antihypertensive treatment.

However, this was not the end of the Tecumseh study; Alan Weder wrote a successful proposal for another five years of field work. It was an entirely new project and Alan moved the study in a direction in which I had no expertise. Consequently I am unable to give you an account of these activities.

## Bibliography

1. Julius S, Jamerson K, Mejia A, Krause L, Schork N, Jones K: The association of borderline hypertension with target organ changes and higher coronary risk. Tecumseh Blood Pressure Study. JAMA 264:354-358, 1990.

2. Julius S, Krause L, Schork NJ, Mejia AD, Jones KA, van de Ven C, Johnson EH, Sekkarie MA, Kjeldsen SE, Petrin J, Schmouder R, Gupta R, Ferraro J, Nazzaro P, Weissfeld J: Hyperkinetic borderline hypertension in Tecumseh, Michigan. J Hypertension 9:77-84, 1991.

3. Palatini P, Julius S: Heart rate and the cardiovascular risk. J Hypertens 15:1-15, 1997.

4. Palatini P, Julius S: Association of tachycardia with morbidity and mortality: pathophysiological considerations. J Human Hypertens 11(suppl 1):S19-S27, 1997.

5. Palatini P, Benetos A, Grassi G, Julius S, Kjeldsen SE, Mancia G, Narkiewicz K, Parati G, Pessina AC, Ruilope LM, Zanchetti A: Identification and management of the hypertensive patient with elevated heart rate: statement of a European Society of

Hypertension Consensus Meeting. J Hypertens 24:603-610, April 2006.

6. Julius S, Mejia A, Jones K, Krause L, Schork N, van de Ven C, Johnson E, Petrin J, Sekkarie MA, Kjeldsen SE, Schmouder R, Gupta R, Ferraro J, Nazzaro P, Weissfeld J: "White coat" versus "sustained" borderline hypertension in Tecumseh, Michigan. Hypertension 16:617-623, 1990.

7. Nesbitt SD, Amerena JV, Grant E, Jamerson KA, Lu H, Weder A, Julius S: Home blood pressure as a predictor of future blood pressure stability in borderline hypertension. The Tecumseh Study. Am J Hypertens 10:1270-1280, 1997.

8. Julius S, McGinn NF, Harburg E, Hoobler SW: Comparison of various clinical measurements of blood pressure with the self-determination technique in normotensive college males. J Chronic Dis 17:391-396, 1964.

9. Julius S, Ellis CN, Pascual AV, Matice M, Hansson L, Hunyor SN, Sandler LN: Home blood pressure determination: value in borderline ("labile") hypertension. JAMA 229:663-666, 1974.

10. Mejia A, Julius S: Practical utility of blood pressure readings obtained by self-determination. J Hypertension 7(Suppl 3):S53-S57, 1989.

**Other Tecumseh Papers**

Amerena J, Nesbitt S, Krause L, Grant E, Lu H, and Julius S. Trends in left ventricular function over three years in the Tecumseh study. Blood Pressure. 1997; 6:262-268.

Gudbrandsson T, Julius S, Jamerson K, Smith S, Krause L, and Schork N. Recreational exercise and cardiovascular status in the rural community of Tecumseh, Michigan. Blood Pressure. 1994; 3:178-184.

Gudbrandsson T, Julius S, Krause L, Jamerson K, Randall O, Schork N, and Weder A. Correlates of the estimated arterial compliance in the population of Tecumseh, Michigan. Blood Pressure. 1992; 1:27-34.

Jamerson K, Schork N, and Julius S. Effect of home blood pressure and gender on estimates of the familial aggregation of blood pressure; The Tecumseh blood pressure study. Hypertension. 1992; 20(3):314-318.

Julius S, Jones K, Schork N, Johnson E, Krause L, Nazzaro P, and Zemva A. Independence of pressure reactivity from pressure levels in Tecumseh, Michigan. Hypertension. 1991; 17(Suppl. III):III-12-III-21.

Julius S, Majahalma S, Nesbitt S, Grant E, Kaciroti N, Ombao H, Vriz O, Valentini M, Amerena J, and Gleiberman L. A "gender blind" relationship of lean body mass and blood pressure in the Tecumseh Study. Am. J. Hypertens. 2002; 15:258-263.

Kneisley J, Schork N, and Julius S. Predictors of blood pressure and hypertension in Tecumseh, Michigan. Clin. Exp. Hypertension - Theory and Practice. 1990; A12(5):693-708.

Marcus R, Krause L, Weder AB, Dominguez-Mejia A, Schork NJ, and Julius S. Sex-specific determinants of increased left ventricular mass in the Tecumseh blood pressure study. Circulation. 1994; 90:928-936.

Mejia AD, Julius S, Jones KA, Schork NJ, and Kneisley J. The Tecumseh blood pressure study: Normative data on BP self-determination. Arch Int. Med. 1990; 150(6):1209-1213.

Palatini P, Amerena J, Nesbitt S, Valentini M, Majahalme S, Krause L, Tikhonoff V, and Julius S. Heritability of left atrial size in the the Tecumseh population. European Journal of Clinical Investigation. 2002; 32(7):467-471.

Palatini, P, Krause L, Amerena J, Nesbitt S, Majahalme S, Tikhonoff V, Valentini M, and Julius S. Genetic contribution to the variance in left ventricular mass: The Tecumseh Offspring Study. J. Hypertens. 2001; 19:1217-1222.

Palatini P, Majahalme S, Amerena J, Nesbitt S, Vriz O, Michieletto M, Krause L, and Julius S. Determinants of left ventricular structure and mass in young subjects with sympathetic over-activity. The Tecumseh offspring study. J. Hypertens. 2000; 18:769-775.

Palatini P, Vriz O, Nesbitt S, Amerena J, Majahalme S, Valentini M, and Julius S. Parental hyperdynamic circulation predicts insulin resistance in offspring. The Tecumseh Study. Hypertension. 1999; 33:769-774.

Shahab ST, Budbrandsson T, Jamerson KA, and Julius S. Isolated "home hypertension" in Tecumseh, Michigan. Croatian Medical Journal.

1993; 34( 4):325-331.

Smith S, Julius S, Jamerson K, Amerena J, and Schork N. Hemotocrit levels and physiologic factors in relationship to cardiovascular risk in Tecumseh, Michigan. Hypertension. 1994; 12(4):455-462.

Vriz O, Nesbitt S, Krause L, Majahalme S, Lu H, and Julius S. Smoking is associated with higher cardiovascular risk in young women than in men: The Tecumseh Blood Pressure Study. J. Hypertens. 1997; 15:127-134.

Weder, AB, Schork, NJ, and Julius S. Linkage of MN locus and erythrocyte lithium-sodium countertransport in Tecumseh, Michigan. Hypertension. 1991; 17:977-981.

Weder AB, Schork, NJ, Krause L, and Julius S. Red blood cell lithium-sodium countertransport in the Tecumseh blood pressure study. Hypertension. 1991; 17:652-660.

Young EA, Nesse RM, Weder A, and Julius S. Anxiety and cardiovascular reactivity in the Tecumseh population. J. Hypertens. 1998; 16:1727-1733.

# From the Forest Back to the Trees; Studies of the Metabolic Syndrome 1991-2004

As you have very likely noted, the time periods of various chapters in this book overlap a great deal, but that's life, you don't neatly finish one job and then start another. In fact, you are a lucky guy if you have only a few research projects. For a while the National Institutes of Health supported large "program projects" to encourage cooperation among investigators and I have seen directors of such enterprises trying to cope with numerous diverse research lines. Some were successful, some not, but all ended spending plenty of time to keep up with less familiar fields while their own research suffered. If I had to do that I'd go nuts. My magic number turned out to be three projects at a time. As the Tecumseh study unfolded between the years 1990 and 2000, we continued to analyze the data and reported a number of new observations. Nevertheless, I just could not stop thinking about the close association of borderline hypertension with other cardiovascular risk factors in Tecumseh.

Most fascinating were the associations of high blood pressure with high plasma insulin levels and with overweight. This was a new field to me and I started to read everything I could get my hands on to better understand the regulation of insulin and how this might relate to blood pressure. Our observation that a relationship between insulin and blood pressure levels exists also within the normal ranges of both blood pressure and insulin suggested this was a physiological relationship and that a common mechanism might regulate both these variables. There also had to be some purpose to this common regulation of two diverse variables.

Oops, having said that, I am getting into the illicit field of teleology, a way of thinking that ought not to be used in science. But to me it makes a lot of sense that most processes in the body make some sense. Explanations in endocrinology literature did not resonate with me. Some diabetologists proposed that the excess of insulin was a primary event which caused retention of fluids, which eventually led to a delayed elevation of the blood pressure. There were some direct arguments against that theory. People with insulinomas, tumors that secrete an excess of insulin do not have a higher prevalence of hypertension than the general population. Furthermore, infusion of insulin into humans causes potent vasodilatation and does not increase the blood pressure. In response colleagues from the endocrine side of the fence would point out that chronic insulin infusion causes hypertension in rats. True, but the same infusion has no effect on the blood pressure in dogs. And if you extend the discussion into animal experimentation you are soon involved in the absurd dialogue whether humans are more like rats than like dogs. But to me the most potent counterargument was the Tecumseh observation of a likely physiologic (regulatory) relationship between these two variables. So why would the body want to regulate a circulatory variable (blood pressure) from an endocrine organ hidden in the retroperitoneal space? If you want something to regulate circulation, you had better place it into the highly distensible blood vessels where even the smallest changes in either pressure or volume can be sensed.

Scientists are territorial creatures and, not surprisingly, people interested in obesity strongly believed that excessive eating is the unifying mechanism which ties the blood pressure, insulin resistance, and obesity together. But insulin resistance, the cause of high insulin in the blood, has been seen also in lean patients with hypertension. Then you start discussing how lean is a lean person - not a very constructive dialogue. Furthermore, Tecumseh records have shown that blood pressure elevation precedes weight gain. If overweight later causes higher blood pressure and insulin resistance, what had caused the early blood pressure elevation?

Being a scientist, I am territorial too and I tend to stick to my roots. Just as we proved that 37% of borderline hypertensives in Tecumseh had high cardiac output, fast heart rate and high plasma norepinephrine levels, that they come from a familial background of hypertension and they had an early blood pressure elevation, I was not about to change horses! Frankly, biased as I might have been, to me the question was how, by which mechanism, could sympathetic overactivity cause insulin resistance? But more importantly, why would the sympathetics do such a crazy thing? Or maybe I was asking the wrong question. One ought to trust the body, it usually knows what it is doing and maybe there is a purpose to the simultaneous increase of insulin resistance and blood pressure?

In Ann Arbor, I frequently interacted with James Neel, the father of modern genetics in the USA. We'd meet in the hospital cafeteria and Jim often spoke about his concept that some "disease of civilization" may reflect natural selection. If a gene offered an early survival advantage it was likely to accumulate in the population. But genes that offered some advantage during periods of famine in the hunter-gatherer era might be detrimental in modern times. The rapidity of social evolution would have outstripped the usefulness of "thrifty" (glucose preserving) genes and modern humans with such genes would become susceptible to diabetes. Neel's concept was easily applicable to genes for a more efficient "defense reaction" which likely were useful in times of hardship but became harmful in contemporary civilization.

Defense reaction is an emotional and complex physiologic response to anticipated danger. High cardiac output and a higher blood pressure prime the circulation for quick and efficient exercise performance. A person in trouble must respond to the situation with physical activity and psychologists prefer to call the defense reaction as the "fight or flight" response. You have two choices but to execute either of them you will have to exercise. Hematological changes also make sense within the context of the defense reaction; the temporary decrease of plasma volume associated with sympathetic stimulation increases the relative concentration of red blood cells so that if you bleed the remaining blood has a better oxygen carrying capacity. Furthermore, the relative platelet concentration also increases which improves the first line defense against bleeding; the ability of platelets to aggregate and initiate coagulation in the wound.

You will remember that we found all these signs in borderline hypertension and it does not take too much of a flight of fancy to conjure up how this temporarily appropriate response may in the long term become detrimental. Heightened blood pressure harms arteries, high cardiac output overtaxes the heart, tachycardia injures coronary arteries and enhanced coagulation favors thrombosis inside blood vessels. Not a good thing to have! But why would the body bother to create insulin resistance at the same time? And if the sympathetic overactivity during the defense reaction really causes insulin resistance, how does it do it; what is the mechanism?

I thought I had the answer for the "why" question. Let's start with the fact that in contradistinction to muscles and the liver, which need insulin to process glucose, the brain does not need insulin to utilize the glucose. Add to this the other fact that the brain needs glucose to function normally and you will start getting the picture. Even a short, half a minute decrease of glucose throws the brain into a disarray; the person first gets confused, next losses consciousness and eventually may have violent convulsions. So to the brain a steady supply of blood glucose is an absolute must, a priority of highest order. But there is a bit of competition for the blood glucose in our bodies; skeletal muscles also use glucose for fuel. There are a lot of muscles in the body, they account for approximately 50% of the total body weight and when they work hard they need plenty of energy. Poor brain, clever but small, is not a match to muscles and if muscles were permitted to eat up the glucose in the blood the brain would suffer. However, muscles have also an alternate source of energy. Under the influence of insulin, they accumulate glucose from the blood in the form of glycogen and store it for later use. So during the defense reaction when it anticipates a need for exercise, the brain increases the sympathetic tone to induce an acute insulin resistance and the muscles are forced to rely on internal sources of energy. Remember, the brain does not need insulin, it can extract glucose directly from the blood! The end result is that everything functions as it should, each organ gets what it needs, or to put it differently, in danger you are ready to act but you also can think your way out of trouble.

Nice story, logical, but is it true? Is there any evidence for acute insulin resistance during the defense reaction? Not exactly! Because the defense reaction is an emotional response, it would be extremely difficult and unethical to threaten people and then measure insulin sensitivity. But in our reviews of literature,[1,2] we found sufficient evidence that the type of sympathetic overactivity seen in the defense reaction can affect insulin sensitivity. There are two naturally occurring agonists for transmission of sympathetic signals to peripheral organs; epinephrine released by the adrenal gland and norepinephrine released at sympathetic nerve endings. These messengers elicit different responses and the overall picture of defense reaction is "epinephrine-like." An infusion of epinephrine increases both the cardiac output and heart rate, raises the BP and dilates small arteries, a picture compatible with stimulation of cardiac and vascular beta adrenergic receptors (Norepinephrine increase the blood pressure by stimulating arteriolar alpha 1 receptors.) Infusion of epinephrine in humans elicits an acute insulin resistance which can be abolished by a blockade of beta adrenergic receptors. If sympathetic activation has the potential for it, then it is very likely that during the defense reaction it will cause insulin resistance. So the body is not stupid and, within the frame of acute defense reaction, the insulin resistance seems to be an appropriate part of the integrated response.

As we discussed various issues among ourselves, some aspects of the insulin resistance defense reaction story bothered me. Yes it made sense from the defense reaction standpoint, but logical as this might be, the observation did not quite square with some other facts. If the beta receptor-mediated insulin resistance is responsible for the observed glucose/insulin abnormalities in hypertension, then treatment with beta adrenergic blockers ought to improve the situation. But far from being useful, treatment of hypertension with beta blockers actually causes insulin resistance. Furthermore, we had clear evidence that beta adrenergic receptor responsiveness decreases in the course of hypertension. This should protect them against insulin resistance and yet the higher the blood pressure, the worse the degree of insulin resistance.

So here we were, we had a nice theory about the purpose of insulin resistance, the body acted as the theory predicted, but that particular mechanism could not explain chronic insulin resistance in hypertension. Nevertheless, our data on association of sympathetic overactivity with borderline hypertension and insulin resistance were too strong; there simply had to exist yet another mechanism for that association. I got obsessed by the issue to the point of asking everybody willing to listen and reciting to myself the same question: "Why would the hemodynamic condition of hypertension be associated with the metabolic condition of insulin resistance?" I will admit that I often talk to myself, and this oft-repeated question sometimes spilled into a loud monologue. But let me break the scientific monotony with a not totally related old time medical school prank. Why? Because it just came to me and because being serious is bad for your health.

You'd ask someone whether he talks to himself and if the answer was "no" you'd say: "You are right, no use talking to a fool."

If he said "yes" you'd hit him with: "Sure, only a fool understands a fool!"

You may not find this particularly funny but it did refresh me enough to move on.

Having convinced myself that a stimulation of beta adrenergic receptors could not be responsible for chronic insulin resistance, I focused on "other" mechanisms that might tie the sympathetic overactivity with chronic and steadily worsening insulin resistance in hypertension? The question would just not go away. Not surprising then that I brought it up during a phone chat with Mladen Vranic, a friend from student times in Zagreb. Mladen is professor of physiology at the University of Toronto and a world renowned expert in the pathophysiology of diabetes. As soon as we were past the preliminary "how are you's," I started to talk about my favorite topic: why would a hemodynamic condition of hypertension be associated with the metabolic condition of diabetes? Could there be some kind of a hemodynamic reason for the development of insulin resistance?

Mladen said that most diabetologists view insulin resistance as a genetic aberration (polymorphism) of the insulin receptor which interferes either with the binding of insulin to its receptor or with the passage of the insulin "signal" from the cell surface to second messengers within the cell. Not surprisingly, another large group of basic scientists investigated possible aberrations in "downstream" post-receptor signaling within the cell. But a rather small group believed that in some patients, insulin resistance may reflect a pre-receptor problem, an inadequate delivery of insulin and glucose to skeletal muscle cells. Now, that was very interesting! Mladen gave me a few names and off we went searching through the literature.

At about that time we got a great reinforcement. Ove Andersson from the Sahlgrenska University Hospital in Goteborg took a sabbatical leave to join us in Ann Arbor. He intended to take a fellowship with us many years earlier but it had not worked out. In some way this delay was good for us. Now we got two mature Goteborg-trained hypertension specialists, Thorkel Gudbrandsson and Ove Andersson. Ove, a pleasant, thoughtful and knowledgeable person did not waste his time. He joined us in the review of literature and later in the active execution of Ken Jamerson's research protocol about insulin resistance. On the whole, Ove was probably more useful to me than I was to him but I do believe his stay in Ann Arbor gave him the needed time to "recharge batteries' and sort out his future priorities. After Ove returned to Goteborg, he was promoted to the position of full professor of Medicine and he is presently a leading hypertension researcher and educator in Sweden.

In a first review of the literature,[3] we proposed the hypothesis that chronic hypertension causes alterations in microcirculation which tend to restrict the nutritional blood flow to skeletal muscles. Consequently, as less glucose and insulin are delivered to muscle cells, less glucose can be cleared in one passage of blood through the muscles. Before we move on, let me define two terms I used above. Under "nutritional blood" flow I mean the flow that reaches individual muscle fibers through a network of capillaries from whence the plasma diffuses to the cells. Even some experts fail to realize that this has nothing to do with the total muscle flow; it reflects only the distribution of the flow within the muscle. The blood flow in the muscles can be directed to flow through shunts and bypass the muscle cells. And the term "restriction of microcirculation" describes reduction of the number of open capillaries in response to a long standing blood pressure elevation. As already discussed in previous chapters, persistent blood pressure elevation alters the vascular wall and small arteries (arterioles) become hyper responsive to vasoconstriction. Assume now that the vasoconstriction is so strong that the lumen of some arterioles close and that some of the downstream capillaries are not filled with blood.

A schematic presentation of such an anatomic situation is given in Figure 1 below. The figure shows a horizontal cross section through skeletal muscles. The large circles represent the muscle bundles of little pistons which contract at our command. The small circles are capillaries, the tiniest branches of the arterial blood tree. Their walls are porous and capillaries bring oxygen and nutrients to muscles. Note the substantially lower number of capillaries in insulin resistant patients. We call this condition vascular rarefaction and Figure 1 remarkably resembles histological specimens from hypertensive and normotensive subjects.

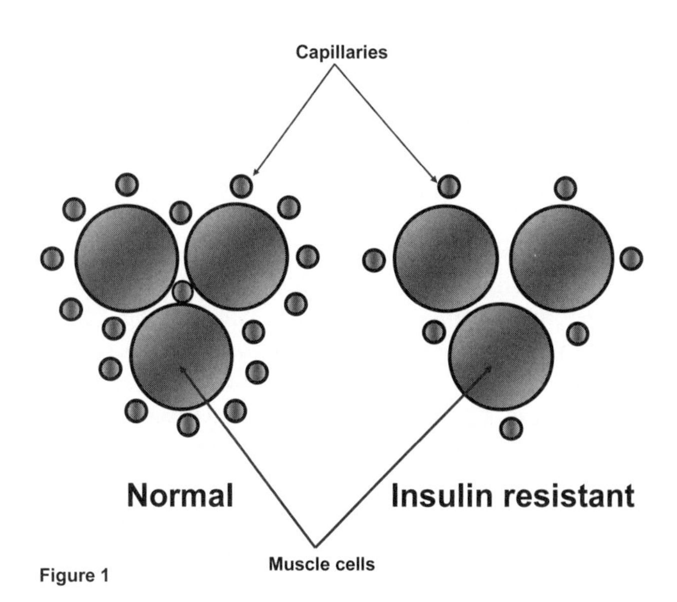


Figure 1

In the first review paper, we strongly hinted that sympathetic stimulation might decrease the nutritional blood flow in skeletal muscles and thereby elicit insulin resistance. In the second paper,[2] we went further and wrote: *"A hypothesis ought to rest on sufficient supporting data and its validity ought to lend itself to experimental verification. We believe our hypothesis meets both criteria. After outlining the supporting evidence we will propose tests which could prove or disprove the hypothesis."* I will admit that before we submitted the paper we had already prepared the protocol of the first experiment to test our hypothesis. Yes, there is a bit of egotism in science. Right or wrong we were going to be the first ones to find out!

The review pointed out that the prevalence of insulin resistance increases with aging and with hypertension and we quoted Conway's early work about insufficient vasodilatation in both conditions. Vascular rarefaction was found in the conjunctiva of patients with hypertension, in autopsies of intestines, and in a number of experimental models of hypertension. Most importantly Heinrich et al. documented vascular rarefaction in the skeletal muscles of patients with hypertension. Unfortunately they published this in the infrequently read German medical journal Klinische Wochenschrift. A relationship between capillary density and insulin sensitivity was reported by Lithell in normotensive Swedish subjects. In the United States Baron showed that people with insulin resistance respond to a glucose load with a blunted increase of skeletal muscle blood flow. He also showed that vasoconstriction causes decreased insulin sensitivity. Furthermore, the Tecumseh study results supported the hemodynamic concept of insulin resistance; there was a strong negative correlation of maximal forearm blood flow with fasting plasma insulin and a positive correlation between insulin and minimal forearm vascular resistance. This correlation between signs of restricted muscle microcirculation (skeletal muscles are the major anatomical component of the forearm) and a marker of insulin resistance (insulin levels) were fully compatible with the hemodynamic hypothesis of insulin resistance. Moreover insulin resistance was found in lean hypertensives as well as in weight matched hypertensive and normotensive subjects suggesting that the association is not necessarily a consequence of overweight in hypertension. This is only a part of the evidence we presented but I do not want to bore you with details. Suffice it to say that the two reviews convinced us we were on the right track. In the next step we started to implement the protocol to test whether changes of the blood flow affect glucose utilization in the human forearm.

The idea was reasonably simple, we'd infuse insulin into the brachial artery, place a catheter also into the vein, measure the forearm blood flow and then use Fick's formula to calculate glucose utilization in the forearm. Adolph Eugene Fick, a nineteenth century German physiologist, was a mathematical wizard with a wide variety of interests ranging from developing the first contact lens to postulating the laws of diffusion and developing a method for measuring cardiac output. He established the principle that if you know the amount of a substance dissolved in the blood and then measure the concentration of that substance in the arterial and venous blood, you can calculate the amount of blood flow in a unit of time. He used oxygen consumption and the difference in oxygen content in the arterial and venous blood samples to calculate cardiac output. In a simplified way Fick's principle predicts that if the oxygen consumption is equal then a smaller arteriovenous difference means that the blood flow was fast and less oxygen was extracted in a unit of time. However, if you can measure the blood flow through an organ and are capable assessing the difference of the concentration of a substance in the arterial and venous blood from that organ, you can turn Fick's formula around to calculate how much of the substance the organ utilized. In our case we measured the forearm blood flow by plethysmography and multiplied it by the arterio-venous difference to calculate oxygen and glucose uptake in the forearm.

Technically this was pretty easy to do. We had the plethysmography set-up for measurements of the forearm blood flow, drawing the venous blood is a simple routine and placing an indwelling catheter into the artery was greatly simplified. In the late 1960's our invasive hemodynamic studies were considered a "state-of-the-art" breakthrough and placing a brachial artery catheter was a big deal. In the 1990's every intern knew how to puncture arteries and disposable needles for threading catheters into arteries were commonplace. However, drawing the venous blood was not as simple as we thought. Normally one places the venous catheter upstream but we had to thread it downstream into the deeper veins to make sure that the sampled blood came from forearm muscles. However, Ken Jamerson mastered the new technique and could easily place the catheter where others failed.

The general idea was to reduce the blood flow through the forearm and see whether this reduces the insulin-stimulated glucose uptake. As we discussed the options there emerged two possibilities; we could have injected various vasoconstrictors into the artery or we could elicit a reflex vasoconstriction in the forearm. Eventually we used both methods but I insisted on starting with reflex vasoconstriction. With infusion of constrictors you always run into the question whether the infused doses were smaller or larger than the range of blood concentrations of that substance seen under normal circumstance. So in the first series of experiments we decided to go back to inflation of blood pressure cuffs on the subject's thighs. As you might remember from Chapter 6 (picture on page 159), we used that trick to study the regulation of renin release in humans. You inflate the cuff to 40 mmHg and this traps a considerable amount of blood in the leg veins but permits the inflow of arterial blood into the extremity. Because the venous return is impeded, the central venous pressure decreases, and this is sensed as a diminished stretch of cardiopulmonary mechanoreceptors and a reflex vasoconstriction ensues.

Ken Jamerson was in charge of the entire project and before undertaking to produce a new device he rummaged through various drawers of the old Kresge building lab and found the set of blood pressure cuffs Wolfgang Kiowski had assembled 15 years previously. Despite its primitive looks- pediatric blood pressure cuffs patched together with tape and connected with protruding tubes- Wolfgang's device was in perfect working order!

Next on the agenda was to determine the right amount of insulin for infusion into the forearm artery. The amount should be sufficient to increase the glucose uptake in the forearm but small enough not to spill into the systemic circulation and decrease glucose levels in the entire body. After reading the literature, Ken proposed to use the dose described by investigators in Pisa, Italy. The infusion rate was calculated for each individual taking into account the forearm blood flow, the forearm volume and expressing the dose as micro units per hour. To avoid adherence of insulin to the tubing, insulin was diluted in a saline solution to which 1 percent of human serum albumin was added.

Eventually Ken developed the protocol for a two hour experiment consisting of 30 minutes of baseline rest, and 90 minutes of insulin infusion (60 minutes to achieve a steady glucose utilization level and 30 minutes for measurements during the cuff inflation). Arterial and venous blood for determination of glucose and oxygen levels was drawn at 20 minute intervals throughout the first sixty minutes and at ten minute intervals after the cuff inflation. At each time point we measured the forearm blood flow with plethysmography. I am telling you these details to explain that this was a very complex study. Multiple dilutions had to be made to prepare the insulin infusion and for each of them the person in charge read out loud what he was doing. The more dilutions you have the larger is the possibility of an error, and sterile glucose for infusion was available to deal with a possible overdose of insulin. We thought we had an airtight procedure. Ove Andersson was one of the first volunteers for the study. Sure enough, in the middle of the procedure Ove became uncomfortable and soon was on the verge of a hypoglycemic coma. A glucose injection followed by glucose infusion took care of the problem. Ove was remarkably pleasant about it but everybody understood how serious the situation was. We tightened the procedure and nothing similar ever happened again.

The results of the study on 14 healthy young volunteers were published in 1993 in Hypertension.[3] The insulin infusion considerably increased the forearm blood flow from 3.8 milliliter/deciliter of forearm volume per minute to 5.0. As an aside this is yet another piece of evidence that insulin in humans is a strong vasodilator and that it could not possibly induce hypertension through a direct vascular effect. Remember the debate, in regards to blood pressure response to insulin humans are apparently more akin to dogs than to rats! And it makes sense that the blood flow should increase; as the infusion of insulin actively stores glucose in to the muscles the local metabolic rate increases and additional flow is needed to remove the excess heat. The body knows what it is doing.

The thigh cuff inflation elicited a significant (18%) decrease of the forearm blood flow. We got the vasoconstriction we desired. Figure 2 shows the results.

Over the initial forty minutes of insulin infusion plasma glucose utilization increased steeply and thereafter leveled off. We then continued to infuse insulin at the same rate and inflated the cuffs for thirty minutes. After inflation of the cuff the glucose utilization declined, and after 20 minutes the decrease was statistically significant. In the last 10 minutes of inflation reflex vasoconstriction diminished and the glucose utilization started to increase again. Throughout the entire experiment the forearm oxygen utilization remained stable. This discrepancy between glucose and oxygen utilization further supported the hemodynamic explanation of insulin resistance. When the forearm blood flow decreases, the diffusion distance between fewer open capillaries and the metabolically active cells increases. The change of distance is likely to impede the delivery of larger molecules such as glucose and insulin whereas freely diffusible oxygen will have no difficulties finding its way to the cell.

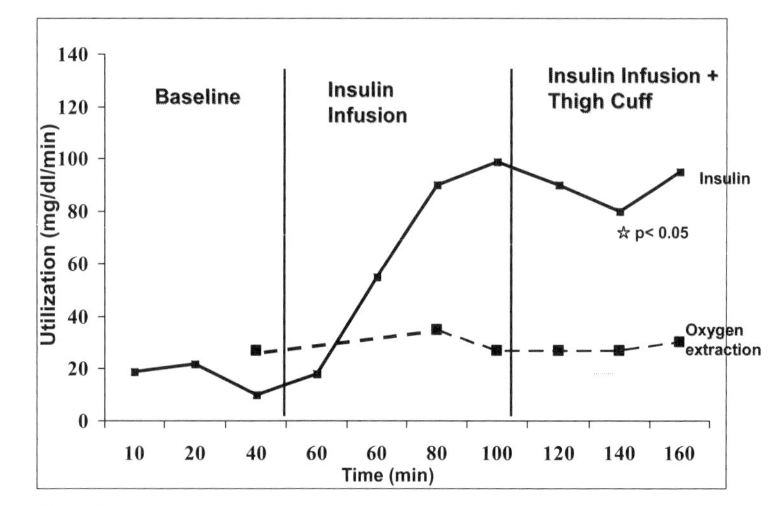


Figure 2

Jamerson, et al Hypertension 1993 21:618-623

This paper did not resolve all issues. A reflex increase of sympathetic tone is likely to activate both alpha and beta adrenergic receptors and acute beta receptor stimulation decreases the glucose metabolism. That effect is apparently independent of the hemodynamic action and is caused by post beta receptor increase of the second messenger (cyclic AMP) which in turn directly inhibits glucose storage. So we could not resolve whether the alpha adrenergic vasoconstriction or beta adrenergic stimulation (or a combination of both) were responsible for the decreased glucose utilization. In fact, to this day Ken Jamerson and I view these results differently, he believes that beta receptors played a role and I stick with the hemodynamic explanation. But one thing is sure, had the decrease of the forearm flow not affected the glucose utilization our hypothesis would be dead. Or as we said in the paper *"...the basic concept espoused in our recent overview of the possible hemodynamic determinants of the association of between insulin resistance, sympathetic overactivity and hypertension would have been proven unfeasible."* Got to give it to us; we knew how to translate simple words into the venerable highfaluting scientific lingo!

Encouraged by these findings, we tested whether similar effects could be elicited with norepinephrine infusion.[4] Norepinephrine predominantly stimulates alpha adrenergic receptors and induces vasoconstriction. To evaluate the effect of the degree of blood flow reduction a smaller dose of norepinephrine was given to one group and a higher dose to the other group. The reduction of forearm blood flow with norepinephrine decreased the glucose uptake by 14% with the smaller dose and by 42% with the larger dose. This was consistent with the hemodynamic concept of insulin resistance but the new study also raised new questions. When we compared the decrease of the forearm blood flow in the previous thigh cuff study with the decrease in the group that received the small dose of norepinephrine, Ken proved to be a genius; he chose the right norepinephrine dose and the degree of the blood flow decrease in the thigh cuff group (18%) was similar to the decrease with the low dose norepinephrine infusion (23%). However the blunting of glucose uptake was larger with thigh cuff (23%) than with the low dose of norepinephrine (14%)! To Ken, who has a very good background in biochemistry, this suggested that a natural reflex vasoconstriction activated beta receptors to increase the intracellular cyclic AMP which blunted the glucose uptake by a biochemical mechanism. To me with my background in physiology, a hemodynamic explanation was more likely. We had a dose-response relationship between the degree of vasoconstriction and the degree of suppression of glucose uptake in the forearm. What more did we need? Well, it was not that simple. By focusing on the vasoconstriction, I ignored the general rule that important functions in the body are usually subserved by different mechanisms. Why shouldn't the body blunt the glucose uptake by activating simultaneously two different mechanisms? In fact, a simultaneous activation would mean that the suppression of glucose uptake was important to the body.

Both Ken and I had fallen victims to the "Rashomon effect" or the "subjectivity of perception" so well illustrated in Kurosawa's famous film. Two observers of an event can produce substantially different but equally plausible accounts of it. So Ken and I had what we call in science an "honest disagreement." Too bad we did not resolve the issue; the answer could have been easily obtained by injecting norepinephrine before and after infusion of a beta blocker into the brachial artery. We started to plan such an experiment, had a number of sessions to discuss the design, but each of us had different priorities and the protocol was never completed.

In my defense, let me say that I was convinced that even if it were important in an acute situation, the beta adrenergic mechanism of decreased glucose uptake could not explain what happens in chronic hypertension. Frequently stimulated beta adrenergic receptors tend to "down regulate" and become less responsive to sympathetic stimulation. Already in 1975 we had shown decreased beta adrenergic responsiveness of the heart rate, stroke volume and vascular resistance in borderline hypertension,[5] a finding that was well documented by other investigators. By contrast, the alpha adrenergic vasoconstriction is enhanced in hypertension in part because these receptors do not down-regulate and in part because pressure-induced structural changes in the arteriolar wall enhance the vasoconstriction. I had a hunch that extra-vascular (metabolic) beta receptors might also be down-regulated in hypertension but this idea was not yet tested. A few years later Sverre Eric Kjeldsen proved me right. Let me get that straight; Sverre did not set out to investigate my hunch, he had a long standing interest in that area and he designed his own protocol to investigate the question.[6]

Sverre recruited by advertisement 40 healthy volunteers for the study. He infused epinephrine into the brachial artery to increase the arterial blood concentration to levels seen during naturally occurring stresses. At such doses epinephrine predominantly stimulates the beta adrenergic receptors. Sverre measured the arterial concentration of epinephrine because epinephrine is taken up by tissues and he wanted to assess the highest (arterial) levels to which various organs were exposed. The epinephrine infusion caused significant increase of glucose and insulin whereas phosphate levels decreased. These results strongly suggested that epinephrine released during various forms of daily stresses may significantly affect the glucose and insulin metabolism. This was yet another important confirmation that sympathetic activation can affect glucose metabolism.

Among 40 volunteers Sverre found 13 subjects with perfectly normal blood pressure and 13 individuals who clearly had hypertension. As Figure 3 shows, hypertensive subjects had a lesser increase of heart rate, glucose and phosphate. Thus, not only a circulatory response but also metabolic beta adrenergic responses were suppressed in hypertension. Since beta receptors were less sensitive, it was unlikely that they would mediate the sympathetically-induced glucose intolerance in hypertension. This leaves us only with the other, hemodynamic, explanation of the association.

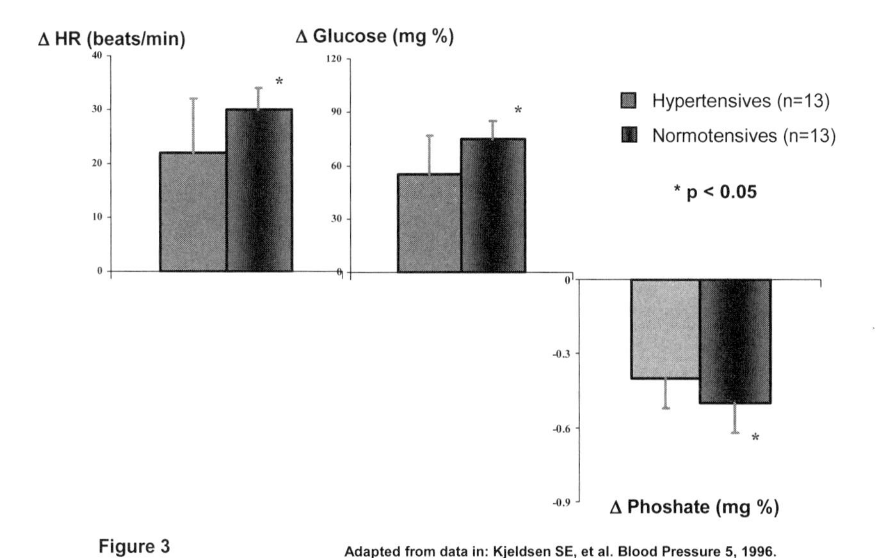


Figure 3

Adapted from data in: Kjeldsen SE, et al. Blood Pressure 5, 1996.

Anatomic evidence supports the hemodynamic explanation of insulin resistance. Vascular rarefaction has been found in all conditions associated with higher incidence of insulin resistance: aging, obesity, diabetes and hypertension. Conversely, physical training improves capillarization in skeletal muscles and increases insulin sensitivity. There is, however, a hitch to the interpretation of these observations. When the metabolic activity in skeletal muscles decreases (as in decreased glucose utilization), the nutritional blood flow decreases and vice versa; when the metabolism increases (as in exercise-induced increased insulin sensitivity), the blood supply increases. So what if the anatomical evidence was secondary to a primary metabolic alteration? And that is where our thigh cuff and norepinephrine infusion experiments come in; we created primary decreases of blood flow and a decreased insulin sensitivity followed. The only other "experiment" similar to ours is the uniformly observed increase in insulin resistance during chronic treatment of hypertension with beta blockers. Beta blockers reduce cardiac output, increase the vascular resistance and by blocking peripheral beta adrenergic vasodilatation they enhance the alpha adrenergic vasoconstriction.

Overall, I believe we sufficiently demonstrated that increased sympathetic tone can elicit insulin resistance by decreasing the nutritional blood flow to skeletal muscles.

To me what we had done was good enough and we started to investigate the relationship between sympathetic overactivity and obesity.

By now I am sure you figured out what we did. Yes, as at any other point in my career, before I moved into a field I first thoroughly reviewed the literature. I'd done this with Tony Schork in 1971 before I decided to pursue borderline hypertension,[7] again with him in 1978 prior to designing the Tecumseh study,[8] and more recently before studies of insulin resistance in hypertension.[1,2] I recommend this to other investigators and here are the Julius rules for such an enterprise. A) Be thorough and review everything in the field. B) It is not a sin to start reading the literature with a bias; after all there is a reason why the field interests you. C) However, hiding opposing views is an inexcusable sin. You must deal with them and explain to yourself and later to the readers, why you decided not to accept these findings or alternatively, if the other side is right -forget about your hypothesis. D) In these times when some people do not view the meta-analysis as a method but consider it to be a science capable of adjudicating controversies, you should remain openly qualitative! Don't weigh apples and oranges to decide from whence the wind blows; you are reading the literature to evaluate whether your hypothesis makes sense. If you find a solid basis for your view, even if the evidence is meager, move on. Don't be overwhelmed by minor objections to your thesis. Let me be a bit crude and say something about academic life. Lazy people always fail but your advancement will not depend on how hard you worked or on your intelligence; it will depend on how you handled the ocean of data. I have seen too many super-objective people frozen into inactivity by constantly weighing pros and cons of any planned research. Even sadder are people who read everything but fail to generate new ideas; I call them music critics, people who always know what is wrong but have never written a symphony.

Luckily when I decided to look into the relationship between sympathetic overactivity and overweight, Paolo Palatini took a sabbatical leave from his professorial position at the University of

Padova and came to work with me. Paolo, with whom I had earlier worked on tachycardia in hypertension, is the epitome of a good scientist; he was curious, objective, focused and willing to work long hours. But not on weekends when he either jumped on his bike for a 80-100 miles round-trip expedition from Ann Arbor or joined as a willing, lively and equal participant in social activities organized by our younger fellows. One weekend I joined Paolo to bicycle through some little-known but interesting areas around Ann Arbor. Of course it was neither an eighty miles trip nor was it at Paolo's usual pace, but I thought it would be worth his while seeing some of the surrounding green spaces, visiting architecturally interesting small communities and stopping at hidden gourmet places. It was fun but there wasn't a single place or route he had not already visited.

Paolo's solitary voyages became a local legend not so much for distances covered but for his choice of places to visit. He went to Detroit via what he considered the most interesting route, but there are some areas of Detroit a solitary traveler ought to avoid. Upon return he could not comprehend our safety concerns; as far as he was concerned everything was just fine. Well, he learned the hard way during a trek to Toledo, Ohio. Having already cycled forty plus miles Paolo stopped at a gas station to buy a Coke in the convenience section. When he came out the bicycle was gone!

Stunned, he sought help from the gas station attendant. There was no need to explain he came to the station with a bicycle; Paolo was clad in professional racer's paraphernalia.

"Did you see me leaning my bicycle against your pump?"

"Sure did!"

"But it is gone now!"

"Yup."

"Do you think it might have been stolen?" Paolo wondered.

"Very likely."

"Did you see someone taking the bike?"

"Nope, I was serving you" said the attendant and quickly changed the topic, "Aren't you insured?"

"No!"

"Well, you should have been!"

Paolo wasn't in a hurry to be thankful for the gratuitous advice, he had a more immediate concern. How would he go back to Ann Arbor? Luckily some teenage girls stopped for coke, took pity on him, squeezed him into their raggedy banged up old vehicle and dropped him off at the bus station. From there Paolo triangulated via Detroit back to Ann Arbor. Palatini was not deterred; he got a less showy piece of equipment and continued his expeditions.

We found only one picture of Paolo which shows him (left side) receiving from me the golden lapel pin with the division logo.

**Paolo Palatini receiving division pin from Stevo Julius**

We handed the pin to people who spent more than six months in the hypertension division and suggested they should wear it at large international meetings. Then visitors to Ann Arbor who did not overlap could recognize one another and chat about old times. Paolo richly deserved the pin; he worked hard and intellectually contributed to everything he was involved with. The picture was taken during a party in Paolo's apartment and well illustrates the general mood in Ann Arbor; we knew how to relax and did not take ourselves too seriously.

**The Hypertension Division logo**

Together with Mariaconsuelo (Consuelo) Valentini, a lively and "can do" Italian fellow, Paolo and I set out to review the literature about sympathetic overactivity, overweight and hypertension. The starting point was my conviction that the often reported decreased cardiovascular beta adrenergic responsiveness might be a generalized response to chronic excessive sympathetic overdrive. Sverre Eric Kjeledsen had already demonstrated that hypertensives respond with a lesser increase of glucose to infusion of the beta adrenergic agonist epinephrine (Chapter 6, Figure 3). Earlier Sverre tied the decreased beta adrenergic responsiveness to sympathetic overactivity; platelet norepinephrine content, a possible indicator of the prevailing sympathetic tone, negatively correlated with beta adrenoreceptor responsiveness in hypertension.[9] This proved two important points; that in hypertension the subnormal beta adrenergic responsiveness is generalized also to metabolic receptors and that this very likely represents a physiologic downregulation of receptors. The phenomenon of "receptor downregulation" is yet another example of a negative feedback; in its balancing act to limit extremes, the body counteracts excessive sympathetic stimulation with decreased receptor responsiveness.

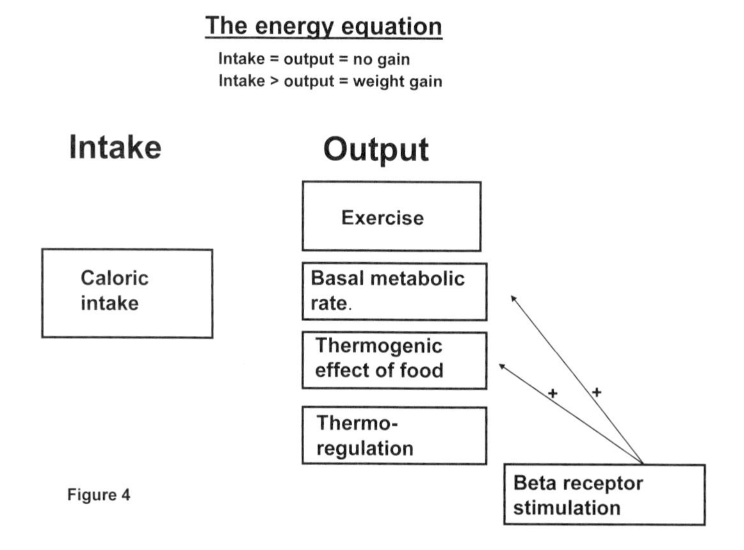


Figure 4

It took only a small leap of imagination from these observations to suggest that a downregulation of metabolic beta adrenergic receptors might favor development of obesity. To follow that reasoning one must first understand the energy equation shown above in Figure 4.

The human body is extremely efficient and it stores excessive calories in the form of fat. This had been helpful to our predecessors who gorged when food was plentiful, then used the accumulated fat as a source of fuel during famines. The urge to overeat has been passed down through the generations. We've got the appetite and rarely can restrain it. To maintain weight the body must precisely match the caloric intake with caloric expenditures, even a small mismatch may over a prolonged time cause a huge increase of weight. If you eat 3500 calories of energy more than you burn, you gain one pound of weight. A moderately active, 180 lb person needs a 3000 calories intake per day to maintain a consistent weight. If that person is in calorie balance and then ingests every week one extra order of French fries (500 calories) over a year he will gain 7.5 pounds. To offset this weight gain that person could walk at a pace of 3.5 miles per hour, two hours a week, and if he prefers a shorter activity he could either run at a pace of 5 miles per hour or leisurely swim one hour a week.

There are two ways to perceive this; that the body is very efficient or that an order of fries will take you a long way. Nevertheless, besides curtailing the food intake, exercising is the only other way to consciously maintain the energy balance. The body can increase the basal metabolic rate or the thermogenic effect of food but these parts of the equation are out of our control and are to a large degree governed by the sympathetic nervous system via beta receptors. And these effects are substantial; in a sedentary person the basal metabolic rate (energy expenditure to maintain basic functions of the body) represents about seventy percent of total daily energy expenditure and the thermogenic effect of food (the caloric cost of processing the food) accounts for another ten percent. Assume now that due to decreased beta receptor sensitivity patients with hypertension are less capable of burning off calories, let's say by 10 percent, and you can see that such patients are destined to gain weight. The evidence that this might happen was very strong and in the abstract of our review paper[10] we wrote the following: "*Both the blood pressure elevation and the gain of weight may reflect a primary increase in sympathetic tone. It is well known that in a milieu of increased sympathetic tone, the beta-adrenergic responsiveness decreases. Sympathetic overactivity and decreased cardiovascular beta-adrenergic responsiveness have been described in hypertension. Beta-adrenergic receptors mediate increases in energy expenditure. If these metabolic receptors were downregulated in hypertension, the ability of hypertensive patients to dissipate calories would decrease and they would gain more weight. The possible relationship of decreased beta-adrenergic responsiveness to weight in hypertension can be experimentally tested. Such research may contribute to an explanation of why patients with hypertension can rarely lose weight."* This time around we finally succeeded to explain our thoughts to the scientific community in plain English. And again in addition to proposing the hypothesis we suggested that it could be tested. So let me now show you some of the evidence that led us to think the hypothesis was reasonable.

As a first step we checked whether we could reproduce from our Tecumseh data an interesting finding reported by the famous Framingham prospective study. It is well known that weight gain is associated with an increase of blood pressure but the Framingham authors have shown that higher baseline blood pressure predicts an excessive future gain of weight. I do not always remember what I read, but that paper left a lasting impression. I liked their idea so much that it became the basis of our review and a motto of its title: "Overweight and Hypertension; a 2-Way Street?"

We utilized previously recorded childhood blood pressure and body stature measurements of people who at an average of 32 years of age participated in the Tecumseh blood pressure study. To assess childhood obesity we used skinfold thickness rather than weight since at that age widely variable growth rates influence the total body weight. We then divided the childhood blood pressure distribution into quintiles and compared the skinfold thickness of the highest and the lowest blood pressure group. Despite a 21 mmHg difference in systolic blood pressure the average skinfold thickness of the two groups was similar (Figure 5). However, at 22 years of age the skinfold thickness of the high childhood blood pressure group was significantly thicker than in the low blood pressure group. The increase of skinfold thickness in the high childhood pressure group exceeded the increase in the low pressure group by 72%. This aspect of the blood pressure-overweight relationship has been roundly ignored; medical literature is replete with reports that gain in weight increases blood pressure.

But the other side of the coin, that higher starting blood pressure results in a larger gain of weight, is even more interesting. We hypothesized that both the early blood pressure increase and the later gain of weight are mediated by sympathetic over activity. When I give a lecture on the topic, I usually at this point get hit with the question: "If the blood pressure is not due to overweight, why then does a decrease of weight lower the blood pressure?" When I don't have time to explain, my quick and admittedly flippant answer is that thiazide diuretics lower the blood pressure but that does not mean hypertension was caused by a lack of thiazides in the blood.

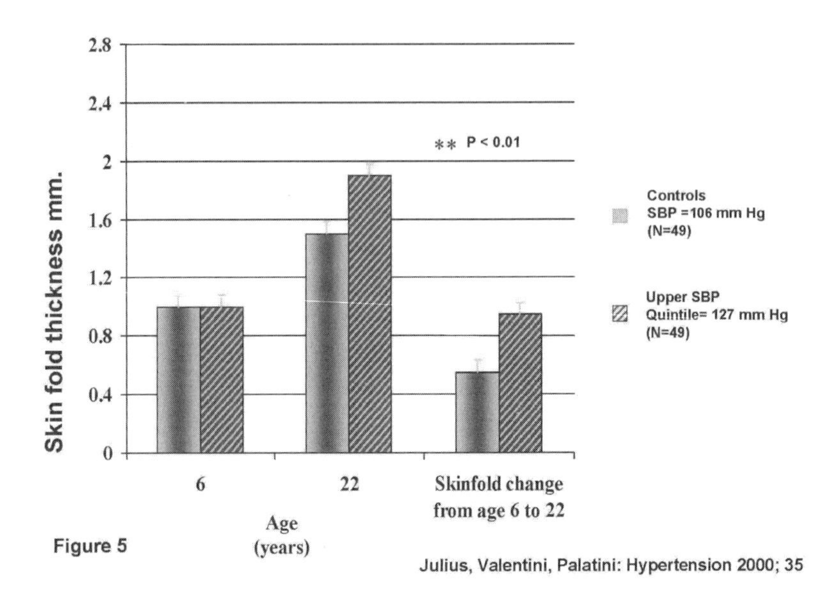


Figure 5

The real answer is that linear thinking "A causes B which in turn causes C" is not applicable to circular events. Circulation is indeed a circle and the starting point of the chain reaction that triggers and accelerates hypertension may be different in different people. But finding out where the process starts may not be as relevant as we think. The circle not only means that the abnormality may start at any point but also that the process can be interrupted at any point of the circle. Yes, if a single cause of a specific type of hypertension is discovered than a cure aiming at removing that cause is indicated. But so far we have not done too well in that respect and it pays to be modest. Incidentally this is a good point for me to reassert that the role of sympathetic over activity is very interesting, but not even in my wildest dreams did I ever think that every patient with hypertension has increased sympathetic tone. In fact, Alan Weder identified in Tecumseh a different phenotype of blood pressure elevation characterized by elevated Li-Na counter-transport in red blood cells[11] and he also demonstrated the linkage of the higher counter-transport with two well-localized genetic markers known to be associated with hypertension.[12] That phenotype did not have any signs of increased sympathetic activity.

A downregulation of beta adrenergic receptors was the center piece of our scheme to tie sympathetic overactivity with gain of weight in hypertension and another analysis of Tecumseh data indirectly supported the hypothesis.[13]

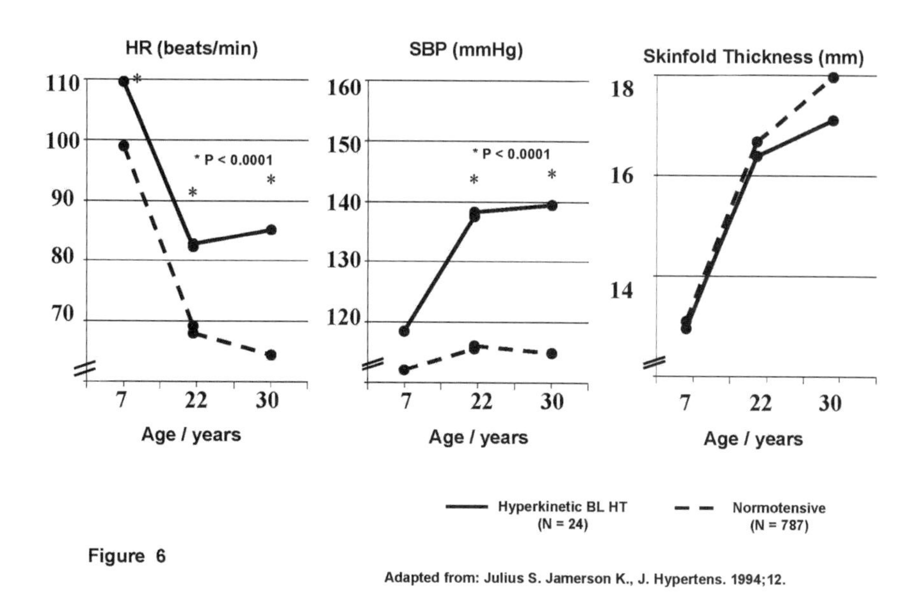


Figure 6

Adapted from: Julius S. Jamerson K., J. Hypertens. 1994;12.

Figure 6 above shows the hemodynamic and skinfold history of subjects with "pure hyperkinetic" borderline hypertension defined by mixture analysis of cardiac index values. By "pure" I mean that we eliminated from the analysis subjects classified by another mixture analysis as having increased Li-Na countertransport. Note in Figure 6 (left panel) that, compared to a large group of normotensive individuals, hyperkinetics (selected on the basis of a high cardiac index at 30 years of age) had fast heart rates as children, young adults and at the time when the classification was made. They developed systolic borderline hypertension at 20 years of age (middle panel) but the right panel shows they never were overweight! At 30 years of age, the average weight of normotensives was 74 Kg and in hyperkinetic borderline hypertensives it was 75 Kg. Interestingly, insulin values were significantly elevated in these lean hyperkinetic adults (18 microunits/ mL versus 12 microunits/ mL in normotensives, $p<.001$).

To us this was pretty convincing evidence that despite persistent sympathetic overactivity the cardiac beta adrenergic receptor did not down regulate in hyperkinetic hypertension. If this were also the case for metabolic beta receptors involved in the regulation of the basal metabolic rate and postprandial thermogenesis, hyperkinetic hypertensives would burn off excess calories in a normal fashion and would not gain weight. So these findings were, as we like to say, "consistent" with our general hypothesis. To prove the hypothesis we had to directly assess whether metabolic responses to beta adrenergic stimulation were suppressed in obese patients with hypertension.

In these experiments we utilized isoproterenol, a very potent and selective stimulant (agonist) of beta adrenergic receptors. It took three different morning sessions scheduled at least one week apart within a 6 weeks period to complete the protocol. In the first session we explained the protocol in detail. If subjects agreed to participate, we performed a physical exam, drew fasting blood samples and gave them bottles to collect 24 hour urine samples for measurement of norepinephrine. During the second visit, we quickly injected through an intravenous catheter increasing doses of isoproterenol at ten minute intervals and measured the heart rate response by electrocardiogram. Finally during the third session the subjects breathed through a ventilated hood into a calorimetry machine to measure oxygen consumption and CO2 production for calculation of energy expenditure. We slowly infused isoproterenol and increased the infusion dose in 30 minute intervals. Calorimetric measurements were taken at baseline and at the end of each 30 minutes infusion period.

You might wonder why we needed to inject isoproterenol two different ways on two different days. The first quick injection served to test the responsiveness of cardiac beta adrenergic receptors. On the second occasion, prolonged infusions increased the energy expenditure which increases the heart rate and cardiac output to subserve increased metabolic needs. Changes in heart rate under this condition would reflect differences in energy expenditure and would not be a specific test of cardiac beta receptor sensitivity.

Consuelo was excellent at networking and she involved in the execution of the protocol two junior division staff members. Both John Bisognano and Robert Brook went through the Ann Arbor internal medicine residency program and both took fellowships in hypertension.

Robert Brook trained further at Northwestern University Hospital in Chicago and returned to Ann Arbor where he is Assistant Professor of medicine in the division of cardiovascular medicine. His clinical focus is on dyslipidemia, but Robert has an inquisitive mind and he acquired an encyclopedic knowledge of all aspect of hypertension and hemodynamics. Presently he is studying the mechanism of the blood pressure increasing effect of air pollution. After a year of fellowship in hypertension, John Bisognano completed training in Cardiology at the University of Colorado in Denver, and then returned to the hypertension division as an Assistant Professor. This pleasant and dynamic man and master of dry humor eventually left Ann Arbor and is now Associate Professor of Medicine in the University of Rochester School of Medicine and director of cardiac rehabilitation and clinical preventive programs in the Strong Health Center. We keep in touch and I never miss the chance of chatting with him.

The isoproterenol infusion protocol was lengthy and the recruitment was slow. Eventually we completed the study in 25 normotensive and only in 13 hypertensive subjects but the results were strong and our manuscript was accepted for publication.[14]

Patients with hypertension responded to swift injection of isoproterenol with a suppressed increase of the heart rate. This was the third time we found an impaired heart rate response to beta adrenergic stimulation in people with higher BP values. In the first report in 1975,[5] an infusion of the same fixed dose of isoproterenol induced a significantly smaller increase of heart rate in borderline hypertensives than in control subjects. However there were legitimate objections to our design; we injected the same dose in everybody but the borderline hypertensives were much heavier. It was conceivable that if the doses were adjusted for body weight the result would be different. Twenty-two years later Kjeldsen infused epinephrine in doses adjusted to the body weight[6] and again the heart rate response was suppressed in the hypertension group. Eight years after Kjeldsen we injected isoproterenol doses adjusted for body surface area. As Figure 7 shows the heart rate increased less in hypertensives at all doses and at higher concentrations of isoproterenol the difference was statistically significant. Such a remarkable reproducibility comes close to declaring a physiological "law" that the same dose of a beta adrenergic agonist will always elicit a smaller than normal heart rate increase in patients with hypertension. But what does it mean? Hypertensives have higher resting heart rates.

Could resting tachycardia limit the heart's ability to further increase the number of heart beats?

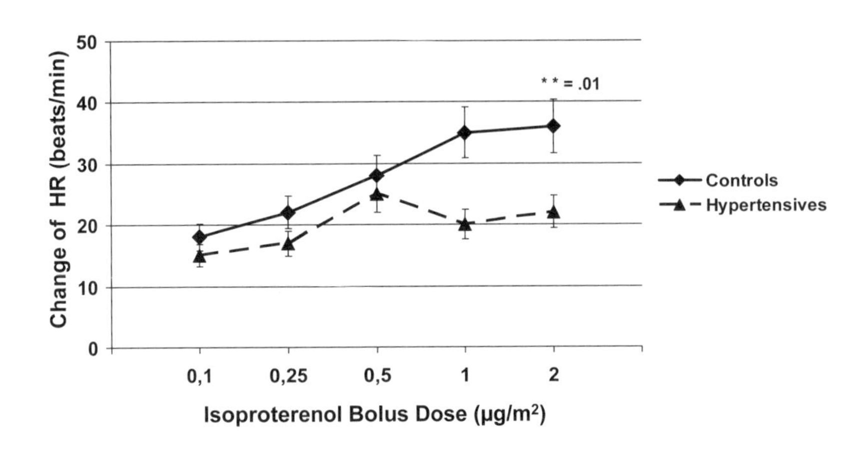


Figure 7

Valentini et al J. Hypertension 2004;22,

We did not think so in part because the new study revealed other abnormalities in the circulation. Figure 8 shows the blood pressure responses to the highest infused dose of isoproterenol at day 3. The systolic blood pressure and pulse pressure (difference between the systolic and diastolic pressure) decreased less in the hypertension group. Apparently in hypertensives the beta adrenergic stimulation elicited a diminished ejection force, less blood was squeezed with a lesser force into circulation which caused a lesser distention of the arterial tree.

Okay, I am trying to create some tension here and you are supposed to pant, hold on to your chair, and be barely able to restrain yourself from taking a peek at the next graph with results of energy expenditure measurements. But then again, you might be just plain bored with this overdose of science. So let's move on.

Lo and behold everything came out as anticipated. (Figure 9) Patients with hypertension utilized less oxygen during the continuous

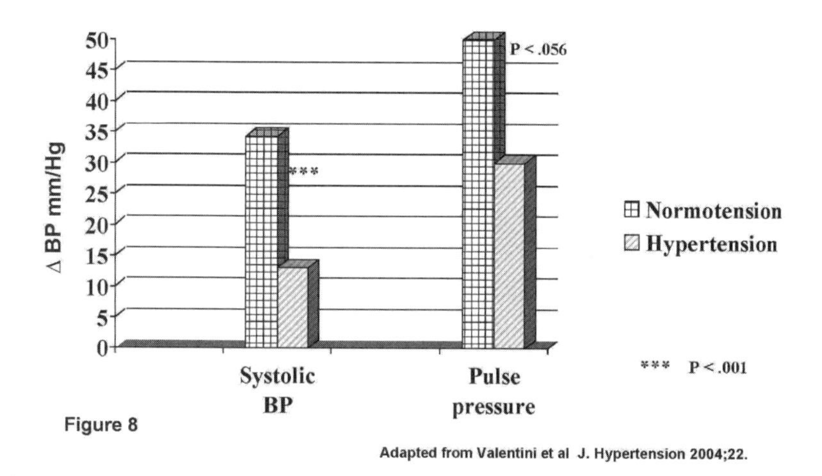


Figure 8

Adapted from Valentini et al J. Hypertension 2004;22.

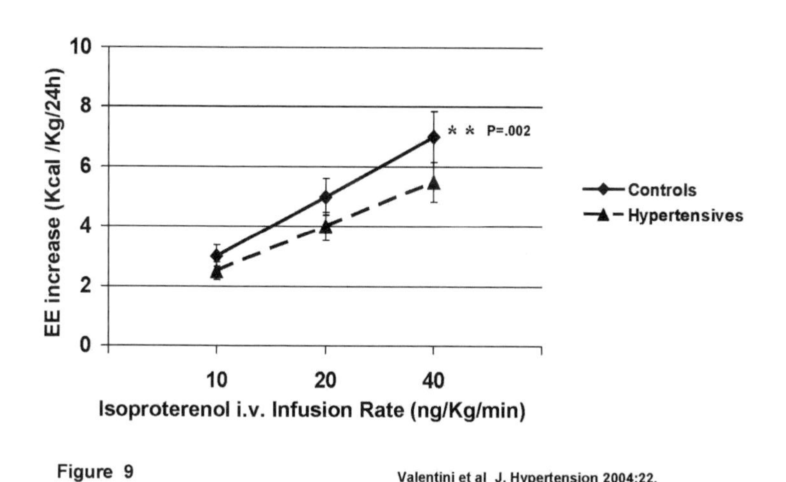


Figure 9

Valentini et al J. Hypertension 2004;22.

isoproterenol infusion and their energy expenditure was suppressed. The tendency for a lesser increase in energy expenditure was seen already at the first infusion dose and the difference between the two curves widened so that at the highest dose the difference became statistically significant. Statistical criteria for declaring significance are very stringent when small groups are compared and despite a meager number of patients in this experiment the "p" value was 0.02. This means that there was only a 2% probability that this result was obtained by chance. Or to put it differently; if 100 well designed studies investigated the same topic, 98 would confirm the finding and two might not. Not bad! Furthermore, it was reassuring that the trend for lesser response in hypertension was seen already at lower isoproterenol doses. Remember that after 30 minutes of infusion we cranked up the dose to the higher level. Metabolic changes are slow and it is conceivable that after a longer lasting infusion the difference in energy expenditure would be evident also at lower doses of isoproterenol.

Our findings were only a "proof of principle;" there was sufficient evidence of a decrease of energy expenditure in patients with hypertension. Despite precise measurement what is shown in the graph must be considered as a qualitative difference. A whole slew of new studies would be required to assess whether this difference is quantitatively sufficient to account for the increase of weight in hypertension. One would like to know how frequently patients and controls were exposed to increased stimulation and whether the degree of stimulation we created in the laboratory mirrored variations seen in real life. This might be difficult to do but there is a simpler way to prove or disprove the point. One could test whether the energy expenditure response to sympathetic stimulation predicts the subsequent increase of weight.

I am not saying this pro forma, I sincerely believe that one finding begets another exploration and wish I could pursue the issue further. In the past I would have designed the next study to clarify the outstanding issues but I will not pursue the topic further. I've been drawn in a different research direction and at some point I have to acknowledge the geriatric reality.

To my great delight, the isoproterenol infusion study closed another gap in our hypothesis. We always assumed that decreased beta adrenergic responsiveness is a consequence of increased sympathetic tone in hypertension. It made sense, molecular biologists have shown details of the process of receptor downregulation; upon excessive stimulation the receptors actually move from the surface to the inner part of the cell where they get chewed up, and the total number of receptors available to respond to the stimulation decreases. This molecular process is also physiologically plausible as the body constantly strives to prevent excesses by various negative feedback mechanisms. But we had no data to convincingly connect the high sympathetic tone with the decrease of beta receptor responsiveness in hypertension. Now take a look at Figure 10 to understand my enthusiasm.

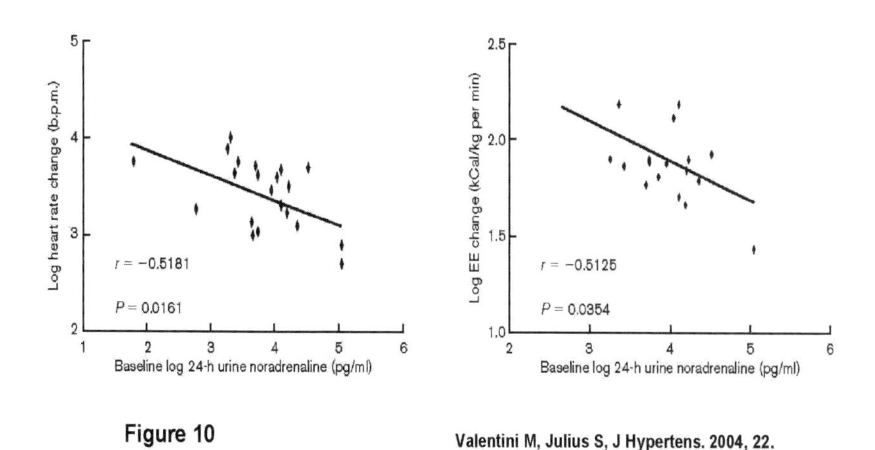


**Figure 10** **Valentini M, Julius S, J Hypertens. 2004, 22.**

The graph shows the relationship between the 24 hour excretion of norepinephrine in the urine and the heart rate response to the highest dose of quick (bolus) injection of isoproterenol (left panel) and the relationship of urinary norepinephrine to the energy expenditure at the highest dose of isoproterenol infusion (right panel) in the entire study population. Both relationships were negative; the higher was norepinephrine in the 24 hour urine collection, the lower were the heart rate and the energy responses to isoproterenol. Thus, the higher the average sympathetic tone over 24 hours, the lower the beta receptor responsiveness. These relationships were statistically significant and reasonably robust but that does not tell the whole story. It is fascinating that we collected samples for urinary norepinephrine at the entry into the study and tested the responsiveness to isoproterenol weeks later on two different occasions separated by at least a week. Plasma norepinephrine which we drew on the day of the procedure also negatively correlated with heart rate and with energy expenditure responses. Plasma norepinephrine oscillates a great deal, two samples drawn in the same individual ten minutes apart, can differ one from another by 50%. Nevertheless, correlation of plasma norepinephrine to heart rate response was highly significant (p=0.004 – a probability of four in one thousand!). The correlation between plasma norepinephrine and energy expenditure was also negative but did not reach statistical significance (p = 0.13). Only a heck of a strong relationship can be detected in samples taken weeks apart. And the fact that the negative relationship prevailed with two different methods for assessment of sympathetic activity is very reassuring. I think we proved the point!

There is yet another facet to this story. Current approaches to the management of the ever-increasing epidemics of obesity have failed. Most appetite suppressing drugs have toxic side effects and none are very effective. Well-designed studies in which teams of physicians, nurses and dietitians persistently gave life style change advice to participants proved that obese patients can lose weight. But in longer studies after the initial success the weight started to increase again. In clinical practice it is nearly impossible to manage overweight. I remember individual successes, the exceptional patient that turned his (hers) life around, but most memories are about the two way embarrassment of interacting with patients who failed to lose weight. They understood my advice, tried their best and were unsuccessful. I was supposed to urge them to do better next time but deep down I knew it was to no avail. The longer you know the patient the harder it gets; after all these are adults and they ask for help, not for a lecture on unobtainable virtues.

We now know that weight gain in hypertension is a physiologic imperative and we understand the mechanism. I hope a solution to the problem is not far away. Many modern drugs are rooted in the understanding of underlying mechanisms. So far all drugs that decrease the sympathetic tone have unpleasant side effects and we need better ones. This might be too much to ask; high sympathetic tone and increased alertness are closely associated and the other side of that coin is that decreasing the sympathetic tone causes- somnolence. And you don't wish to sleep through most of your time on this planet. It is more realistic to expect that molecular biologists will come up with some method of cajoling the beta adrenergic receptor not to down-regulate.

I hope by now you detected how much fun it was to first identify the problem in the Tecumseh field study and then return to the laboratory to study the underlying mechanism. I worked with great guys, and everything was well planned. Protocols were precisely executed and, of course, it helps that the studies were positive. Both the insulin resistance and the obesity studies were the first of a kind and as a friend of mine used to say: "Nobody ever remembers who the second man to discover America was." We all have big egos and some of us admit it.

Moreover, the Tecumseh study and the two latest investigations of pathophysiology made me change my ways. For years I used to pussyfoot around every topic by throwing into a presentation all sorts of habitual scientific disclaimers: "On the one hand this means this and that, but more studies are needed." And I was a grand master of ambiguous adjectives; "this suggests that...," "our findings are potentially..." and so it went. In describing the impact of sympathetic over activity, borderline or white coat hypertension and tachycardia I invariably stated that these conditions "are not innocuous." I particularly loved the "not innocuous" phrase; it just sounded so good! The time has come to be straight and tell the double negative: "these are not 'not innocuous' conditions; these conditions are bad for our patients!

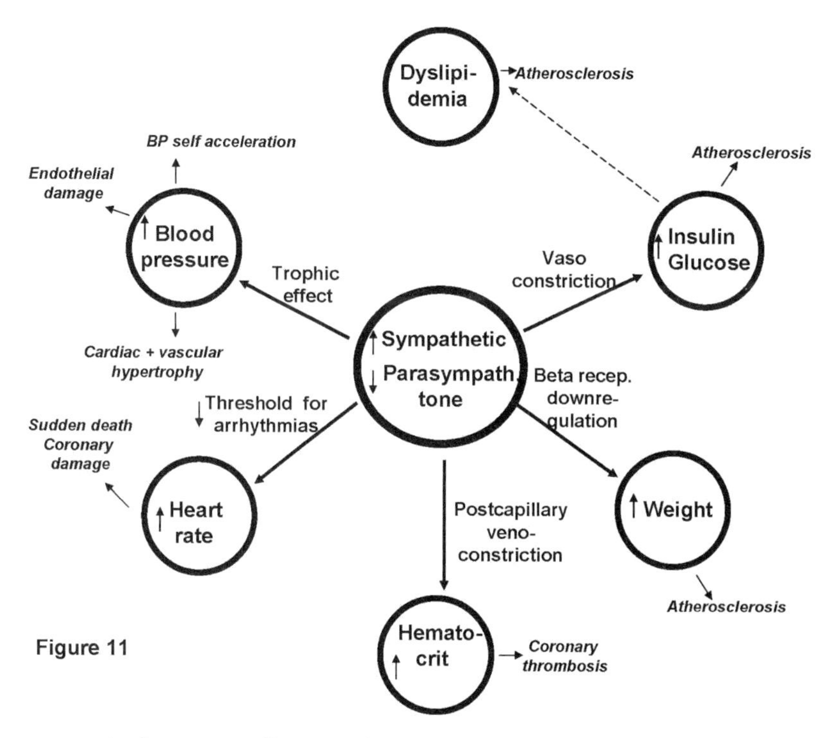

Figure 11

And more studies are always needed, but since it is unlikely that I will do them I had better tell you what this all adds up to. In Figure 11 below, I placed the abnormal autonomic tone squarely in the middle of all bad things and that is where it should be. However, when I say that the autonomic imbalance "causes" various other abnormalities in the graph, I am not stating that this is the primary problem. I neither know nor need to know whether the imbalance preceded the insulin resistance or vice versa. Regardless how it started, the basic message is that the autonomic abnormality is bad for the body, that we know the mechanism by which it negatively affects other abnormalities and that these additional abnormalities further damage the body.

The central circle in Figure 11 shows the autonomic imbalance; sympathetic tone is increased and the parasympathetic tone is decreased. To our knowledge the decreased parasympathetic tone is relevant only for cardiac arrhythmias and sudden death; in all other abnormalities the increased sympathetic tone is the major culprit.

The figure is pretty self-explanatory. Arrows show the mechanism by which the autonomic imbalance causes hemodynamic and metabolic abnormalities. Next to the circles (in italics), I show the consequences of each abnormality. Quite a horror story and I am not making this up; there is solid evidence in the literature that the intermediate mechanisms shown in circles are strongly associated with diverse negative outcomes.

I hope you agree with me that there is no such a thing as "just a bit of nervousness" and "just a touch of high blood pressure." If they are left unattended both conditions can shorten your life span. The Emperor has no clothes and time has come to say so.

**Bibliography**

1. Julius S, Gudbrandsson T, Jamerson K, Shahab ST, Andersson O: Hypothesis. The hemodynamic link between insulin resistance and hypertension. J Hypertension 9(11):983-986, 1991.

2. Julius S, Gudbrandsson T, Jamerson K, Andersson O: The interconnection between sympathetics, microcirculation and insulin resistance in hypertension. Blood Pressure 1: 9-19, 1992.

3. Jamerson KA, Julius S, Gudbrandsson T, Andersson O, Brant DO: Reflex sympathetic activation induces acute insulin resistance in the human forearm. Hypertension 21, 5:618-623, 1993.

4. 4. Jamerson K, Smith SD, Amerena JV, Grant E, Julius S: Vasoconstriction with norepinephrine causes less forearm insulin resistance than a reflex sympathetic vasoconstriction. Hypertension 23(6):1006-1011, 1994.

5. Julius S, Randall OS, Esler MD, Kashima T, Ellis CN, Bennett J: Altered cardiac responsiveness and regulation in the normal cardiac output type of borderline hypertension. Circ Res 36-37 (Suppl. I):I-199-I-207, 1975.

6. Kjeldsen SE, Moan A, Petrin J, Weder AB, Julius S: Effects of increased arterial epinephrine on insulin, glucose and phosphate. Blood Pressure 5:25-31, 1996.

7. Julius S, Schork MA: Borderline hypertension - a critical review. J Chronic Dis 23:723-754, 1971.

8. Julius S and Schork MA: "Predictors of hypertension", in: HM Perry, Jr. and WM Smith (eds.), Mild Hypertension: To Treat or Not to Treat. Annals of the New York Academy of Sciences, New York Academy of Sciences Vol. 304, pp. 38-52, 1978.

9. Kjeldsen SE, Zweifler AJ, Petrin J, Weder AB, Julius S: Sympathetic nervous system involvement in essential hypertension: increased platelet noradrenaline coincides with decreased ß-adrenoreceptor responsiveness, Blood Pressure 3:164-171, 1994.

10. Julius S, Valentini M, Palatini P: Overweight and hypertension. A 2-way street? Hypertension 35:807-813, 2000.

11. Weder AB, Schork NJ, Krause L, Julius S: Red blood cell lithium-sodium countertransport in the Tecumseh blood pressure study. Hypertension 17:652-660, 1991.

12. Weder AB, Schork NJ, Julius S: Linkage of MN locus and erythrocyte lithium-sodium countertransport in Tecumseh, Michigan. Hypertension 17:977-981, 1991.

13. Julius S, Jamerson K: Sympathetics, insulin resistance and coronary risk in hypertension: the 'chicken and egg' question. J Hypertens 12:495-502, 1994.

14. Valentini M, Julius S, Palatini P, Brook RD, Bard RL, Bisognano JD, Kaciroti N: Attenuation of haemodynamic, metabolic and energy expenditure responses to isoproterenol in patients with hypertension. J Hypertens. 22:1999-2006, July 2004.

# The TROPHY Trial: A Dream Come True 2000 - 2007

After twenty five years of service as the division Chief, in 1999 I handed over the leadership of the Hypertension Division to Alan Weder. I got rid of administrative responsibilities hoping to get more free time but this did not materialize. I got involved in large multi-center trials and my work load actually increased. It started with my old buddy, Len Hansson, who had already organized a number of studies in Scandinavia and was about to mount the large international HOT trial.[1]

Hypertension is a marvelous topic for investigation in as much as the blood pressure relates in a linear fashion to negative outcomes and that both the pressure and, unfortunately, outcomes are easy to measure. Worldwide multiple epidemiological studies proved that the higher the pressure, the more frequent are strokes, heart attacks, heart failure, kidney failure and a whole host of other disasters. That relationship is quite precise; you can tell the percentage increase of strokes or heart attacks for each millimeter of increase in blood pressure. And the outcomes are solid; it is darn easy to tell whether a patient is alive or dead and whether he had a stroke or heart attack.

Compare this to some other disease, let's say arthritis. The symptoms: pain, swelling of a joint, and decreased joint mobility, are difficult to assess and it is almost impossible to compare symptoms in different individuals. We can tell precisely how much a pill lowered the blood pressure; they have to deal with subjective reports.

Add to this yet another unique property of hypertension and you will understand why the effect of treatment is better documented in hypertension than in any other field of medicine. Medicine generally prefers etiological over symptomatic treatment; in a case of infection the right antibiotic rids the patient of the invading microorganism and cures him whereas controlling the symptom of fever relieves the discomfort but doesn't affect the outcome. However, in hypertension, treating the symptom of high blood pressure improves the outcome and saves lives.

Mind you, the knowledge that decreasing blood pressure is beneficial is fairly recent. Before the advent of effective antihypertensive drugs, some physicians argued that high blood pressure was needed to overcome the increased vascular resistance and they predicted that decreasing the pressure would impair the perfusion of vital organs. In 1970 when the first Veterans Administration Trial proved that blood pressure reduction dramatically reduces morbidity and mortality in hypertension, things turned around and the skeptics quietly withdrew. Withdrew but did not go away; all along the 50 year old trail of antihypertensive treatment skeptics kept setting up ambushes. Would blood pressure lowering be good in milder forms of hypertension? A series of trials said yes! Would it be helpful to lower blood pressure in elderly individuals? My old colleague Tony Amery organized the "Sys Europe" study and Len Hansson did a study of the elderly in Scandinavia and both said yes. Is it worth lowering blood pressure in isolated systolic hypertension of the elderly? The SHEP trial in the USA said yes.

In the mid 1990's when Len invited me to the HOT (Hypertension Optimal Treatment) trial, a new group of skeptics raised the specter of the "J" curve. They suggested that blood pressure lowering reduces mortality up to a certain low pressure point beyond which a further decrease of pressure actually increases mortality. No problem with that; obviously if you drop the blood pressure to zero the patient will die but the question of how low is too low has not been resolved. Some J curve proponents suggested that in patients with coronary heart disease lowering diastolic blood pressure below 95 mmHg. was harmful, whereas most opinion leaders and all guidelines agreed that that the blood pressure ought to be decreased to below 90 mmHg in all patients. Eventually the guidelines were accepted, practicing physicians became more aggressive and morbidity and mortality in populations steadily decreased. The new J curve controversy threatened to reverse this encouraging trend. Thus it was scientifically very important to determine whether we were too aggressive or the skeptics were too conservative and Len Hansson together with Alberto Zanchetti succeeded in mounting a world-wide large scale trial to resolve the issue. The study was funded by the Swedish pharmaceutical company Astra. I accepted with pleasure the offer to be the USA National Coordinator and serve on the Executive Committee of the HOT study. Patients were randomly assigned to one of the three study arms; diastolic blood pressure goal a) <90, b) <85, and

c) <80 mmHg. Eventually 19,193 patients from 26 countries were enrolled into the study.

I have nothing but good memories of the HOT times. Completing outcome studies is not as easy as it seems. In many aspects such studies are huge towers of Babel; you must assure that all participants regardless of their language and local circumstances "sing from the same sheet" and constant supervision and quality control are needed. Besides complex organizational and operational issues, outcome trials are hobbled yet by another problem. In animal research if one approach doesn't work you are free to change the study design. Not so with longitudinal investigation in humans, at least not yet. When I say that we are "not yet" free to change the design, I refer to the troubling recent trend of altering the definition of endpoints or changing "*statistical power*" in the middle of the trial. Presently we just say "tsk tsk" when something like that happens. But I recently read the prospectus of a company which offers to design for you "feasibility driven" and "flexible" protocols. Should such protocols become the norm, they will open the proverbial Pandora's Box, nobody will know whether to believe the results and we will have another fight on our hands. As thing stood with the HOT trial and still stand now, you had to declare the protocol in advance, indicate what sort of statistics you'd use, and you were stuck with that for years to come. And this is how it should be, but the other side of that coin is that you remain blinded and you might be in for surprises at the end of the study. In HOT, the surprise was that instead of the planned 5 mmHg between group diastolic blood pressure difference, the actual difference was around 2 mmHg. Yet the results proved that lowering the blood pressure below 90 was safe and strongly indicated that a stepwise decrease of the blood pressure elicited a stepwise decline in morbidity and mortality.

Strikingly, all outcome trials report that the worse a patient's prognosis is, the larger the effect of the antihypertensive treatment. This is counterintuitive; a person with a well localized small tumor responds to treatment better than someone with a bad large cancer. The apparent paradox reflects statistical norms; the fewer events occur, the less is it possible to demonstrate that an observed difference is statistically significant. So the groups with a highest incidence of events are always the "winners." And in HOT the best results were achieved in high risk diabetics.

Len's HOT trial had a huge impact on medical practice, the fears that aggressive treatment might be harmful had been alleviated and contemporary treatment guidelines are now particularly aggressive in high risk patients. There is no such thing as full victory in science, I am sure some people are still not convinced, but thanks to Len we nowadays rarely hear about the J curve.

In the next two outcome studies we asked a different question: if blood pressure is lowered to the same degree with two different antihypertensive drugs, would the outcomes be the same or not? As I keep reiterating in this book from the very beginning hypertension is intertwined with other cardiovascular risk factors. Various drugs have a different effect on these associated cardiovascular risk factors. Diuretics increase insulin resistance when taken in higher doses. Beta blockers definitely negatively affect insulin resistance in hypertension. Most vasodilators, including some calcium antagonists, elicit compensatory (counteracting) increases of sympathetic activity. Conversely, drugs that interfere with the renin angiotensin system also decrease the sympathetic activity and also diminish the pro-hypertrophic and endothelium damaging effects of sympathetic and angiotensin over-activity. Furthermore, there were some inklings that these drugs could decrease insulin resistance. So there were good theoretic reasons to expect different results with different antihypertensive agents and comparative studies of two or more drugs started to proliferate.

After giving a lecture in Indianapolis, I was relaxing at the home of my good friend, Myron Weinberger, when his telephone rang; somebody wanted to talk to me! Whenever I am on the road, my wife gets a phone number to call if needed. But Susan never called me; only upon my return would I learn that she navigated alone through all sorts of family mini emergencies. What could it be now? I readied myself for bad news and grabbed the receiver. It was not my wife but two unknown men from the Merck Company. They called to ask whether I'd be willing to join the leadership of the LIFE study. I asked them to send me the protocol. I liked the study design; it took in high risk patients who had left ventricular hypertrophy and randomized them to either treatment with the angiotensin blocking agent (ARB) losartan or the beta blocker atenolol. The basic idea of the LIFE (Losartan Intervention For Endpoint reduction) study[2] was that by blocking the pro hypertrophic properties of angiotensin, losartan might induce a better regression of cardiac hypertrophy and this, in turn, would be more beneficial to patients than simple blood pressure lowering with atenolol. This was a reasonable hypothesis well worth investigating.

The principal mover of the study was Bjorn Dahlof from Goteborg, a trainee of Len Hansson. He developed the protocol in close cooperation with the world class cardiologist, Richard Devereux, from Cornell University in New York. My task was to supervise the USA operation and interact with American investigators.

Bjorn Dahlof and his mentor, Len Hansson, had parted ways and Len was not invited to join the LIFE study in any capacity. There are two sides to each story; Bjorn probably felt he had sufficiently matured to step out of his mentor's shadow and Len was crushed by what he considered as thankless behavior of his former pupil. As I have learned with my own fellows; separation is bound to occur at some time point and it is never pleasant. But this one between two presumably cool Swedes was charged with emotions and I was sorry to see it. And here, dear reader, I will offer a bit of gratuitous advice; in situations like this decide what you wish to do but don't act as an arbiter or a mediator. You are ill equipped to understand the personal pain of both sides and you couldn't possibly judge what is going on. And when it comes to mediation let me pass on a piece of wisdom given to me by Norman Mayer, a Professor of psychology at the University of Michigan and a world expert in industrial conflict resolution.

"I understand your batting average is pretty good. How do you go about resolving a conflict?" I wondered.

'Simple," said Norman, "I first go to the place to assess the situation. If it is resolvable and there is enough good will on both sides, I am ready to arbitrate. But if that is not the case, I stay out of it. Trying to arbitrate in such situations only deepens the wounds."

"And that is why I have a high batting average," said the wise guy.

I accepted the LIFE job and called Len to inform him of my decision. He was hurt but our friendship was too strong not to survive this challenge. A few years earlier Len and I had landed on the opposite sides of another issue, but we leveled one with another and our friendship continued. I had decided to run for the office of President of the International Society of Hypertension. Len was committed to another candidate, told me so and predicted I would lose the contest. I felt I should stick to my guns, and as he predicted I lost. No hard feelings!

LIFE was a large multinational study totaling 9,193 patients seen in 945 research centers in Scandinavia, England and the USA. After seven years, the study was un-blinded and the results were exciting[2]. During the trial blood pressure of the two groups was practically identical and in the losartan arm of the study there was a significant reduction of events as well as a larger degree of reversal of left ventricular mass. So, a) patients with ventricular hypertrophy had more events, b) reversal of left ventricular hypertrophy was more substantial in the losartan group, and c) events were decreased in the losartan group; ergo losartan reduces excessive events by reversing ventricular hypertrophy. Logical, isn't it! Not so fast! Losartan significantly reduced strokes but did not have a salutary effect on cardiac mortality, myocardial infarction and congestive heart failure! So the favorable reduction of the cardiac mass in the losartan group was only a marker of a favorable effect on the distant brain but did not directly affect proximal cardiac parameters.

To me another extremely interesting finding was the substantially lower rate of new onset diabetes in the losartan group. This supported my concept of a hemodynamic component of insulin resistance diabetes. Patients treated with the vaso-constrictive beta blocker had more new onset diabetes than the group that received the vaso-dilating losartan.

Next in line was the VALUE (Valsartan Antihypertensive Long-term Use Evaluation) trial. In the spring of 1997, I was in Barcelona, Spain working in my hotel room on the next day's lecture when the telephone rang. Peter Dumovic from Novartis informed me they had in principle decided to organize a large outcome study comparing their angiotensin blocking agent valsartan to another antihypertensive medication. Could I within a few days prepare an overview of all ongoing outcome trials in hypertension and suggest to which other antihypertensive agent they should compare valsartan? I knew how industry works, and Peter didn't have to apologize for the short deadline. Years may have been lost in internal deliberations but once they reach the decision everything becomes very urgent and presto, preferably yesterday, they need the study protocol. I was scheduled to go from Spain to Croatia and having a few lonely hotel days in front of me, I accepted the challenge. The gist of my 5 page report was that all ongoing trials compared "old" antihypertensive drugs (diuretics and beta blockers) to "new" second generation antihypertensive medications. Doing another variation of these trials would not make sense and the time had come to compare "new" versus "new" drugs. In my opinion, the most interesting and logical would be a comparison of valsartan with a calcium channel blocker.

The principle was accepted and together with a scientific advisory board assembled by Novartis I led the development of the protocol. We worked hard, the protocol was ready for implementation within six weeks, I became the Study Chair, Sverre Eric Kjeldsen, the Associate Study Chair, and the advisory board morphed into the Executive Committee of the study. I enjoyed leading the study. We had an excellent executive committee (HR Brunner, L Hansson, SE Kjeldsen, J Laragh, GT McInnes, MA Schork , M Weber and A Zanchetti), and an outstanding Steering Committee representing national coordinators of 32 countries from five continents around the world. The recruitment was completed within 2 years and 15,245 patients were enrolled in this double blinded trial.

The study compared outcomes in groups treated either with a calcium antagonist (amlodipine) or an angiotensin receptor blocker (valsartan). We enrolled patients at high risk of complications and a stepwise treatment scheme of increasing doses in each study arm to achieve the desired blood pressure goal (<140 and <90 mmHg.). If mono therapy was not successful other drugs to reach the goal were added to the primary drugs.[3]

Day to day management of the study was directed by an Operations Committee of people from Novartis who managed the logistics, along with Sverre Kjeldsen and me. We closely monitored the progress and focused on country by country blood pressure control, regularity of scheduled return visits and the number of patients lost to follow up. Statistics about progress of the whole study, comparative graphs of the status in various countries, as well as comments on specific acute issues were regularly published in a study bulletin. Once a year there was an investigator meeting during which we also held Steering Committee and Executive Committee meetings.

Throughout the study I learned to appreciate Sverre Kjeldsen even more than before. We are very different people and we complement one another. I am forgetful and Sverre has a vise-like memory; I am temperamental, Sverre is calm and methodical; I tend to talk a lot, Sverre speaks up only when it is truly necessary. When I felt I might have over reacted, I took advice from Sverre and when I missed a deadline Sverre reminded me in his unemotional methodical fashion. Sverre also had an incredible grasp of technical details from previous studies, something that my brain unjustifiably but systematically files into the cerebral equivalent of a waste basket. And when it comes to results of other studies and the knowledge of relevant medical literature,

Sverre is top shelf together with only one other person with a similar treasury of facts: Alberto Zanchetti. Add to this Sverre's willingness to take on any responsibility delegated to him, and you can understand how valuable he was to the study. His penchant for work often brought out the best in me; having "dumped" a few jobs on him I'd realize I was getting close to overdoing it and I'd force myself to take a fair share of the work load.

My main role was to be the motivator and cheerleader of the study. I also worked hard on getting input from leaders, seeking consensus on issues and creating a collegial atmosphere. I think I succeeded, VALUE investigators learned to know one another and I still continue to keep in touch with many of them.

Finally after seven years when we broke the study code, the results proved very interesting and important. As usual there was a great deal of pressure to publish the results as soon as possible. We aimed to present the study during the June 2004 meeting of the International Society of Hypertension in Paris, and were left with a mere three months to submit the paper to Lancet, correct it if it was accepted, and hope that Lancet would publish it in time to coincide with the oral presentation in Paris. To facilitate this Sverre and I, while still blinded, developed an outline of the paper fashioned according to the more or less standard form of reporting such large trials. By doing so, we clearly prioritized the statistical analysis. As soon as these analyses became available, a small group of primary writers and statisticians "holed up" in a French hotel near Nice and within three days wrote the first drafts of two papers. The writing group consisted of two primary statisticians (T Hua, AM Schork) and three primary writers (S Julius, SE Kjeldsen, M Weber). The group produced drafts of two papers which were soon presented to the Executive Committee. After a day of intense discussion and a line-by-line correction of the text, the Executive Committee substantially modified the primary paper. Never before have I witnessed such an intense, constructive and goal oriented discussion. Punches were not held back, tempers flared, but the end product was much better than the draft. We stuck to the deadline, the Lancet proved tough but cooperative and the primary[4] and companion[5] papers were published one day before the presentation in Paris.

VALUE showed that in high risk patients it is important to promptly lower blood pressure. Due to the up-titration scheme and doses chosen for comparison, the blood pressure came down quicker in the amlodipine-based group than in the valsartan-based group and it took 6 months for the blood pressure of the two groups to converge. Within the first six months, there was an excess of strokes in the valsartan group but after the blood pressure difference between the two groups decreased the amlodipine group "caught up" and the cumulative stroke events in both groups were equal. In his companion paper using a sophisticated post hoc analysis,[5] Michael Weber elegantly argued that regardless of which compound was used a prompt blood pressure lowering offers added protection to patients.

The other very important message related to the fact that VALUE was a "roll over study;" a huge majority of patients (92%) was previously treated and for safety reasons we decided not to burden them with a "washout" placebo period. Instead we gave them active treatment immediately after they enrolled into the study. Knowing their blood pressure at enrollment, we could look at the blood pressure control rates (percentage with blood pressures <140 and <90 mmHg) with the usual treatment and compare them to rates achieved in the study. The organized and disciplined goal-oriented approach to treatment in both arms of VALUE dramatically increased the blood pressure control rates. The control rates of the systolic blood pressure stood at 22% at the entry into the study. At the end of the study the systolic blood pressure control rate almost tripled to 60%. At the end of the study the rate of control of the diastolic pressure was 90%.

The findings above provided important guidance for clinical practice but in VALUE there were also interesting outcome differences between the two treatment regimens. Towards the end of the study there was a steady increase of heart failure in the amlodipine group but the difference did not reach statistical significance. However, rates of new onset diabetes were clearly and significantly increased in the amlodipine group (16.4% amlodipine and 13.1% valsartan $p < 0.0001$).

As you must by now realize, I was enthused with the new onset diabetes finding. In the LIFE study we found a lower incidence of new onset diabetes in the angiotensin receptor blocker (ARB) than in the beta blocker group. This was compatible with the hemodynamic concept of insulin resistance and diabetes. However, atenolol by itself worsens insulin resistance and it was not clear whether the results reflected only the negative effect of atenolol or also a positive influence of losartan. VALUE resolved the question; a difference was also seen when the angiotensin blocker valsartan was compared to amlodipine, a drug that has no negative effects on the glucose mechanism. But nothing is simple; in VALUE if the blood pressure was not controlled we mandated adding diuretics in two steps and if the blood pressure goal still had not been reached an addition of other drugs was permitted. What if the differences we found reflected a difference in doses of added diuretics or if a mixture of other drugs that have a negative effect on glucose metabolism changed the picture?

At this point I am as darned eager to finish writing this book as you might be keen on completing reading it, but I must yet take another, hopefully short, detour. Most outcome trials investigate the overall efficacy of a treatment. In the simplest words this is sort of an economical input/output equation; if I give X amount of pills to Y number of people, what do I get for it? How many lives will that investment save? Here it does not matter how the pill works and it is unimportant whether the patient took the pill or not; the overall effect is what counts. Why would one want such a crude measure of success? Well, let's assume that a drug is 100% efficacious but has many nasty side effects and only 10 % of patients will agree to keep taking it. For that 10% the success will be phenomenal, but in the general patient population the drug will be a failure. So it has long been agreed that all trials must use the intention to treat (ITT) approach: everybody that signed in for the study counts regardless whether he was in the study for one day or seven years. All patients who later drop out of the study also count. And that is how we analyzed the outcomes in the HOT, LIFE and VALUE trials. However, the ITT approach is not suitable for telling whether two antihypertensive drugs that equally lower blood pressure, but through different mechanisms, could have different effects on some other variable, let's say blood sugar. For that you must analyze results in patients who actually took the two study drugs and did not take any other drugs.

Therefore, I reanalyzed the VALUE data to focus only on patients who were taking only one of the two primary drugs (monotherapy) at the end of the up titration period of 6 months. From then on, the data were analyzed in two ways, a) by the ITT approach counting throughout the entire study every patient who took monotherapy at six months and, b) by "censoring" (removing from the analysis) patients if and when they later started to receive other drugs.[6] Forty six percent of VALUE study participants (N=7080 patients) were on monotherapy at six months (valsartan = 3623, amlodipine = 3817 individuals). Despite the fact that that there were substantially fewer people in the monotherapy analysis (which decreases the chance of proving significance), in the new analysis we still found a robust and significantly lower incidence of new onset diabetes in the valsartan group (p 0.012 censored, p 0.034 ITT) than in the amlodipine group. This answered the question whether previous reports of a difference meant that ARB's positively affect glucose regulation or that beta blockers/diuretics have a negative effect on insulin resistance. Since it has been repeatedly shown that calcium antagonists are metabolically "neutral" (devoid of any influence on insulin resistance), the monotherapy findings proved that ARB's in their own right have a salutary effect on glucose metabolism.

In VALUE, we compared two vasodilators and yet the ARB valsartan group had less new onset diabetes than the group receiving the calcium antagonist amlodipine. How could this fit into the hemodynamic theory of insulin resistance? The answer is that to have a positive effect on glucose metabolism the vasodilatation must take place in skeletal muscles and it must improve the nutritional (capillary) muscle blood flow. Calcium antagonists cause flushing and it is very likely that they shunt the blood away from skeletal muscles. Furthermore, dihydropyridine calcium antagonists cause tachycardia, a reliable sign that they activate a compensatory increase of the sympathetic tone. As I explained earlier, increased sympathetic activity elicits insulin resistance. So, as far as our concept of the association of hypertension with insulin resistance is concerned, the monotherapy VALUE results were good news.

I really loved working on large and long lasting outcome projects. Why? Maybe I can explain this if I give you a verbal auto portrait. I view myself as an incurable people watcher, someone who loves company and who is capable of learning from interactions with others. Long ago in former Yugoslavia understanding how other people functioned was a survival tool, in the USA it became a pleasurable and only marginally useful habit. I am truly happy if I understand the inner workings of a person and succeed in circumnavigating his sensitivities to integrate him into the group. This is utterly crazy. You'd think after what I had to go through during World War II and in postwar communist Yugoslavia I'd become leery of others but I actually like people. In the three large trials described, I met a number of highly intelligent and interesting people be they academicians, scientists from pharmaceutical industry or people responsible for the logistics of the study. This is probably the right place for a comment on the role of industry in scientific studies. Nowadays, academicians working with the

Industry are examined by conflictoscopes operated by outsiders who salivate to uncover ethical impurities. The relationship with the Industry is indeed difficult. If you work for a University you must bring in a grant to secure some free time for outside activity and, presto, you are in an apparent conflict. A person paid by the industry! No doubt there are certain companies that will try to push you in a pre specified direction. Luckily they are rare and pretty transparent. You don't have to worry; if you ask relevant questions they will realize who you are and will not invite you to the next planning meeting. I was involved in three mega trials and can say without reservation that in all of them companies were represented by highly ethical and very knowledgeable scientists. Dag Elmfeldt in HOT, Jonathan Edelman in LIFE, as well as Gillian Pincus and Francis Plat in VALUE were a pleasure to work with and they were uniformly more watchful of ethical issues than the primary investigators.

This was also true for technical personnel assigned to these studies; they were always helpful and more skilful than the investigators in resolving logistic issues. Only in one of the three large trials did we have to interact with an impossible individual. I will call him Iago; an insecure person with overblown ego, a character assassin and an accomplished winner of backstabbing skirmishes in his Company. He was a competent and useful enforcer of study logistics but couldn't limit himself to that role. Despite a woeful lack of scientific knowledge, he held firm opinions about things he could not comprehend; a classical example of active ignorance. Lack of knowledge can be an asset to a person willing to learn but if ignorance resides in a gravely immature and recklessly impolite individual the outcome is disastrous. Like Don Basilio in Rossini's Barber of Seville, he'd plant rumors and if he started to badmouth someone in the Company you knew that person would be removed within weeks. He was shameless; amid a meeting he'd whisper something into a higher up's ear and like Iago he viscerally enjoyed the process. Since his machinations did not affect the scientific leadership of the study team, his outbursts became an entertaining side show. Nothing is entirely bad; I learned that operas and dramas indeed mirror life, and that once in while you can run into a true personification of an Iago or Don Basilio.

One of the nicest features of outcome studies was the chance to meet former Ann Arborites. There were six of us in the HOT study: Hansson, Ibsen, Kjeldsen, Jamerson, Ramirez and Julius. In the LIFE study, I again worked with Sverre Kjeldsen and Hans Ibsen. And in VALUE there was a whole gang of us: John Amerena, Len Hansson,

Ken Jamerson, Sverre Kjeldsen, Silja Majahalme, Tony Schork and myself. We enjoyed meeting and while not being cliquish we'd always find time to chat.

I also had a bit of a hidden agenda in joining outcome trials. I longed to get on with the last item of the 1969 "master plan," a study of pharmacological treatment of borderline hypertension. I hoped that a curriculum vitae showing I held leadership positions in three large outcome trials might become handy when I applied for a grant to complete my dream study. The major concept in designing the 1969 master plan (Figure 1) was my conviction that high blood pressure per se, just because it is high, very early alters the function and structure of cardiovascular organs. I was particularly fascinated by Folkow's concept that structural changes in resistance vessels become "amplifiers" of hypertension and I entered into the graph the dream of early treatment (Figure 1, circle).

If such amplifiers really exist, then early blood pressure lowering to prevent arteriolar restructuring might interrupt the vicious cycle of hypertension. And yes, I was young and optimistic enough to presume that with early treatment we might be able to prevent the later development of hypertension!

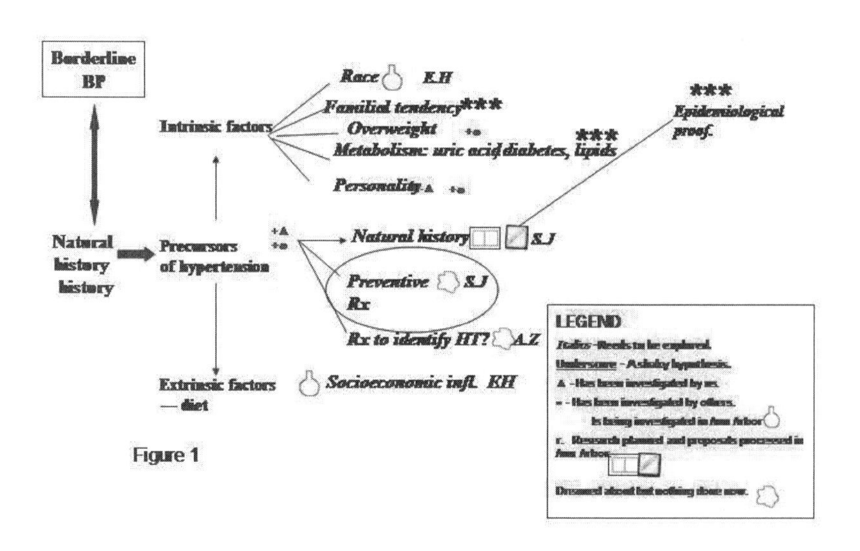


Figure 1

About twenty five years after I stuck the idea into the "master plan," I felt the time was right to seriously consider a "primary prevention" of hypertension study. To use the presently very popular global warming phrase, there was a "climate change" in the field of prehypertension and I decide to write up the protocol for the early treatment study. These days you need a guarantee that the leadership would consist of enthusiastic people who would sincerely care about proper and timely execution of the study. Also people who knew me well enough not to hesitate telling me if something was wrong. In the course of the study Shawna Nesbitt moved to Southwestern Medical School in Dallas, Texas and that transition worked out very well. The overall budget was modest but the Company reserved sufficient resources for the core study function. We held annual meetings with investigators and nurses, issued progress bulletins to research centers, and sent regular newsletters to study participants. Anne Keezer, the wise, energetic and enthusiastic study leader, soon delegated the supervision to Melissa Grozinski, a pleasant person who coordinated all activities with ease and precision. Dr. Eric Michelson, a supporter of TROPHY, was the scientific liaison with the Company. I should better say "Companies" as in the course of the trial, Astra Merck morphed to Astra Ltd and finally to AstraZeneca. Transition periods within companies are usually tense; new leadership arrives, budgets change, projects are abandoned and in the name of downsizing people lose jobs. Surprisingly, nobody touched TROPHY and we never felt endangered. I had no input into selection of clinic centers but somebody did an excellent job; this was the most dedicated group of investigators I ever worked with. They knew TROPHY was different from other trials and were eager to participate in it.

Figure 2 shows the TROPHY design.[10] The main outcome in the study was development of clinical hypertension which was defined as having blood pressure > 140 and/ or 90 mmHg at three different clinic visits. Subjects qualifying for the study were randomly assigned either to the placebo group or to the candesartan treatment group. After randomization, the participants were seen at three month intervals until the end of the study. If a person developed hypertension, antihypertensive treatment was initiated. If they did not develop hypertension, participants in the placebo group continued to receive placebo for four years. The active group received a single daily dose of

16 mg of the angiotensin receptor blocker candesartan for two years and after the second year they were converted to placebo.

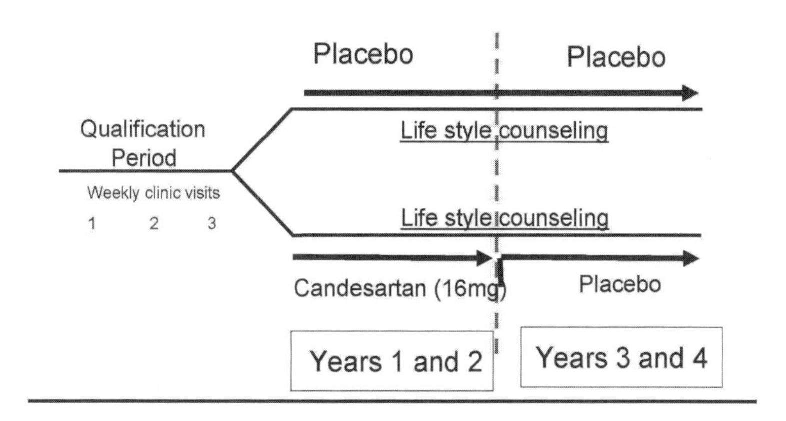


**At Visit 1:** **$\leq$ 159/85-99 mm Hg or 130-159/ $\leq$ 99 mm Hg**

**Avg. of 3 Visits:** **$\leq$ 139/85-89 mm Hg or 130-139/ $\leq$ 89 mm Hg**

**Figure 2** **Julius et al; N Engl J Med 354, 2006**

The study focused on two different but related issues. In the first two years, the question was whether patients with minimal blood pressure elevation could be successfully and safely treated with antihypertensive agents. This was not a forgone conclusion; it was quite possible that the antihypertensive pill would lower the blood pressure too much. Too low a blood pressure could induce unpleasant symptoms and at worst it might cause serious accidents. By the same token it was also possible that candesartan would not affect the blood pressure in prehypertension. It seems illogical but it is also true: the closer you get to normal blood pressure the harder it is to decrease it further. The body is set to protect itself from hypotension and it activates a host of mechanisms to counteract blood pressure lowering.

In the second part of the study when we switched the candesartan group to placebo the question was whether, akin to what was shown in spontaneously hypertensive rats, a short period of treatment would prevent or postpone the development of hypertension.

Nearly 20% of the population has marginal blood pressure elevation but knowing that they are out there doesn't mean you'll find them. People with prehypertension have no symptoms and rarely use medical services. It took us about a year to recruit 772 youngish (mean age 48.5 years) participants (391 in the candesartan and 381 in the placebo groups). In 2006, nearly six years after the study started, we stood ready to un-blind the data and complete analyses.

In the first two years, the blood pressure in the actively treated candesartan group was -10/-4 mmHg lower than in the placebo group, a highly significant between group difference. To our delight, the antihypertensive treatment was very well tolerated. We had a hint that blood pressure lowering with candesartan might have been safe, otherwise, the Data Safety Monitoring Board (DSMB) would have stopped the study. Nevertheless, we were not quite sure and to understand our quandary you need to know how a DSMB works. Experts in a DSMB operate independently from the study leadership and periodically examine adverse events in a blinded fashion. At least that is the idea. In reality, they are given the data as drug A versus drug B and if necessary they can open the envelope with the code. However, all antihypertensive drugs have specific side effects; calcium antagonists cause flushing and swelling of the legs, beta blockers cause sensitivity to cold weather, angiotensin converting enzyme inhibitors cause dry cough. It is pretty easy do decipher the drugs but nevertheless we expect the DSMB to break the code or stop the study only when it is absolutely necessary. This is an excellent rule. Experts tend to form strong opinions about drugs. My professor of infectious diseases in Zagreb had seen a case of bone marrow damage with aminopyrine and swore never to use the drug again. The incidence of this side effect is one in 300,000 cases but the professor could use alternative drugs and it was his right to do so. If he were to sit on a DSMB, he'd stop the study at the first sign of trouble.

Our DSMB generally was unlikely to overreact and could have tolerated a minor imbalance in adverse events without expressing any concerns. But even the smallest difference would be a great deterrent to the use of antihypertensive drugs in prehypertension. Many patients with established hypertension generally feel good and after a while they stop taking pills, despite the fact that it has been proven over and over again that taking drugs saves lives! It is doubtful whether people who know they have only a minor blood pressure elevation would be willing to cope with side effects. However, in TROPHY the four years incidence rate of serious adverse events was 5.9% in the placebo and 3.5% in the candesartan group. We resisted the urge to brag about the lesser percentage of adverse events in the candesartan group, not having more troubles with candesartan was good enough.

Figure 3 shows the cumulative incidence of new onset hypertension in TROPHY. At the two year point there was a 26% absolute risk reduction (AR) of new hypertension in the candesartan group. However, in outcome trials it is customary to calculate also the relative risk reduction (RR) which measures the magnitude of the absolute risk reduction (in our case due to candesartan) compared to the risk in the placebo group. The RR reduction was 66%. Absolute and relative risk reductions of such magnitude are rarely seen in large outcome trials.

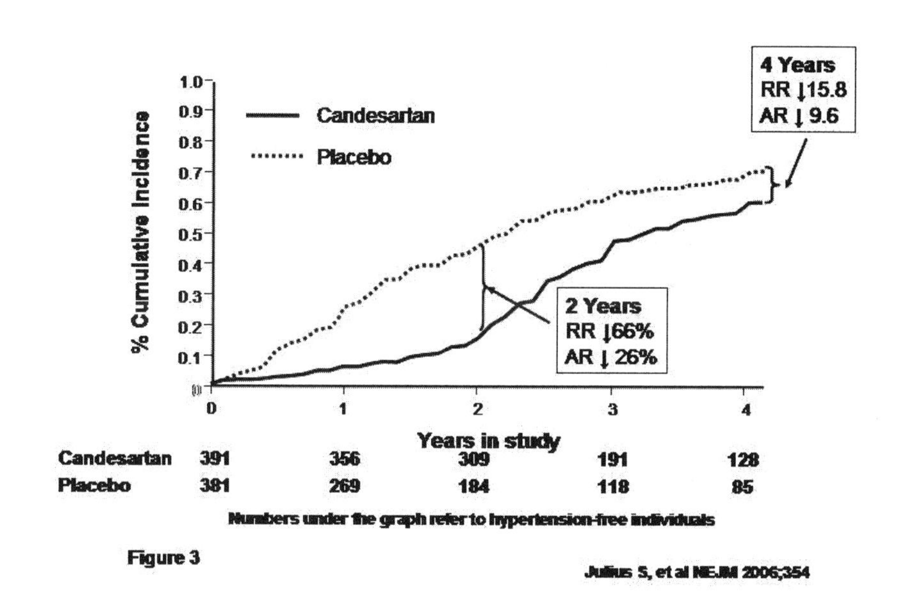


Figure 3

The findings from the second half of the TROPHY study, after we discontinued candesartan treatment, were downright exciting. Initially, one year after we stopped candesartan, there was a substantial increase of new hypertension but thereafter the curve flattened, and at the end of the study the incidence of hypertension in the previously candesartan treated group remained significantly lower than in the placebo group, an absolute difference of 9.6% and relative risk reduction of 15.8%. There is a huge difference between a statistically significant finding and a clinically important result. The design of TROPHY was unique and having no statistical precedents to rely on we declared in advance how we intended to analyze the data.[11] We developed a variety of statistical approaches while clearly indicating that the study was powered to detect at the end of the study a 30% relative risk reduction in the group that initially received treatment with candesartan. In addition we were very careful not to overreach even if the results were as positive as we expected. Here is the quote: "*The TROPHY study seeks only the proof of principle that early pharmacological treatment of prehypertension might delay or prevent development of clinical (stage 1) hypertension. If the TROPHY results confirm this hypothesis, further studies would be warranted to investigate how this affects the compliance with treatment and cardiovascular outcomes in hypertension.*"

Well, two years after we discontinued candesartan there was still a clear reduction of hypertension in the candesartan group but to use scientific jargon the finding was not as "robust" as we expected. The TROPHY soon became an instance of the proverbial "glass half empty or half full" situation. (Almost literarily "half" since we expected a 28% reduction and found 15.8%!). Internally we were quite happy and to this day I believe, but cannot prove, that some of the effect reflected a "vascular healing" and that we succeeded to interrupt the arteriolar blood pressure amplifier in hypertension.

I understand the phenomenon of newness; if you publish a confirmation of previous findings, your scientific contribution will be small but nobody will challenge your findings. I call this the "left handed, color blind and overweight Slavic hypertensives" technique; showing that a drug works also in a small hitherto not studied group. I expected some flack to our truly new findings. To prevent the tempest we reiterated in the title of the paper ("Feasibility of Treating Prehypertension with an Angiotensin Receptor Blocker.") that we sought only a proof of the principle that it is possible to treat prehypertension with antihypertensive drugs. In the abstract, the part that everybody reads we concluded: *"Treatment of prehypertension with candesartan appeared to be well tolerated and reduced the risk of incident hypertension during the study period. Thus, treatment of prehypertension appears to be feasible"* Later in the discussion we stated: "*Our study also indicates that the effect of active treatment on delaying the onset of hypertension can extend to up to two years after the discontinuation of treatment. However, the absolute reduction of 9.8 percent in incident hypertension in the study at four years was modest.*

*Although the observations in this study indicate that candesartan may ameliorate blood pressure in persons with prehypertension, we do not advocate treatment of the 25 million people with prehypertension. We are unaware of any ongoing prospective trials in prehypertension, and hope that the present results will stimulate further research. The public health implications of such research are potentially large. Further studies are needed to answer a number of questions.* And after that we enumerated a number of open issues.

The call for new studies was not a scientific cop-out. I sincerely hope we opened the doors for future research in this promising area. Wouldn't it be beautiful if somewhere in the future it were proven that, akin to results in spontaneously hypertensive rats, pharmacological pretreatment might prevent the presently inevitable increase of blood pressure in untreated hypertension?

Our paper was accepted for publication in the prestigious New England Journal of Medicine and the Editors invited an editorial comment to our paper. On the balance this was a negative review but the piece was written in a measured tone. The title: "Pharmacotherapy for Prehypertension- Mission Accomplished?" was a bit needling. We never claimed TROPHY was the end of the story nor did I land on an airplane carrier to proclaim victory. The writer, Dr Heribert Schunkert, was a German cardiologist whom I never met in hypertension circles. However, he raised a valid objection. By our definition, a subject had new onset hypertension if he had three elevated blood pressure readings throughout the trial. We thought this was a reasonable definition and that in real life practicing physicians would initiate treatment under such circumstances. Dr. Schunkert pointed out that, in essence, by suppressing blood pressure with active treatment and then requiring the individual to have three elevated readings in the second half of the study we artificially delayed the onset of hypertension in the previously candesartan treated group. Or to give Schunkert's example; if a subject in a placebo group had two high blood pressure readings in the first half of the study and then had only one additional high reading in the second half, he'd have hypertension. However, a person treated with candesartan, having had suppressed blood pressure readings in the first half of the study, would require at least three return clinic visits before the diagnosis on hypertension could be declared. So we might have artificially delayed the onset of hypertension in the candesartan group. A similar concern about "carry over" values from one half to the other half of the study was raised in the article by Drs. Persel and Baker published in the American Journal of Hypertension (2006 vol. 19).

The rules of the game did not permit changes of the study protocol and we had to stick to the original definition of hypertension in the first paper.[11] However, in response to critiques, we reanalyzed our results using the Seventh Report of the Joint National Committee on Prevention, Detection, Evaluation, and Treatment of High Blood Pressure (JNC 7) definition of hypertension. This definition has been accepted in clinical practice and it specifies that stage 1 hypertension can be declared only in patients who had elevated blood pressure reading at two <u>successive</u> clinic visits. The results of this analysis[12] are shown in Figure 4.

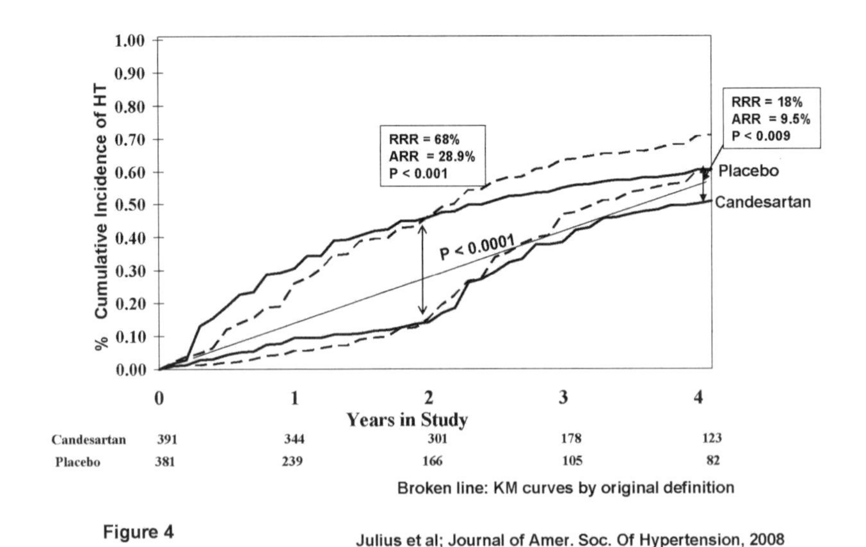


Figure 4

Broken lines show the cumulative incidence of hypertension with the old and solid lines with the new definition. The overall incidence of new hypertension in the reanalysis was 11% lower than in the original paper. Despite this lower incidence of new hypertension both graphs show a very similar trend with the old and new definitions.

With the new definition the absolute and relative risk reduction of hypertension in the candesartan group at two years and at the end of the study was practically identical to the original report and the significance of the difference was also the same with both definitions of hypertension.

This reanalysis addressed the blood pressure measurement "carry over" issue raised after the publication of the first report. Because it is stipulated that hypertension can be declared only if there were two successive high readings none of the non successive repeated blood pressure elevations (even if there were three or more such readings) could be carried over from the first to the second half of the study. More importantly, in this reanalysis the definition was equal to criteria used to initiate treatment with blood pressure lowering drugs in clinical practice. With this in mind, it is of interest to take a look at the median time to development of hypertension, which gives the time frame within which 50% of subjects in each group reach hypertension. This was 2.3 years in the placebo group and 4 years in the candesartan group. Thus using the routine "gold standard" for diagnosing hypertension, fifty percent of previously candesartan treated individuals would not develop hypertension up to two years after discontinuation of the drug. I am saying up to two years since we did not follow the patients longer than two years after discontinuation of candesartan.

This does not answer the question whether the delay of hypertension after discontinuation was due to a positive vascular healing effect of the treatment. However the practical meaning of this finding is undisputable; if you treat a prehypertensive with candesartan for two years and then discontinue treatment he has a 50% chance of not developing hypertension within two years. Regardless whether it reflects vascular healing or not, this finding calls for further research. Here are some questions to investigate: We know that patients dislike life long treatment of hypertension; would treatment followed by a "treatment vacation" improve long term compliance with medication? Would such an intermittent regimen continuously keep the patient's blood pressure in a prehypertension to stage 1 hypertension range and prevent long term acceleration of hypertension? How does this regimen compare to the presently recommended life style modification for management of prehypertension? Finally there is the issue of cost effectiveness. Physicians are not economists and they ought to be committed to improving patient's health regardless of the cost. Nevertheless, doctors do understand the simple principle that if two treatment regimens have exactly the same effect the cheaper one should be used. I would not be surprised if long term blood pressure control were better and the aggregate cost of pills, physician visits, hospital stays and loss of work proved to be lesser with an intermittent treatment regimen.

The current national guidelines recommend life style modification for management of prehypertension. In TROPHY, we vigorously encouraged life style modification in both treatment groups. Life style measures are successful in dedicated individuals but in the majority of cases they fail. We certainly must continue insisting on life style changes but we also ought to be realistic as to what can be achieved with them.

Figure 5 compares the absolute risk reductions in TROPHY and in the dietary intervention Trial of Hypertension Prevention (TOHP).

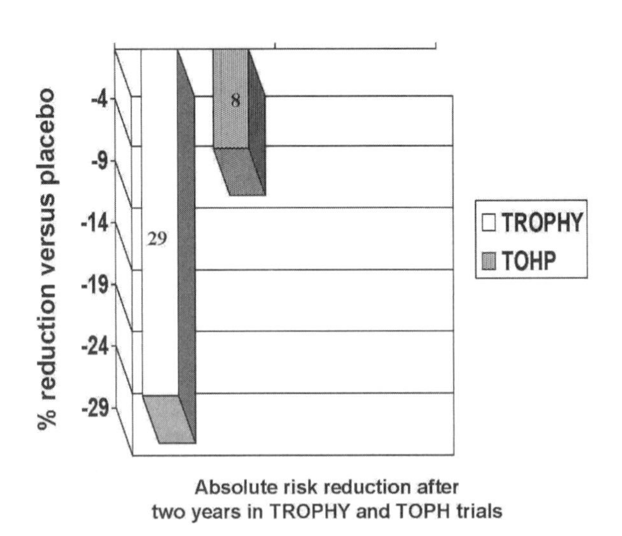


Figure 5

TROPHY results using the JNC 7 definition of hypertension are shown. Clearly the suppression of hypertension with candesartan was much more substantial than with the dietary treatment. In real life this difference is likely to be even larger since throughout the TOHP trial dedicated expert teams continuously monitored individual patients and reinforced the need to comply with life style modification.

TROPHY is a dream come true and I am proud of it. Despite occasionally vigorous critiques, by and large our findings were well received. The horse is out of the barn; new groups of investigators are planning TROPHY- like trials and that is the biggest possible compliment to our work. I am not directly involved but I know of one research proposal to the NIH and of one protocol internally planned by NIH. An imaginative study has been launched in Japan. Many colleagues have approached Industry but unfortunately they still meet with the same resistance I encountered before I came upon the good people in the Astra Merck Company. I hope my colleagues persevere. Prehypertension is an important field and a great opportunity for further research.

There is a need to better define the type of patients who are most likely to benefit from treatment, and to explore which drug regimen might result in better results. If drug treatment works in rats why should we not explore the effect of drugs in humans? In spontaneous hypertension rats there is “a window of opportunity” for prevention of hypertension; early blood pressure lowering ameliorated future hypertension but in “middle aged” rats preventative treatment failed. Let’s hope future research will teach us whether such a window also exists in humans.

Nothing will come to pass if we do not openly hammer in the message that what we are doing now is not sufficient. While we preach restraint the whole nation is steadily gaining weight at an alarming pace and the incidence of prehypertension has slightly increased. We cannot resolve a society-wide problem by blaming the patient. To influence the trend we would have to involve city planners to build remote parking spaces, design more bike lanes, and find ways to facilitate walking. We’d also have to think how to get the right food to people; the DASH diet is very costly and unavailable to a large proportion of the population. Maybe high gasoline prices will remind our teenagers that they have legs! In the meantime we must face the truth that what we presently recommend is not a plan for action but rather an eloquent invitation to failure.

We should not be deterred by occasional admonitions that we are pill mongers bent on helping the industry to sell drugs and that the nation can’t afford the additional cost of mediation. We heard that argument four decades ago when, out of fear that too many people would be treated, the normal age related systolic blood pressure was defined as 100 plus your age. So 70 years old people would not be treated unless their systolic blood pressure exceeded 170 mmHg. New data left that formula in the dust; we now treat elderly people with systolic blood pressure over 140, the national economy is not in shambles and we have saved many lives. Why not consider designing well documented studies in prehypertension and if the future data warrant it routinely start treating prehypertension? We've become realistic in our approach to people who are close to the end of their life cycle - why not aim at younger folks?

Looking back at forty three years of research and clinical work in Ann Arbor I don't recall too many bad days in the office. I am sure this is wrong; like most people I keep forgetting bad things and tend to scan the past through colorful sunglasses. But thanks to my second chairman, Bill Kelley, I can categorically say I never had three difficult days in a row. At some point Bill half seriously needled me about something in the hypertension division. Instead of shrugging it off, as I usually did, I kept thinking about the unpleasant encounter the entire next day. I felt bad but the actual issue did not bother me; upset by not being able to get rid of bad thoughts I became miserable about being miserable. And then I remembered my father's story about the Austro Hungarian Army. An Austrian solider had the right to complain against an officer but the grievance could at the earliest be filed only three days after the actual event! Right then I decided that if I could not cool off in the next two days or if I ever in the future had three unhappy days in row I'd quit. I swear I would have resigned but it was not necessary; throughout Kelley's tenure and through six chairmen after him I remained a happy camper.

By and by my life style has changed for the better. Four years ago I became an active emeritus professor at the University of Michigan which means that I can work but do not get paid for it. I do not feel exploited by the University; I can pay the personnel from my research funds, I need not fill out forms for absences and I can set my own time schedule. Not bad, but there is a hook to it. Every year my status has to be confirmed by the Chairman of Medicine, the Dean and the Regents of the University. In its wisdom the University has granted me the proverbial "second youth;" just as at the very beginning of my career I now have a renewable and presumably less secure position. It is a smart arrangement; though nobody has asked me to do so, I prepare at the year's end a progress report and if I had nothing to show I wouldn't reapply for the next year.

I continue planning. A whole bunch of TROPHY data awaits analysis and frankly until now, a self-imposed deadline to complete this book has stood in the way. I continue international lecturing but have changed the focus. Instead of talking about any topic in hypertension I now prefer invitations to talk about prehypertension, including the role of the sympathetic overactivity and tachycardia as cardiovascular risk factors. To my own surprise I am getting more aggressive about early drug treatment of prehypertension. I chiefly advocate for more studies but sometimes the real Stevo Julius pops up; the one that knows that clinicians ignore prehypertension, that currently recommended life style modification cannot solve the problem and that treating prehypertension is feasible. It has not been proven that such treatment saves lives but in TROPHY candesartan had a placebo-like side effects profile. I wonder why shouldn't we treat prehypertensives? In hypertension we are so evidence bound that lack of proof that treatment decreases mortality is considered as evidence against using treatment in prehypertension. This compunction is not shared by diabetologists who deal with a chronic disease where, similarly to hypertension, the symptom (blood sugar level) is used to diagnose the disease. They constantly decrease the blood sugar level at which they diagnose diabetes and they don't wait for outcome studies to initiate early treatment.

Next on my agenda is initiating cooperative research with colleagues in my old medical school in Zagreb. For a while it was very difficult for me to connect with my old country. For a good ten years after I left Zagreb, I was considered a political refugee and I did not venture going back. When the dust settled I started to visit as a private citizen either alone or with the family. By and by invitations to speak started coming, and I officially visited my old hospital. Just when everything looked great the tragic fratricidal war broke out in former Yugoslavia. After the war everyone was upset and old friends appeared sadly one dimensional; they kept repeating horror stories and wallowed in the sea of injustice. And there was nothing I could do for them. Luckily that understandable phase is just about over and things are nearly back to normal. Two years ago the School of Medicine in Zagreb elected me as Visiting Professor and I promised this would not be a pro forma appointment. I am currently trying to set up in Croatia Tecumseh-like study. Though it is a rather small country, Croatia has widely different regions; people in the Slavonian plains eat fatty food whereas in Dalmatia, particularly on some islands, a Mediterranean life style prevails. By surveying these populations we would establish a good basis for longitudinal observations. Additionally, a cross sectional comparison of the two contrasting groups might offer new insights into the relationship between hypertension, obesity and dyslipidemia. We've already approached a funding agency and at this point things look good.

The last two years I have reduced my office work to four hours a day. This leaves me enough time for non medical things. For those not believing I've learned to slow down, I am appending this picture of myself and Andy Zweifler during a recent celebration in his house. Yup, we have grown older but we still know how to relax!

**Stevo and Andy**

I undertook to visit all surviving principals from my book "Neither Red Nor Dead." Susan and I visited my classmates from the first bench in Marshal Tito's High School in Zagreb: Milovan Domjan in Trogir and Vinko Guberina in Sibenik. Both are retired and despite not having met them for decades we reconnected instantly. Susan and I also met with my "Angel Protector," Dr. Tode Curuvija. In 1954 after my Father's suicide, Tode protected me and when things got too hot he sent me to practice medicine in Bosnia. During the recent fratricidal war it was Tode's turn to leave town under pressure but he took it in stride. In his 90's, Tode remains a lively optimistic and knowledgeable individual. We also keep touch with Dr. Mehmed Mutevelic with whom I practiced medicine during my "exile" in Gorazde. He has visited us in Ann Arbor and I went twice to Sarajevo to spend some quality time with him and his friends.

Other friendships are equally important and Susan and I use every opportunity to keep in touch with our extended family; the former staff and fellows of the hypertension division in Ann Arbor. Writing this book became a good excuse to meet friends; including one chapter written in Wolgang Kiowski's apartment in a chalet in Swiss Alps, and another in Paolo Palatini's beautiful old villa in Vittorio Veneto in Italy.

Forty years in science has been a blast! And I am not done yet!

## Bibliography

1. Hansson L, Zanchetti A, Carruthers SG, Dahlöff B, Elmfeldt D, Julius S, Ménard J, Rahn KH, Wedel H, Westerling S for the HOT Study Group: Effects of intensive blood-pressure lowering and acetylsalicylic acid in patients with hypertension: principal results of the Hypertension Optimal Treatment (HOT) randomised trial. Lancet 351:1755-1762, 1998.

2. Dahlöf B, Devereux RB, Kjeldsen SE, Julius S, Beevers G, de Faire U, Fyhrquist F, Ibsen H, Kristiansson K, Lederballe-Pedersen O, Lindholm LH, Nieminen MS, Omvik P, Oparil S, Wedel H for the LIFE study group: Cardiovascular morbidity and mortality in the Losartan Intervention For Endpoint reduction in hypertension study (LIFE): a randomized trial against atenolol. Lancet 359:995-1003, 2002.

3. Mann J, Julius S for the VALUE Trial Group: The Valsartan Antihypertensive Long-term Use Evaluation (VALUE) trial of cardiovascular events in hypertension. Rationale and design. Blood Pressure 7:176-183, 1998.

4. Julius S, Kjeldsen, SE, Weber M, Brunner HR, Ekman S, Hansson L, Hua T, Laragh J, McInnes GT, Mitchell L, Plat F, Schork MA, Smith B, Zanchetti A, for the VALUE trial group: Outcomes in hypertensive patients at high cardiovascular risk treated with regimens based on valsartan or amlodipine: the VALUE randomized trial. Lancet. 363:2022-31, June 2004.

5. Weber MA, Julius S, Kjeldsen SE, Brunner HR, Ekman S, Hansson L, Hua T, Laragh JH, McInnes GT, Mitchell L, Plat F, Schork MA, Smith B, Zanchetti A: Blood pressure dependent and independent effects of antihypertensive treatment on clinical events in the VALUE Trial. Lancet. 363:2049-51, June 2004.

6. Julius S, Weber MA, Kjeldsen SE, McInnes GT, Zanchetti A, Brunner HR, Laragh J, Schork MA, Hua TA, Amerena J, Balazovjech I, Cassel G, Herczeg B, Koylan N, Magometschnigg D, Majahalme S, Martinez F, Oigman W, Seabra Gomes R, Zhu JR. The valsartan antihypertensive long-term use evaluation (VALUE) trial. Outcomes in patients receiving monotherapy. Hypertens 48:385-391, March 2006.

7. Julius S, Jamerson K, Mejia A, Krause L, Schork N, Jones K: The association of borderline hypertension with target organ changes and higher coronary risk. Tecumseh Blood Pressure Study. JAMA 264:354-358, 1990.

8. Julius S, Mejia A, Jones K, Krause L, Schork N, van de Ven C, Johnson E, Petrin J, Sekkarie MA, Kjeldsen SE, Schmouder R, Gupta R, Ferraro J, Nazzaro P, Weissfeld J: "White coat" versus "sustained" borderline hypertension in Tecumseh, Michigan. Hypertension 16:617-623, 1990.

9. Egan B, Panis R, Hinderliter A, Schork N, Julius S: Mechanism of increased alpha-adrenergic vasoconstriction in human essential hypertension. J Clin Invest 80:812-817, 1987.

10. Julius S, Nesbitt SD, Egan BM, Weber MA, Michelson EL, Kaciroti N, Black HR, Grimm RH Jr, Messerli FH, Oparil S, Schork MA for the Trial of Preventing Hypertension (TROPHY) Study Investigators: Feasibility of treating prehypertension with

an angiotensin-receptor blocker. N Engl J Med 354:1685-97, April 2006.

11. Julius S, Nesbitt S, Egan B, Kaciroti N, Schork MA, Grozinski M, Michelson E, for the TROPHY study group: Trial of preventing hypertension, Design and 2-year progress report. Hypertension. 44:146-151, August 2004.

12. Julius S, Kaciroti N, Egan BM, Nesbitt S, Michelson EL for the Trial of Preventing Hypertension (TROPHY) Investigators. TROPHY Study: Outcomes based on the Seventh Report of the Joint National Committee on Hypertension definition of hypertension. Journal of the American Society of Hypertension 2(1) 39-43, 2008

# Autonomic Nervous System and Circulation; A Sketch of Underlying Physiology

Since this book deals with the autonomic nervous control of the circulation let's first start with the physiology of the control system and later proceed with a description of the circulation.

The autonomic nervous system (ANS) alters the function of various organs independently of our volition. We can consciously order our muscles to contract in a certain way and they will do as told. But we have no control over how much blood the heart should pump into the vessels, how much urine the kidney should produce, and whether our stomach should contract or relax. Whereas this system works independently of our consciousness (therefore "autonomic"), it readily responds to emotions. Sexual arousal, embarrassment, anger, or fear will raise your blood pressure, triple your heart rate and, depending on the circumstances, make your skin pale, flushed or sweaty. Because of this connection to the state of mind, it is believed that our inner disposition can subconsciously change the function of the ANS. Scientists interested in this *psychosomatic* connection suggest that if you were able to reach deep inside yourself, and becalm your mental disarray, you might set your body to function in a normal way. Easier said than done!

The *neurons* of the ANS emanate from a little area at the bottom of the brain *(brain stem).* The system is divided into two opposing but interrelated branches; the *sympathetic and parasympathetic* nerves. They are opposed because each of them works in a different direction. Take the simple example of the number of heart beats per minute. The beats are generated by an automatic center in the *atrium* of the heart (cardiac *pacemaker)* which, if left alone, will make the heart beat at a steady pace of about 110 beats per minute. The sympathetic nerves tell the pacemaker to speed up and the parasympathetic order it to slow down. In humans, the resting heart rate is set around 60 beats per minute chiefly through the inhibitory action of parasympathetic nerves which slow the pacemaker from 110 to 60 beats. But the parasympathetic nerves cannot tell the pacemaker to speed up beyond 110 beats; that is the domain of the sympathetics. Thank the sympathetic nerves for raising your maximal rate to about 190 beats during physical exercise; you need those extra beats to deliver more blood to hard working muscles. Otherwise your exercise capacity would be greatly reduced. Athletes speak about the "second wind," the sudden burst of energy during a prolonged and exhaustive exercise. There it is; the sympathetic activity at its helping best!

The two opposing branches of the ANS are also interrelated. Their relationship is organized in a reciprocal fashion; if one part increases its function the other will decrease it. That, of course, makes perfect sense, otherwise everything would come to a screeching halt. In the case of the cardiac pacemaker, the two opposing systems would reduce the heart rate to a very narrow range. By the way, the reciprocal control of the heart rate also acts as fall back system; a kind of insurance to preserve the cardiac function when one branch fails to do its job. Nowadays we have at our disposal drugs that can entirely block the effect of the sympathetic nerves on the heart. These so called *beta blockers* are used to treat hypertension, heart failure and heart attacks. In patients receiving beta blockers, the heart rate accelerating effect of the sympathetic nervous system is completely abolished. Nevertheless, such patients can increase the heart rate through a withdrawal of the parasympathetic inhibition of the pacemaker. Their exercise capacity is decreased but not seriously limited; they might not be able to run a marathon but they can meet the exercise demands of daily life.

The degree of function of the two branches of the ANS is called *tone* and depends on the frequency of electrical impulses traveling through the nerves. These electrical impulses are not directly transmitted to various organs. Instead, each nerve type releases a different *messenger* - a chemical that tells the organ what to do. Sympathetic nerve endings release *norepinephrine* whereas the parasympathetic branch releases *acetylcholine.* These messengers bind to *receptors* on the cell surface and thereby initiate the response in various organs. Generally, the response of the organ depends on the amount of the messenger released but there is another important "but." The response also depends on the distribution of various receptors in the organ. An organ with a larger density of receptors will respond more than a less endowed organ. Furthermore, the response also depends on the position of the receptor within the cell and whether the receptor is occupied or free to respond. The sum of these factors determines an organ's *"sensitivity"* in responding to the tone. As if this is not complicated enough, there are different receptor types in various organs and these are usually functionally opposed one to another. In arterioles, enhanced sympathetic tone activates both alpha and beta *adrenergic receptors* with very different results; alpha stimulation leads to *constriction* whereas stimulation of beta receptors causes *dilatation* of arterioles. This interaction between functionally opposed adrenergic receptors is just a variation of the tendency to provide balance in the body. We do not quite know how this works, or better said I don't know how it works. Just imagine, if one receptor tells the blood vessels to constrict and another to dilate the end results will be a big fat zero. That is not the way to run a body! Part of the answer might be that norepineprine released at the nerve endings has a greater affinity to bind to alpha than to beta receptors. It is conceivable that beta receptor stimulation occurs only when large amounts of norepinephrine are released. On the other hand beta receptors readily respond to *epinephrine* a hormone released *by adrenal glands.* So now we have two types of receptors and two different hormones to stimulate them. It is all part of a body's balancing act to keep everything in the middle and prevent extremes.

In short, how an organ responds to stimulation will depend on the balance of sympathetic and parasympathetic tone, the sensitivity of receptors, the distribution of various types of receptors within an organ, and how the brain distributes the autonomic tone to various organs.

Let's now talk about the basics of the circulation of the blood in the body.

The heart is a four chambered pump in the middle of the circulation. Its left and right sides are separated but operate in synchrony so that both sides contract and relax at the same time. Each side has a smaller entrance chamber with a thin wall called the *atrium* and a larger chamber with a thick muscular wall called the *ventricle.* The pumping action of the heart is divided into two phases. In *diastole,* heart muscles are relaxed and the ventricles are filling up with blood. In *systole,* the ventricular muscles shorten (contract) and develop pressure to propel the blood from the ventricles into the circulation. A special set of valves points the blood flow in the desired direction. During systole, valves located at the atrio-ventricular border close while the ones on the border to large exit arteries open up. During the diastole, the atrial valves open whereas the valves to the large arteries close. The right ventricle ejects the blood through the pulmonary artery into the lungs, while the left ventricle pumps the blood into the *aorta* from whence it is distributed via a network of *arteries* throughout the entire body. The highest pressure developed during the systole is called *systolic pressure*. The lowest value to which the pressure in blood vessels falls before the next systole occurs, while the heart is filling with blood, is called *diastolic pressure*.

The right ventricle has the easier job. The pulmonary circuit is short and the right ventricle develops about half of the pressure normally generated by the left ventricle. The left ventricle receives the oxygenated blood from the lungs via the left atrium and ejects it through the aorta into arteries. Systolic blood pressure stretches the elastic wall of the aorta and during diastole - akin to releasing the neck of an inflated balloon - the aorta recoils and helps propel the blood towards peripheral tissues. This passive storage of cardiac energy in the arterial wall is very important. When we age and the elasticity of the arteries diminishes, the heart must work much harder. The pressure generated by the left ventricle is pretty much equally distributed through the larger arteries. We call them *conduit arteries* as their major function is to conduct the blood to various parts of the body.

When the blood flow hits the network of *arterioles*, things start happening. Because arterioles have thick walls and small *lumens,* they resist the flow much more than larger vessels. Consequently, most of the energy created by a heart beat is "spent" in arterioles and downstream the blood pressure decreases dramatically. The arterioles are also "traffic cops" of the circulation; the degree to which they constrict or relax determines how much blood will flow to which region. Downstream from arterioles the vascular tree branches further into a network of ever smaller vessels and the blood pressure continues to decrease. Eventually when the blood reaches very small vessels with

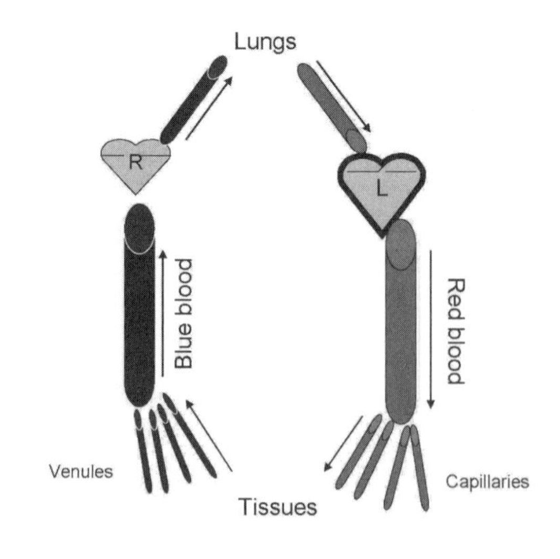


Figure 1

very thin walls (*capillaries*), the pressure falls to just a few millimeters, the vessels are narrow, and their walls are porous. So tapered are the capillaries that the small red blood cells pass through them only in a single file. This is where gas, nutrition, and heat exchange between the blood and tissues takes place. In capillaries, the red blood cells release oxygen to tissues, exchange it for carbon dioxide and the blood turns from red to dark blue. The dark blood drains into small *venules* from whence it returns through a system of ever larger *veins* to the right side of the heart. Remember; in arteries red blood flows from large to small vessels and in veins blue blood flows from small to larger vessels. There is another thing to remember about veins; they have thin walls and are very distensible. Very small changes in the venous pressure cause a large increase in their diameter and they also serve as a reservoir for blood. In our jargon, in contradistinction to large conduit arteries, the big veins are called the *capacitance* vessels to denote their capacity to store the blood. Once the venous blood reaches the right atrium, it flows into the right ventricle which ejects it into the lungs. In the lungs, the red blood cells release carbon dioxide, exchange it for oxygen, and the oxygenated blood returns via the pulmonary vein to the left side of the heart.

The heart can pump more blood either by increasing the number of beats (*heart rate*) or by increasing the volume of single beats (*stroke volume*). Sometimes the two mechanisms combine but there is a catch to this; when the heart rate increases, the period between two heart beats shortens and less time is left to fill the ventricle which, in turn, decreases the stroke volume. The heart can solve this problem by increasing the strength of contraction to squeeze out more blood with each stroke. In our highfalutin jargon, variations in the strength of cardiac ejections are called *inotropy.* Sympathetic stimulation increases cardiac inotropy (positive inotropy). The heart has yet another very neat intrinsic mechanism to increase inotropy. By intrinsic I mean that the heart can increase the force without sympathetic stimulation, and in the absence of other hormones with positive inotropic properties. In the late $19^{th}$ century, a German (Otto Frank) and a British (Ernest Starling) physiologist determined that the force of contraction is proportional to the initial length of the cardiac muscle. The more a cardiac muscle (*myocardium*) is stretched the stronger is its contraction! Whereas this vaguely resembles an inflated rubber balloon, the increased inotropy in response to myocardial stretch is an active process; the cells develop more energy and actively increase the ejection force. Just to convince you how haughty is our nomenclature, let me introduce the word "*chronotropy*" which simply means that the frequency of heart beats can increase or decrease (positive or negative chronotropy). The word, however, is important since sometimes positive chronotropic and inotropic stimuli are intertwined but sometimes they are not. Sympathetic stimulation increases both the frequency of the heart beats and the strength of the cardiac contraction. However, the parasympathetic branch of the ANS can also speed the heart but will do so without noticeably affecting the cardiac strength.

All important functions in the human body are controlled by numerous redundant mechanisms. The human heart is an excellent example of this general physiological principle of multiple controls of a single function. As mentioned earlier, the cardiac ejection force depends very much on the degree to which the myocardial fibers are stretched. Every time we stand up there is a huge change of the distribution of the blood inside distensible veins. In the upright posture the force of gravity pools blood into the legs. With all that blood trapped in the legs, it would make no sense for the sympathetic nervous system to increase cardiac inotropy and chronotropy without securing an adequate venous return to stretch the myocardium. And sure enough, the sympathetic stimulation goes also to the veins, stiffens their walls, decreases their diameter, and shifts the blood from legs towards the heart. A clever solution, if you cannot fill up the container you had better reduce its size. The sympathetic nervous system intervenes in a similar manner during large bleedings or dehydration. Good, but how does the sympathetic nervous system learn that you stood up, that you lost some blood or that you were dehydrated? To explain this we have to say a few words about *stretch receptors* in the circulation. Most of the mechanisms regulate the circulation in a fashion similar to a thermostat. Technically this is called a *negative feedback* (Figure 2).

When the room temperature reaches a high level, the thermostat turns the heat down and vice versa; a low temperature causes the thermostat to increase the heat. Central to this arrangement is a sensor - a device that can detect changes of the regulated variable. To regulate the amount of blood returning to the heart, something somewhere must be able to sense the volume of the blood inside blood vessels.

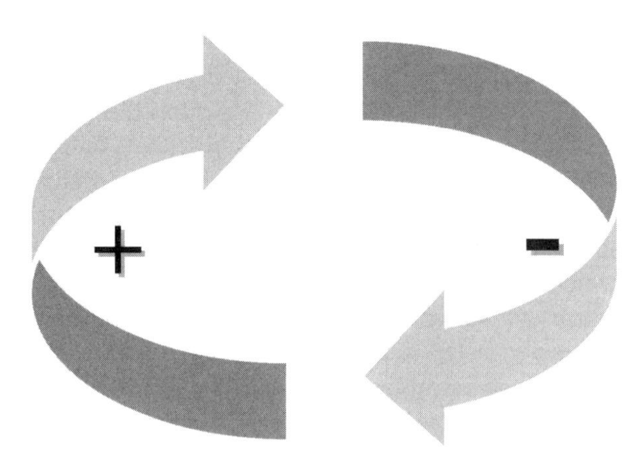


**Figure 2**

If you were an engineer designing such a device, would you put it in the stiff arterial side or in the very distensible venous part of the circulation? Obviously it ought to be on the venous side. And if you were concerned about measuring the volume of blood returning to the heart, you'd put the sensor in the heart. And that is exactly how it works. At the entry to the heart, both in the left and the right atrium there are specialized stretch receptors which sense whether the atrial wall has been more or less distended. In the upright posture or with bleeding, the nerve endings of these atrial cells are less distorted and they will send a signal to the brain to increase the sympathetic outflow to the heart and veins. In reverse situations, such as fluid overload, the system works to decrease the sympathetic tone to the heart and veins. Incidentally, the atrial stretch receptors regulate also the sympathetic tone to the kidneys and to the arterial side of the circulation.

The body has at its disposal also a dual system of stretch receptors to directly regulate the blood pressure level; one located in the arch of the aorta and the other in the left and right carotid artery. Both these *baroreceptors* systems sense how much each beat of the heart distends the arteries. If the blood pressure is excessive the distortion of the aortic stretch-sensitive receptors increases and they send a signal to the brain to decrease the sympathetic tone and thereby lower the blood pressure. Conversely, if the blood pressure is too low the aortic baroreceptors will tell the brain to increase the sympathetic outflow. If left alone, the cardiovascular regulation center in the brain stem would increase the blood pressure to very high levels and both the arterial and cardiopulmonary receptors continuously inhibit the activity of that center. So technically speaking the baroreceptor does not send a signal to the brain to increase the sympathetic tone; rather the baroreceptor decreases its inhibitory influence on the brain. So when we wish to sound scientific, we speak about inhibition and de-inhibition of the cardiovascular center in the brain stem. Things are complicated enough and I will not use the term de-inhibition but if you decide to study baroreceptors further you might run into "inhibition" and "de-inhibition" terms in the scientific literature.

Interestingly, the aortic and carotid baroreceptors appear to operate around different "set points." The aortic baroreceptor in normal people is set to keep the blood pressure around the 120/80 mmHg level. Carotid arteries provide the blood to the brain and their baroreceptors operate around a lower set point than the aortic ones. Apparently the main function of the carotid baroreceptors is to protect the brain from too low blood pressure levels.

These baroreceptor set points can quickly change; it has been shown in experimental hypertension in animals that baroreceptors are reset to higher values within days after hypertension evolves. Instead of fighting the high blood pressure, the baroreceptors become a part of the problem and they soon try to return the blood pressure to the higher value.

This is about the right point to summarize the story about the heart and move on.

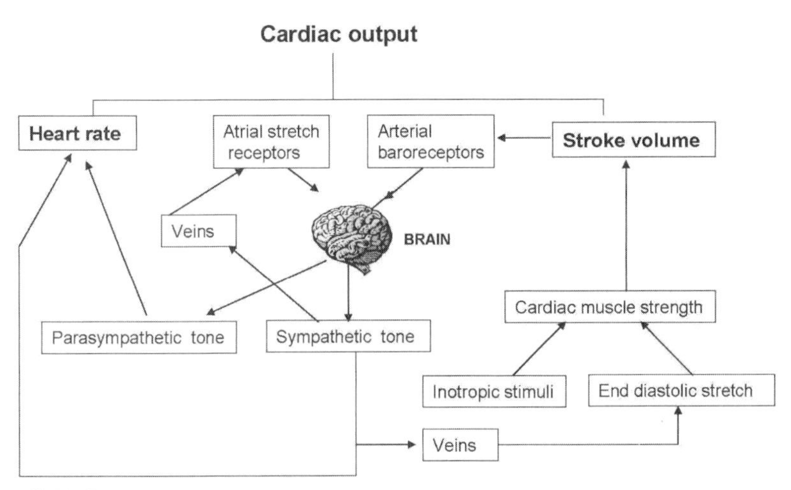


Figure 3

Of course, the above graph is a gross simplification; in reality things are much more complex. For example the graph fails to show that chrono and inotropic response to sympathetic stimulation depend on the sensitivity of cardiac autonomic nervous receptors. The intrinsic amplification of myocardial contractility (strength) does not only depend on the venous return but also on the structure of the heart. The heart stiffens with aging and also in hypertension. In both instances an adequate venous return will elicit a lesser distention of the myocardium (at the end of diastole) and the contractility of the heart will be diminished.

The graph of the cardiac control systems may also leave the wrong impression that the brain is subservient to the atrium; that it waits for cardiac and arterial signals to tell it what to do. Not at all! The brain can override all other controlling systems and it does so quite frequently. A good example of a situation in which the brain will go it alone is the *defense reaction.* The defense reaction, frequently also called the fight or flight reaction, is an emotionally induced circulatory response to anticipated danger. Sensing that the body might need to exercise to cope with imminent danger, the brain increases both the heart rate and the myocardial strength. Driven by the ANS, the cardiac output increases to more than double above the resting state. In parallel, the ANS causes *vasodilation* in skeletal muscles. This siphons off a large amount of blood from other organs to muscles well before the need for exercise (fight or flight) actually arises. During the defense reaction and in many other emotional situations, and for that matter during regular exercise, the brain can override all other systems.

The ANS also tells the arterioles where to send the blood within an organ. When they constrict, the arterioles can shunt the blood away from tissue cells and when they relax (dilate) more blood flows to the cells. This is important to understand since most techniques measure the total blood flow to an organ but give no information about the distribution of the flow within that organ. We believe that some patients with hypertension have high insulin levels because arterioles in their skeletal muscles are constricted in such a way that the blood zooms through the muscle but does not reach the cells on which insulin acts. You'd be surprised how many reputable scientists did not get the point and countered our hypothesis with the argument that in hypertension the total blood flow through the muscles is normal.

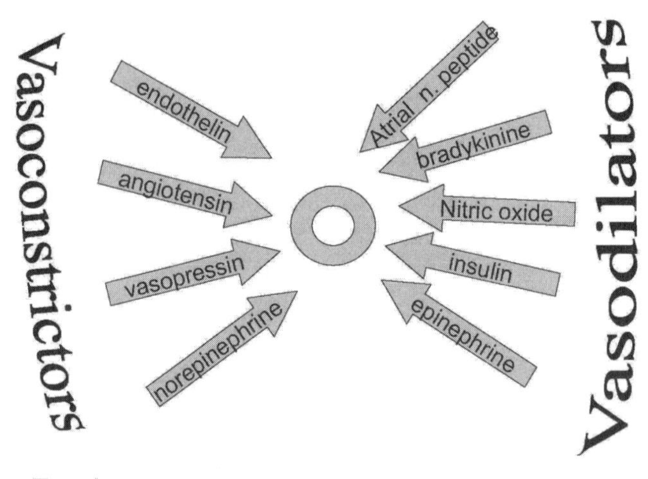


Figure 4

The arterioles are also the major site of vascular resistance. As it flows through large conduit arteries, the blood meets very little resistance but in arterioles the resistance to flow is very high. And at risk of boring you, I will at this point repeat the basic formula of blood pressure: blood pressure = CO x VR. (Blood pressure equals cardiac output time vascular resistance). In other words, the blood pressure can only rise in one of three ways: 1) the cardiac output is high, 2) vascular resistance is increased, and 3) both the output and resistance are high. In established hypertension the vascular resistance is always increased. Since we are dealing with research in hypertension a review of the regulation of vascular resistance is in order (Figure 4).

When it comes to vascular resistance and other important functions, the body doesn't rely on only one mechanism. On the left side of the graph is a partial list of substances that induce arteriolar constriction and thereby increase the vascular resistance; *norepinephrine, angiotensin, vasopressin, endothelin.* Opposed to vasoconstrictors are resistance-lowering vasodilators such as *nitric oxide, bradykinin, atrial natriuretic peptide, epinephrine, and insulin.* Obviously, the human circulation has an enormous arsenal of diverse mechanisms for the regulation of vascular resistance. When it is in the interest of the body to increase the resistance, all pressor substances are organized as fall back systems. Since these constrictors of arterioles use different signaling paths in the cell, if one of them fails to do the trick, the others chip in. Some of these resistance-regulating systems are located on the spot in the inner lining of arterioles (*endothelium*), others are just a bit less close (norepinephrine is secreted from nerve endings at various spots in arterioles) and some are quite distant. The cascade of release of angiotensin, one of the most potent vasoconstrictors, starts in the kidney which secretes an *enzyme* called *renin.* Renin acts on a large protein produced by the liver and cleaves off from that substrate a short polypeptide consisting of 10 amino acids (*angiotensin I*). When angiotensin 1 reaches the lungs, two additional amino acids are cleaved off by the *angiotensin converting enzyme.* The shorter molecule is the potent vasoconstrictor *angiotensin II.* Obviously the response time of this complex system is slower than that of substances released in the arteriolar endothelium.

The cardiac output and vascular resistance are major determinants of the blood pressure and each of them is regulated by multiple and complex mechanisms. It follows that resolving the underlying mechanisms of hypertension will be a difficult task and, more importantly, that it is extremely unlikely that anyone will ever discover a single cause of hypertension. Many controlling mechanisms can go wrong and it is easy to predict that many different subsets of hypertension will be identified in the future. Show me someone that claims he has found the cause of hypertension and I will show you a simpleton.

Looking at the complexity of the system and realizing that each of its components can malfunction, my mentor Bjorn Folkow considered it a minor miracle that hypertension is seen only in 20 percent of the population and wondered why doesn't everybody eventually develop high blood pressure. He said this in jest but he was right; in the eighth decade of life almost everyone has high blood pressure readings.

I chose to study only two aspects of hypertension; the early phase of prehypertension and the role of the ANS. We made considerable progress in understanding that side of the problem but I will never claim that this is all that is to hypertension. Hypertension experts might as well paraphrase the Roman adage that all roads lead to Rome and say that many paths in the body lead to hypertension. So enjoy the book but please keep in mind that I investigated only two aspects of the complex disease of hypertension.

# Technical Vocabulary

**ACEI.** See angiotensin converting enzyme inhibitor

**Acetylcholine.** A neurotransmitter released from parasympathetic nerve endings.

**Adrenal glands.** Small endocrine glands located on the top of both kidneys. Consist of the outer layer called cortex and inner layer called medulla.

Cortex produces mineralo corticoids and gluco corticoids. Mineralo corticoids (aldosterone) cause sodium retention. Gluco corticoids (cortisol and corticosterone) are involved in glucose regulation. They also have anti inflammatory and anti allergic properties.

Medulla secretes the sympathetic neurotransmitter epinephrine also called adrenaline.

**Adrenergic receptors.** These receptors transmit sympathetic signals to cells. They bind with catecholamines (norepinephrine, epinephrine and a number of synthetic substances). Adrenergic receptors are divided into alpha and beta types.

**Agonist.** A substance (natural or synthetic) that binds with a receptor and induces a physiologic response.

**Aldosterone.** The main mineralo corticoid produced by the adrenal cortex. It is released from the cortex by angiotensin. It promotes retention of sodium and bicarbonate, the excretion of potassium and hydrogen ions, and retention of water.

**Anaerobic exercise.** Occurs when the intensity of work exceeds the oxygen supply provided by the circulation. At that point muscles continue to work but they generate metabolic products which later must be oxidized (Oxygen debt).

**Angioplasty.** An intervention to repair a blood vessel either by surgical means or by introduction of a catheter into the vessel.

**Angiotensin.** A hormone produced by renin which is secreted by kidneys. Renin acts on a substrate in the blood (angiotensinogen) and cleaves off the decapeptide angiotensin 1.

**Angiotensin 1.** An inactive peptide consisting of ten aminoacids Angiotensin 1 is converted in the lungs into the active hormone angiotensin 2.

**Angiotensin 2.** An octapeptide released from angiotensin 1by the converting enzyme in the lungs. Angiotensin 2 is a potent vasoconstrictor and releases aldosterone from adrenal glands. The combined effect of direct arteriolar constriction and sodium retention by aldosterone increases the blood pressure.

**Angiotensin converting enzyme.** An enzyme that converts angiotensin 1 into angiotensin 2 when the blood passes through the lungs.

**Angiotensin converting enzyme inhibitors.** Drugs that interfere with the conversion of angiotensin1 into angiotensin 2.Used for treatment of hypertension.

**Angiotensin receptor blockers.** Drugs that inhibit the action of angiotensin by binding to the angiotensin 2 receptors. Used in the treatment of hypertension.

**ANS.** See autonomic nervous system.

**Aorta.** The largest artery in the body. Carries the oxygenated (red) blood from the left ventricle. Branches in the chest and the abdomen to other large arteries.

**Arteries.** Blood vessels carrying the oxygenated (red) blood to various organs. Their walls contain smooth muscle cells which when contracted narrow the arterial lumen (vasoconstriction).

**Arteriography.** A method to radiologically visualize an artery.

**Arterioles.** Small arteries with thick muscular walls. They are the main site of the resistance to flow (vascular resistance).

**Atrial natriuretic peptide.** A hormone released from the atria of the heart. The hormone has two actions a) it increase the excretion of sodium and water by kidneys, b) it dilates the arterioles and arteries.

**Atrio ventricular block.** Occurs when the passage of the signal from the atrial pacemaker to the ventricles is disrupted. Main symptom is a very slow heart rate.

**Atrium (Atria).** Small thin walled chambers at the entry to muscular cardiac ventricles.

**Atropine.** Drug that binds to parasympathetic receptors and blocks the effects of the parasympathetic neurotransmitter acetylcholine.

**Autonomic nervous system.** A system of nerves that controls functions of internal organs and operates without conscious control or sensation. The system has two branches; the sympathetic and parasympathetic nerves.

**Autoregulation.** The intrinsic (local) capability of an organ to maintain a constant blood flow despite changes in perfusion pressure.

**Baroreceptors.** Sensory nerve endings that detect changes in a vessel's diameter. Baroreceptors act as a negative feedback for the autonomic nervous control of control of blood pressure and blood volume.

**Basal metabolic rate.** The resting energy expenditure needed for the basic function of vital organs.

**Beta adrenergic receptors.** Receptors which respond to sympathetic neurotransmitters norepinephrine and epinephrine. **Beta 1 receptors** increase the heart rate and the cardiac output. **Beta 2 receptors** dilate the blood vessels.

**Beta blockers.** Drugs that bind to beta receptors and block the effect of sympathetic neurotransmitters epinephrine. One subtype blocks only cardiac beta 1 receptors and the other subtype antagonizes both beta 1 and beta 2 receptor responses.

**Brachial artery.** A large artery palpable on the anterior aspect of the elbow.

**Brain stem.** The area at the bottom of the brain and adjacent to the spinal cord. This part of the brain regulates the sympathetic and parasympathetic tone.

**Bradykinine.** A potent but short-lived product of an enzymatic interaction in the blood. Causes dilatation of blood vessels and increases capillary permeability.

**Capacitance vessels.** A system of veins capable of holding and storing blood.

**Capillaries.** The smallest branches of the arterial system that pass blood from the arteries into the veins. Capillaries have thin walls which enables exchange of substances between tissues and the blood.

**Cardiac index.** Cardiac output divided by a person's body surface area. Used to compare cardiac output among people of different body size.

**Cardiac output.** The amount of blood (in liters) that the heart expels in one minute.

**Cardiac pacemaker.** Specialized myocardial cells in the right atrium that generate the impulse for the heart rate. The pacemaker rate is increased by sympathetic and decreased by parasympathetic nerves.

**Cardiopulmonary or central blood volume.** The amount of blood in the heart and lungs. Reflects the venous filling of the heart.

**Cephalic vein.** A large superficial vein on the anterior aspect of the elbow.

**Catecholmaines.** Neurotransmitters with similar chemical formulas. Main catecholamines are epinephrine secreted by adrenal glands, norepinephrine released by sympathetic nerve ending and dopamine released by brain cells.

**Chronotrophy.** Describes the timing of cardiac contraction (heart rate); positive chronotropy is an increase and negative chronotropy a decrease of the heart rate.

**Conduit arteries.** Are aorta and other arteries of with a large cross-sectional area.

**Constriction.** The narrowing of the inner diameter of a blood vessel due to contraction of muscles in its wall.

**Control subjects.** A term for normal healthy people that provide a standard of comparison in experimental studies. In drug intervention studies the term control subjects applies to a group of patients which did not receive active treatment.

**Coronary arteries.** Large caliber vessels which provide arterial blood to cardiac muscles.

**Defense reaction.** An autonomic nervous response to anticipated danger consisting of increased blood pressure, heart rate and cardiac output.

**Denervation.** Disruption (by surgical or chemical means) of nervous signals to an area or an organ.

**Diastole.** The period during which the heart relaxes and fills with blood before the next cardiac beat.

**Diastolic blood pressure.** The lowest blood pressure between the two heart beats.

**Diencephalon.** See brain stem.

**Dilation or dilatation.** The widening of the inner diameter of a blood vessel due to relaxation of muscles in the vascular wall.

**Downregulation.** Occurs when a cell or organ has been subjected to prolonged stimulation. Downregulated receptors are less sensitive to stimulation.

**Dye dilution.** A precise method for determination of cardiac output and plasma volume. The dye for plasma volume measurement (Evans blue) is injected into veins and its concentration is assessed in plasma. Indo cyanine green for cardiac output assessment is injected directly in the right atrium. Thereafter the optical density of the blood is measured as the arterial blood passes through a densitometer.

**Dyslipidemia.** A general term for abnormal lipids in the blood. It refers to high values of low density lipoproteins (LDL), cholesterol and triglycerides as well as to low levels of high density cholesterol (HDL).

**Endothelium.** The thin layer of cells that line the interior surface of blood vessels. Endothelial cells secrete potent vasodilating (nitric oxide) and vasoconstricting (endothelin) substances.

**Enzymes.** Proteins that cause chemical changes in other substances in the body, without being changed themselves. See Renin.

**Epinephrine (Adrenaline).** A catecholamine released by adrenal medulla. Predominantly stimulates sympathetic beta receptors but in larger concentrations stimulates also alpha adrenergic receptors.

**Eutrophic inward remodeling.** A response of arterioles to chronic vasoconstriction and high blood pressure levels. The lumen of a remodeled vessel is reduced, the wall thickness is increased but the muscular layer of the wall is not hypertrophic.

**Exudates.** Eye ground changes seen in diabetic and hypertensive patients. They are indicative of vascular damage and reflect leakage of the plasma into the eye ground.

**Fibromuscular dysplasia.** A disease that can cause narrowing of the renal arteries which in turn can cause hypertension.

Hypertension due to fibromuscular dysplasia can be cured by surgery or by balloon catheterization.

**Ganglionic blockade.** See ganglionic blocking agents.

**Ganglionic blocking agents.** Chemical compounds that block transmission of signals through the entire autonomic nervous system. They were initially used for treatment of hypertension but induced intolerable side effects and were soon replaced by better tolerated drugs.

**Heart rate.** The number of heart beats in a minute.

**Hemodynamics.** Analyzes how the blood circulates in the body by measuring cardiac output, blood pressure and by assessing vascular resistance.

**Homeostasis.** The capability of the organism to regulate its internal environment in order to maintain a stable constant condition.

**Hyperglycemia.** An excessive concentration of glucose in the blood. Main symptoms are increased urination and excessive thirst. In extreme cases the loss of fluids through the kidneys can cause dehydration and alteration of the mental status(sleepiness and unconsciousness).

**Hyperkinetic state.** Found in early phases of human hypertension, is characterized by a slight increase of blood pressure and a substantial increase of resting cardiac output and heart rate.

**Hyperosmolar.** Abnormally increased concentration of solutes in body fluids as in dehydration and hyperglycemia.

**Inotropy.** Intrinsic strenghth of contractilty of the cardiac muscle.

**Insulin.** A hormone secreted by pancreatic islet cells. Insulin causes liver and muscle cells to take in glucose and store it in the form of glycogen.

**Insulin resistance.** Decreased sensitivity of peripheral tissues to insulin. Insulin resistant individuals need higher levels of insulin to store glucose in the liver and skeletal mucles.

**Isoproterenol.** A synthetic beta adrenergic agonist which stimulates beta 1, beta 2 and beta 3 receptors.

**Lumen.** Internal diameter (caliber) of a blood vessel.

**Medulla oblongata.** See brain stem.

**Metabolic syndrome.** A combination of medical disorders that increase the risk of developing cardiovascular disease and diabetes. Presently the syndrome is defined as co-occurrence of at least three of the following signs: elevated waist circumference, high triglycerides, low high density lipoporotein cholesterolol, high fasting blood glucose and high blood pressure. Some defitnion also include insulin resistance.

**Messenger.** A molecule that induces biological responses in cells.

**Myocardium.** Contractile muscle cells of the heart.

**Negative feedback.** System consisting of a signal, a sensor and a loop that tends to stabilize the signal. In arterial barroreceptors the signal is the blood pressure, the sensors are the arterial wall stretch receptors and the feedback loop is to the centeral nervous system. If the pressure is high the brain decreases the blood pressure. If the pressure is low the brain increases the pressure.

**Nephrectomey.** Surgical removal of kidneys.

**Neurons.** Electrically excitable cells in the nervous system that process and transmit information.

**Nitric oxide.** A potent but short lived vasodilator released by the vascular endothelium.

**Noradrenaline.** See norepinephrine.

**Norepinephrine.** The neurotransmitter released from sympathetic nerve endings which acts on vacular alpha adrenergic receptors to cause vasoconstriction. Norepinephrine is also release from the adrenal medulla.

**Oxygen consumption.** The amount of oxygen used by the body at rest or during exercise.

**Pancreatic.** From the pancreas. Pancreas is a gland situate behind the stomach. It secretes into the blood hormones insulin and glucagon. Pancreas also secretes a number of enzymes needed to process the food in the guts.

**Parasympathetic.** A branch of the autonomic nervous system releasing the neurotransmitter acetylcholine. Parasympathetic nerves inhibit the cardiac pacemaker and thereby slow the heart rate. Parasympathetic nerves also affect the function of the gastro intestinal system, bronchi and parotid glands.

**Pheochromocytoma.** A tumor of the adrenal medulla causing hypertension by producing huge amounts of epinephrine and norepinephrine.

**Phentolamine.** An injectable alpha adrenergic receptor blocking agant.

**Plethysmography.** (Of the forearm); a method to measure the blood flow in the forearm by obstructing the venous outflow and measuring changes in the volume of the forearm.

**Psychosomatic.** Bodily (physical) symptoms caused by mental or emotional disturbance.

**Receptors.** Proteins in the cell membrane that have affinity to bind with specific molecules (agonists) in order to transmit stimuli into the cell.

**Regression towards the mean.** Statistical term that refers to the likelihood that on repeated measurements initially high scores will be lower and low scores higher.

**Renin.** Enzyme released by the kidneys in response to changes in sodium balance, blood pressure and sympathetic stimulation. See angiotensin, angiotensin 1, and angiotensin 2.

**Retroperitoneal** (space)**.** The area behind the peritoenal membrane at the back of the abdominal cavity.

**Revascularization.** Procedure to provide blood to an organ with inadequate arterial blood supply either by surgery or by balloon catheterization.

**Selection bias.** Statistical term referring to the possibility that an investigated population is not representative of the population at large.

**Sensitivy.** Propensity of an organ to respond to a stimulus.

**Skeletal muscles.** Also called striated muscles, are muscles attached to bones. They are responsible for voluntary movements of the body.

**Sleep apnea.** Inadequate and irregular breathing during sleep.

**Smooth muscles.** Responsible for involuntary contractility of hollow organs, such as blood vessels, the gastrointestinal tract, the bladder, or the uterus.

**Starling mechanism.** Refers to the ability of the heart to change its force of contraction in response to changes in venous return. (Based on the observation that increased stretch of myocardial cells results in larger force of contraction).

**Stoke Adams attacks.** An extreme slowing of the heart or a stoppage of the heart due to an **atrio ventricular block**. Symptoms are dizziness, fainting and convulsion.

**Stretch receptors.** See baroreceptors.

**Stroke volume.** The amount of blood ejected with one heart beat.

**Systole.** The contraction of heart chambers to eject the blood.

**Systolic pressure.** The peak blood pressure in the arteries.

**Tachycardia.** Fast heart rate.

**Teleological, teleology.** A concept that all natural events are shaped by a purpose.

**Tone.** In this book refers to the frequency of impulses in the autonomic nervous system.

**Vascular remodeling.** The change of the structure of arterioles in response to chronic vasoconstriction and blood pressure elevation. See also eutrophic vascular remodeling.

**Vascular resistance.** The resistance that must be overcome to push blood through the circulatory system. Small arteries (arterioles) offer the largest resistance to flow. The total vascular resistance cannot be directly measured and is calculated by dividing the mean arterial pressure with the cardiac output. Regional vascular resistance of an organ or an area of the body is calculated from the mean arterial pressure and the local blood flow.

**Vasodilation, vasodilatation**; Widening of the inner diameter of a blood vessel due to relaxation of muscles in its wall.

**Vasopressin.** A vasoconstricting hormone released by the pituitary gland.

**Veins.** Thin walled blood vessels carrying the oxygen-unsaturated (blue) blood from the periphery to the heart.

**Ventricles.** The muscular thick walled chambers of the heart.

**Venules.** Small veins